Henry William Bristow

The Geology of the Isle of Wight

Henry William Bristow

The Geology of the Isle of Wight

ISBN/EAN: 9783337409043

Printed in Europe, USA, Canada, Australia, Japan

Cover: Foto ©berggeist007 / pixelio.de

More available books at **www.hansebooks.com**

MEMOIRS OF THE GEOLOGICAL SURVEY.

ENGLAND AND WALES.

THE GEOLOGY

OF THE

ISLE OF WIGHT,

BY

HENRY WILLIAM BRISTOW, F.R.S., F.G.S.

SECOND EDITION, REVISED AND ENLARGED,

BY

CLEMENT REID, F.L.S., F.G.S.,

AND

AUBREY STRAHAN, M.A., F.G.S.

PREFACE.

THE onward progress of geological science during more than a quarter of a century since the first edition of the present Memoir appeared has not left the Isle of Wight unaffected. The geological formations on which the beauty of that fair Island so largely depends have been studied in great detail in all parts of the South of England, as well as in foreign countries. The coast-sections of the Isle of Wight have even become subjects of discussion and controversy. When, therefore, the first edition of this Memoir was nearly exhausted, and it became necessary to undertake the preparation of a second edition, I felt that no satisfactory progress could be made in this task until the Map of the Island had been first revised and brought abreast of the present condition of Geology. The publication of the large Ordnance Survey Maps on the scale of six inches to a mile supplied for such a revision a far more accurate and convenient basis than was available at the time when the Island was originally mapped by the Geological Survey.

Accordingly, Mr. Bristow, the Senior Director, to whom science is mainly indebted for the first Survey Map of the Isle of Wight, and for the Memoir descriptive of the structure of the Island, undertook the serious labour of superintending the preparation of new editions, both of Map and Memoir.

In the following Prefatory Note supplied by him he has stated how this work has been carried on under his general supervision. The revision of the Map became in fact a re-survey of the Island, as all the lines were retraced on the ground. It is, however, due to Mr. Bristow to add that the main geological lines remain nearly as he mapped them more than 30 years ago.

In the preparation of the present edition of the Memoir so many and important have been the changes required that the work might not unfairly be described as a new one. The revision alike of Map and Memoir has been made under Mr. Bristow's direction and with his co-operation, by two of the officers of the Survey, Mr. C. Reid, who took the Tertiary area, and Mr. A. Strahan, who had assigned to him the Secondary Rocks. I have also myself personally visited the Island with Messrs. Reid and Strahan, and read over on the ground the proofs of the following chapters. I will here briefly mention some of the more important alterations and additions.

In discussing the relations of the Wealden to the Upper Neocomian Rocks it is shown that these two groups are separated by a sharply-defined lithological demarcation, accompanied by a palæontological break.

In re-mapping the Lower Greensand Mr. Strahan has taken advantage of certain broad lithological characters, which being traceable across the Island, permitted of a convenient subdivision of that formation into groups whose respective limits could be shown on the Map. This subdivision, for which a new scheme of colouring has been adopted, is only intended for the Isle of Wight, where it is of considerable local service. Mr. Strahan found that an upper subgroup of the Lower Greensand, corresponding to the Folkestone Beds, existed on the Island, capable of subdivision into an upper ferruginous and slightly conglomeratic rock, the Carstone, which passes up into the Gault, and a lower sandrock resembling in lithological characters the Folkestone Beds, and passing downwards into ferruginous sands. Another subgroup, exhibiting both the lithological and palæontological features of the Sandgate Beds, has been placed with these underlying sands (the Hythe Beds) under the name of the Ferruginous Sands. The position and extent of the Atherfield Clay remain nearly as in the first edition of the Map.

A few fossils have been added to the small fauna hitherto yielded by the Gault. A line has been engraved on the Map to mark the position of the bold topographical feature formed by the Chert beds of the Upper Greensand in the central parts of the Island.

The subdivisions of the Chalk which can be traced on the ground have now been inserted on the Map. The Chalk-rock is so shown, but the Melbourn-rock, though frequently recognised in place, is not represented on the Map for want of space.

In the preparation of the following Chapters it has been found necessary entirely to re-measure the cliff sections of the Secondary Rocks. This has been done in Compton Bay from the Upper Greensand downwards, in Atherfield Bay from the Chalk-marl downwards, and in Sandown Bay from the Chalk-rock downwards. The total thickness of strata measured at the last-named locality was 1,218 feet. The results of this detailed re-examination are shown graphically in Plate II., which represents the coast-section from Compton Bay to Blackgang, and in Plate III., which contains a series of comparative Vertical Sections showing the varying thickness of the Secondary formations in different parts of the Island and on the adjacent coast of Dorsetshire.

In revising the Tertiary area of the Island, Mr. Reid found that only slight changes were required in the Eocene lines of the Map. In the Sections and Memoir he has somewhat modified the boundaries of the Bracklesham and Barton Beds in conformity with the recent researches of the Rev. Osmond Fisher and Mr. Keeping. The so-called "Upper Bagshot Sands" of the Isle of Wight are not improbably considerably higher than the division of that name in the actual Bagshot district. Hence, until the position of the glass-sands of the Island has been definitely ascertained, it has been thought desirable not to speak of these deposits as "Upper Bagshot," but to revert to the older name of "Headen Hill Sands."

The classification of the Eocene formations into Upper, Middle, and Lower, adopted in the first edition of the Memoir, has been modified. The so-called "fluvio-marine beds" of the Isle of Wight are now classed as Oligocene.

The most important alteration of the Map of the Tertiary part of the Island has been in the tract occupied by the Hamstead (Hempstead) Beds. These strata have been detected by Mr. Reid by means of a boring apparatus over a large area, so that instead of covering a space of only two or three square miles, they really spread over half of the Tertiary district of the Island. They also prove to be of considerably greater thickness than has been supposed, their actual thickness being 260 feet instead of 170 feet. The sections in the Tertiary districts have been re-measured where it was thought desirable. The Chapters on the Tertiary rocks in the present Memoir have been largely extended and in great part re-written.

In the recent re-survey of the Isle of Wight the superficial deposits have been mapped out in detail. They have been arranged in four groups which are based, as far as possible, on chronological order. Excluding the angular flint-gravel of the Chalk Downs, the age of which is doubtful, the oldest group, that of the Plateau Gravels, is shown to be probably as old as, and perhaps contemporaneous with, some of the Glacial deposits of the Midlands. But no conclusive evidence has been obtained in the Isle of Wight of the co-operation of coast-ice or land-ice in the formation of these deposits.

The later groups (Valley Gravels and Alluvia) contain the records of successive stages in the excavation of the present system of valleys. This chapter of geological history possesses a special interest and value from the insular position of the Isle of Wight and the changes that have resulted from the cutting back of the coast-line by the sea. The drainage system of the Island, like that of the South of England generally, has been determined by the great lines of anticlinal and synclinal folds into which the Secondary and Tertiary strata have been thrown. Each main anticline became a line of watershed, but in the subsequent gradual denudation of the general surface of the land the forms and elevations of the topography have resulted, not from these underground movements, but from the relative durability of the rocks. The areas of maximum elevation at the present day are not those where the greatest amount of upheaval took place in past time.

Mr. Strahan's survey of the superficial deposits in the south of the Isle of Wight affords a glimpse of an older and different topography before the Chalk Downs of that region had been reduced to their present limited area. An extensive sheet of river-gravel in the south-west of the Island marks the course of what must at one time have been a considerable stream, taking its rise among the Southern Downs which then stretched southwards into the English Channel. As Mr. Codrington has suggested, this stream flowed westwards and northwards by Freshwater to

Yarmouth. But by the gradual encroachment of the sea its drainage area has been greatly reduced, and at last its valley has actually been reached and cut across by the waves, so that the stream there enters the sea, and the lower part of the valley is left almost dry.

One of the following chapters has been devoted to a description of the nature and position of the various anticlinal and synclinal folds which play so large a part in the geological structure, not only of the Isle of Wight but of the whole of the south-eastern mainland. From the evidence obtainable in the Island we know that these plications of the rocks were produced at some time subsequent to the deposition of the Oligocene strata. Elsewhere we obtain proofs that they were completed before the Pliocene period. The limits of their geological date are thus fixed.

The Appendices include a number of well-sections and borings collected and arranged by Mr. Reid. The fossil lists formerly dispersed through the Memoir have been thrown together into one tabular statement which has been prepared by Messrs. Reid and Strahan with the assistance of Mr. G. Sharman and Mr. E. T. Newton, Palæontologists of the Geological Survey. A geological bibliography, compiled by Mr. Bristow, has been added to the Memoir.

ARCH. GEIKIE,
Director-General.

Geological Survey Office,
    London,
       April 1889.

[Since this preface was written, and while these pages are passing through the press, Mr. Bristow has been removed from us by death. We hoped that he would have lived to see the final publication of this Memoir, in the preparation of which he took so keen an interest. The correction of his "Notice" formed his last piece of scientific work, and in returning it to me only a few weeks before the illness from which he never recovered, he expressed with characteristic courtesy his approval of all that had been done to make this new edition a fitting termination to the labours of his long career in the Geological Survey. We cherish his memory as a loyal and helpful friend and a distinguished colleague.

A. G.

June 24th, 1889.]

NOTICE

(By H. W. Bristow, F.R.S.)

The original survey of the Isle of Wight on the one-inch scale was commenced under the personal superintendence of Sir Henry T. De la Beche in the year 1848, and was carried on at intervals between that year and 1856 by the late Professor Edward Forbes and myself, Mr. W. T. Aveline at the same time completing a portion of the Secondary area between Chale and Dunnose, the whole being under the direction of Professor A. C. Ramsay. During part of the time that the Island was being surveyed assistance was rendered by the late Mr. R. A. C. Godwin-Austen, Mr. Henry Keeping (now of the Woodwardian Museum, Cambridge), and by the Fossil Collectors, Richard Gibbs and John Cotton.

A re-survey of the Island on the six-inch scale instituted by the present Director-General was begun in November 1886, and was completed by the end of the year 1887, the northern or Tertiary half of the Island being mapped by Mr. C. Reid, and the southern or Secondary half by Mr. A. Strahan. This re-survey, reduced to the new one-inch Ordnance Map, was published in 1888. Clean copies of the six-inch Maps have been deposited in the Geological Survey Office for reference, and a duplicate set of these sheets, mounted as a wall-map, was exhibited at the International Geological Congress in 1888, and is now suspended in the Museum of Practical Geology.

The first edition of the present Memoir was published in 1862. It was written by myself by desire of the late Sir Roderick J. Murchison, then Director-General, use being made, when necessary, of the posthumous Memoir on the Fluvio-marine Formation of the Isle of Wight by Professor E. Forbes, in which some of the notes I had made had already appeared. In the preparation of the present edition of the Memoir the authorship of the revision has followed the same general distribution as in the case of the mapping. The account of the Secondary rocks has been revised and enlarged by Mr. Strahan, who, besides examining these rocks in the Isle of Wight, continued the mapping of their subdivisions into the neighbouring coast of Dorsetshire. The comparisons with the Geology of the mainland made in the following account of the Secondary rocks are thus entirely his.

The chapters on the Tertiary rocks have been revised and much enlarged by Mr. Reid. The most important change which he has been able to make in the Map, the great extension he has given to the Hamstead Beds, has been rendered possible by the application of a boring apparatus, whereby no fewer than 358 borings, ranging from 10 to 33 feet in depth, were made in the Tertiary area of the Island.

The lists of fossils have undergone a thorough revision by
Messrs. G. Sharman and E. T. Newton, who have also named the
additional specimens collected during the progress of the re-
survey.

Professor T. Rupert Jones undertook the determination of the
Ostracoda, revised the lists of these crustacea, and furnished
Table V., which gives a synoptical view of their distribution. We
are also indebted to Mr. J. Starkie Gardner for the account of
the Flora of the Bagshot Beds of Alum Bay, and to Mr. Carruthers for looking over the lists (in MS.) of the plants of the
Secondary rocks. Mr. W. Hill kindly undertook the examination
under the microscope of nodules from the Upper Chalk of Whitecliff. Advantage was taken also of the intimate knowledge of
the Geology of the Isle of Wight possessed by Mr. Henry Keeping
to obtain his assistance in revising some of the detailed sections
of the Tertiary strata.

H. W. Bristow.

London, March 30, 1889.

TABLE OF CONTENTS.

CHAPTER VI.

THE GAULT AND UPPER GREENSAND.

THE GAULT :—
  INTRODUCTION - - - - - - - 60
  LANDSLIPS - - - - - - - 60
  DESCRIPTION OF SECTIONS - - - - - 62
  CORRELATION WITH THE MAINLAND - - - - 65

UPPER GREENSAND :—
  INTRODUCTION - - - - - - - 65
  COAST SECTIONS :—
    1. Compton Bay - - - - - - 68
    2. Blackgang to Shanklin - - - - - 68
    3. Culver Cliff - - - - - - 70
  INLAND SECTIONS :—
    1. Along the Central Downs - - - - - 70
    2. Around the Southern Downs - - - - 72

CHAPTER VII.

THE CHALK :—
  INTRODUCTION - - - - - - - 73
  CHLORITIC MARL - - - - - - - 79
  UPPER, MIDDLE, AND LOWER CHALK :—
    1. Compton Bay along the Central Downs to Culver Cliff - 82
    2. The Southern Downs - - - - - 90
  DIVISION OF THE UPPER CHALK INTO ZONES - - 92

CHAPTER VIII.

EOCENE :—
  INTRODUCTION - - - - - - - 94
  READING BEDS - - - - - - - 94
  LONDON CLAY - - - - - - - 97
  LOWER BAGSHOT BEDS - - - - - - 101
  ON THE FLORA OF ALUM BAY, BY J. STARKIE GARDNER - 104

CHAPTER IX.

EOCENE—continued :—
  BRACKLESHAM AND BARTON BEDS :—
    BRACKLESHAM BEDS - - - - - 109
    BARTON CLAY - - - - - - 117
    HEADON HILL SANDS - - - - - 122

CHAPTER X.

OLIGOCENE :—
  INTRODUCTION - - - - - - - 124
  HEADON BEDS - - - - - - - 128

CHAPTER XI.

OLIGOCENE—continued :—
  OSBORNE BEDS - - - - - - - 148
  BEMBRIDGE LIMESTONE - - - - - - 158
  BEMBRIDGE MARLS - - - - - - 170

xii

LIST OF ILLUSTRATIONS.

xiv

THE GEOLOGY

OF THE

ISLE OF WIGHT.

THE GEOLOGY

OF THE

ISLE OF WIGHT.

CHAPTER I.

INTRODUCTION AND TABLE OF STRATA.

THE Isle of Wight is of a lozenge shape, with its longer axis extending nearly east and west from the Foreland to the Needles, a distance of $22\frac{1}{2}$ miles, and its shorter axis nearly north and south from West Cowes to Rocken End, a distance of 13 miles. The northern apex is situated immediately opposite the mouth of Southampton Water. The two northern sides of the Island are nearly parallel with the mainland of Hampshire, from which they are separated by the Solent on the west, and on the east by the sea between Southampton Water and Spithead. The nearest point to the mainland is Cliff End, which is only a mile distant from the bank of shingle and sand on which Hurst Castle is situated; but the Solent is generally from two to three miles in width, while the channel east of Southampton Water reaches a breadth of four miles. The area of the Island, as deduced from the Ordnance Survey, is 155 square miles 370.209 acres, in which are included 122.684 acres of water, 9 square miles 34.076 acres of foreshore, and 434.454 acres of tidal water. It is divided into East and West Medina by the River Medina, which, rising near the southern apex of the Island, runs northwards through a gap in the chalk range, and discharges itself into the sea between East and West Cowes.* A more marked physical division is that produced by a bold range of Chalk Downs, which extends from the Needles to Culver Cliff.† The area lying to the north of this range is occupied by Tertiary strata, and is chiefly characterised by the heavy and clayey nature of the land, and by the numerous woods which cover its surface, especially east of the River Medina. The tract of land south of the chalk range is occupied chiefly by the Lower Greensand, and presents a

---

* The Isle of Wight was called "Meden" in former times. The Roman name for it was VECTIS. In Camden's Saxon Chronicle and Domesday Book and in the oldest records it is written "Wict."—H. W. B.

† Culver Cliff (after the Anglo-Saxon name "culfre," a dove) was probably so named from its being the resort of numerous wild pigeons of a small species (*Columba saxitilis*) which made it their haunt. Pennant states that "these birds make at a certain season most enormous flights; they come daily in vast flocks, as far as the neighbourhood of *Oxford*, to feed on the turnip-fields, and return again to these and *Freshwater Cliffs*, where they pass the night." (Pennant's Journey, p. 151.) Culver Cliff was also famous for a breed of hawks in the time of Queen Elizabeth.—H. W. B.

striking contrast to the north side of the Island in its generally light and loamy soil, and in the absence of woods. In the southern part of the Island also is found a group of hills, capped by outliers of chalk, which rise to a far greater height than any part of the generally low Tertiary district, and in fact form the most elevated tract in the Island.

A considerable part of both the northern and southern parts of the Island is overspread by gravels and Alluvium, the former being of considerable thickness and commercial importance.

The following table gives in descending order the formations shown upon the map :—

Blown Sand		} Recent.
Alluvium		
Peat		
River Terraces (Gravel)		} Pleistocene.
Angular flint-gravel of the Chalk Downs		
Plateau Gravel		
Hamstead Beds		} Oligocene. ["Fluvio-marine" of E. Forbes.*]
Bembridge Marls		
———— Limestone		
Osborne Beds		
Headon Beds		
Headon Hill Sands		
Barton Clay		} Eocene.
Bracklesham Beds		
Lower Bagshot Beds		
London Clay		
Reading Beds		
Chalk-with-flints		} Upper Cretaceous.
Chalk Rock		
Middle and Lower Chalk with Melbourn Rock.		
Chloritic Marl		
Chert Beds } Upper Greensand Sands		
Gault		
Carstone		} Lower Cretaceous.
Sand-rock Series } Lower Greensand or Ferruginous Sands } Upper Neocomian.		
Atherfield Clay		
Wealden Beds with beds of sandstone		

The above formations will be described in ascending order, commencing with the Wealden—the lowest and oldest strata seen in the Isle of Wight.

* The term "Vectian" was proposed for this group by Prof. John Phillips, but has not been generally adopted.

CHAPTER II.

THE WEALDEN BEDS.

INTRODUCTION.

THESE beds rise to the surface on the southern and eastern sides of the Island, where they have been elevated along the anticlinal axes of Brixton and Sandown. The entire area occupied by them is very inconsiderable, not exceeding five square miles; and there is no good section inland. On the coast, however, for six miles from Compton Bay to Atherfield, they are well exhibited in the cliffs (*see* Plates I. and II.), and there is also a tolerably fair exposure of them on the coast in Sandown Bay. The lowest beds exposed in the Island are the variegated Wealden clays and sandstones of Brook Bay. Judging from the section at Swanage, where the whole of the Wealden formation is displayed, there may be about as great a thickness of these beds below the sea-level in the Isle of Wight, as is seen cropping out in the cliffs.

The Wealden Beds include two different but perfectly conformable types, the one consisting of dark-blue or almost black shales, evenly bedded and splitting into thin laminae, together with layers of shelly limestone and ironstone, and very thinly laminated "paper-shales," crowded with the shells of minute ostracoda (Cyprids). Fossils are abundant in this type, though the number of genera is somewhat limited. *Paludina*, *Cyrena* (*Cyclas*), and *Unio* occur in profusion everywhere, and *Vicarya* (= *Cerithium*, *Melania*, *Potamides* of previous writers) is abundant at Atherfield. This type is found invariably at the top of the Wealden formation, immediately under the Lower Greensand, but appears also to be interstratified with the type now to be described.

FIG. 1.   FIG. 2.   FIG. 3.

Cypridea spinigera, Sow.   *Cyrena*.   *Paludina fluviorum*, Sow.

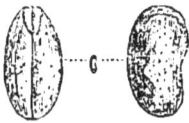  

The other type, under which the Wealden beds appear, is that of red, green, and variegated marls and clays (curiously resembling the Keuper Marl), with numerous included bands of sandstone of variable thickness. The bedding is far from regular, and fossils are comparatively scarce. A large freshwater shell (*Unio valdensis*, Mant.), dirfted wood in great abundance, the remains of fish, and the water-worn bones of terrestrial reptiles are met with throughout the group.

A 2

The correlation of these two groups of the Isle of Wight with the Wealden strata of the mainland has caused some diversity of opinion. Dr. Fitton and the older authors spoke of the upper group only as Wealden and of the lower as Hastings Sand. By the Geological Survey they were both included under the name Wealden, but in 1856 Mr. Godwin Austin[*] stated that the Weald Clay might be seen "to alternate with, and therefore to be synchronous with, the marine Neocomian." Professor Judd [†] also in 1871 stated that he looked upon "the great mass of variegated strata containing only freshwater and terrestrial fossils . . . . as the Wealden proper," and that the upper group or Punfield Beds, as he called them, "may be regarded indifferently either as the highest member of the Wealden in our classification of terrestrial strata, or as a portion of the Neocomian in our grouping of the marine series." This view of their relations was suggested by the intermingling of brackish water or marine forms such as *Cardita*, dwarfed oysters, and the estuarine *Vicarya* with purely freshwater forms such as *Paludina* and *Unio*. But unfortunately, the true base of the Lower Greensand not having been then discovered at Punfield, a large part of this formation, with its highly characteristic fauna, was included in the "Punfield Beds" of Professor Judd, with the result that the fauna of these Punfield Beds was made up partly from the Lower Greensand and partly from the Wealden.

This fact was first ascertained by Mr. Meyer[‡] in the years 1871-72. He observed that the Atherfield Clay with some of its characteristic fossils occurred beneath the fossiliferous zone from which many of the marine Punfield fossils had been obtained, and that the characteristic cypridiferous shales with limestone occurred beneath and nowhere above this marine band. His conclusions were strengthened by observations made by the Geologists' Association[§] in 1882, and have been fully confirmed by the examination that was undertaken for the purpose of the present Memoir. The results and measurements obtained during this examination will be incorporated in the following pages, but it may be stated here that at Punfield, as in the Isle of Wight, the palæonto-logical break between the Wealden and Lower Greensand is complete, and is accompanied by evidence of considerable erosion of the former.

The name of Punfield Beds, therefore, having been applied to strata belonging to two distinct groups, will not be used here. But at the same time it will be convenient to distinguish the beds for which the name was intended from the variegated Wealden type which has been mentioned above. The name Upper Wealden is scarcely suitable, for, though generally found at the top of the Wealden formation, they appear also to be interstra-

---

* *Quart. Journ. Geol. Soc.*, vol. xii. p. 66.
† *Ibid.*, vol. xxvii. p. 207.
‡ *Ibid.*, vol. xxviii. p. 243, and vol. xxix. p. 70.
§ *Proc. Geol. Assoc.*, vol. vii. p. 388.

tified at various horizons in it.* On the other hand the most striking characteristic of the beds is their shaly character, as compared with the almost structureless variegated clays, and the name of Wealden Shales will perhaps be sufficiently distinctive.

The Wealden Beds rise from beneath the Lower Greensand in Brixton and Sandown Bays, on the south-western and south-eastern sides of the Island respectively. In both bays they rise with a steep dip from beneath the rocks which compose the central range of the Island. On receding from this central axis of disturbance the angle of dip grows less, until the beds finally assume a horizontal position, as may be seen near Brook, in Brixton Bay, and in Sandown Bay at the point where the coast-line cuts the Alluvium of Sandown Marsh. Still further south in each of these bays a gentle southerly dip sets in, and the higher beds of the Wealden series pass in succession below the beach. The structure, therefore, is similar at each locality, namely, that of a dome with a steep side to the north.

BROOK AND COMPTON BAY. (*See* Plates II. and III.)

The lowest beds displayed in the Island are those forming the shore near Brook and at Sedmore Point, half a mile south-east of Brook Chine.† At Sedmore Point a bed of sandstone forms the foot of the cliff for about 400 yards. Above it are blue, purple, and deep-red marls, overlain about half-way up the cliff by an impersistent bed of sandstone, with a gravelly band about 18 inches thick, made up of fragments of sandstone with many small bones, at its base. *Cyclas*, *Paludina*, and *Unio* are recorded by Fitton from this bed. The upper part of the cliff consists of purple and blue marls, with light-coloured bands containing much lignite.

Between this Point and Brook Chine the strata have slipped, forming an undercliff, known as Roughland, along the whole length of which (some 500 yards) there is no clear exposure of rock in place, though the extent of the slip shows that the beds must be chiefly clays. As we approach Brook Chine the section becomes clear again. A greenish band may be seen to rise westwards from beneath the beach, and to run along the upper part of the cliff past Brook Chine to a small chine 180 yards south of Brook Chine, where it descends once more to the beach. This bed is easily traced by its colour, and by the fact that it is crowded with large flattened masses of lignite, especially to the south and west of Brook Chine. It shows that the strata form a

* Geology of the Weald (Geological Survey Memoir), and Drew, *Quart. Journ. Geol. Soc.*, vol. xvii. p. 283.
† The local name for the deep fissures or gullies, which are termed chines in the Isle of Wight, is derived from the Anglo-Saxon *cinu*, a cleft. Wyclif speaks of the " chyne of a ston-wall." So also, Spenser—
" Where byting deepe, so deadly it imprest,
That quite it chyned his backe behind the sell."
—Faerie Queene, b. iv., canto 6, xiii.

gentle anticline, the centre being near Brook Chine, the deep-red and variegated marls of which are perhaps the lowest rocks seen in the Island.

The lignite bed described above appears to pass out to sea south of, and therefore below, a similar bed which is seen at Brook Point, but the strata are so variable that it is impossible to speak with certainty. The section at the Point shows upwards of 100 feet of red, purple, and blue clay with impersistent bands of sandstone, underlain by 13 feet of grey clay, the lower part of which contains numerous flattened masses of black shining lignite. This lignite band rests upon a bed of hard sandstone, to which the Point owes its existence. It is a whitish or pale-grey rock, about 6 feet thick, containing fragments of marl and clay, and with iron-pyrites abundantly disseminated through its upper part. It is irregularly stratified, and its surface is undulating and covered with fucoidal and hollow vertical markings.

Below and partly imbedded in this rock lie the scattered trunks of coniferous trees, known as the " Pine Raft." They were first observed by Webster in 1811,* but were more fully described by Mantell in 1846.† The trunks lie prostrate in all directions, broken up into cylindrical fragments. They are covered by thin bark, now in the state of lignite, the wood having been con-verted into a black or greyish calcareous stone,‡ with much iron pyrites. Many of the trees still present traces of woody structure, and the annular rings of growth are clearly perceptible; but they are traversed also by numerous threads of pyrites. The trunks are generally of considerable magnitude, being from one to three feet in diameter; two upwards of twenty feet in length, and of such size as to indicate a height of forty or fifty feet when entire, were noticed by Mantell.

The "Pine Raft" can be seen at low water only. During spring tides it may be observed to rest on variegated marls, but all attempts to trace it eastwards from Brook Point have failed, pro-bably on account of its being of local development only. The purple marls forming the cliff above it are apparently the same beds that have made the great slip of Roughland, and the Pine Raft, if it is continuous, should be found in the cliff near Sedmore Point; but though many large fragments of trunks are lying on the beach, there is no bed in the cliff exactly corresponding to that of Brook Point.

As suggested by Mantell, the trees were probably drifted from a distance, in the same manner as the trunks, brought down by the Mississippi at the present day, are deposited in large rafts in the delta of that river. It is not to be expected, therefore, that

---

* Englefield's Isle of Wight. 1816.

† *Quart. Journ. Geol. Soc.*, vol. ii. p. 91. 1846.

‡ Unlike the trunks in the dirt-beds of the Isle of Portland, which are sili-cified. Professor Way pointed out the probability " that the fossil forest imbedded in the Weald Clay at Brook Point is impregnated with phosphoric acid, instead of carbonic acid, as is generally assumed." *Journ. Roy. Agric. Soc. of England*, vol. ix. p. 82.

the " Pine Raft" is of wide range, or that the horizon at which it occurs should be recognisable when the trees are not present. There is no evidence that any of the trees in this or any other part of the Wealden series grew upon the spots where they are now found.

In the cliffs of this neighbourhood there have been found also the cones to which more special reference is made in the fossil list on p. 258.

Mantell records also the occurrence of *Clathraria Lyellii* as a pebble on the beach of Brook Bay.

The large freshwater shell, *Unio valdensis*, was first observed by Mantell "in the sandy clay beds immediately above the fossil forest" (*op. cit.*, p. 94). It occurs also in some hard irony concretions, which have fallen to the beach on the west side of Sedmore Point.

FIG. 4.

Unio valdensis, Mant.

A large number of reptilian bones also has been obtained from the cliffs. Those on which the species *Iguanodon Seelyi* was founded were obtained by Mr. Hulke in the small chine 180 yards south of Brook Chine.[*] *Ornithopsis Hulkei* also occurred in Brook Bay, and footprints, believed to be those of an Iguanodon, have been found 600 yards to the west of Brook Point, and near Sedmore Point by Mr. Beckles.[†] The prints occurred as casts, attached to a thin bed of hard sandrock on the shore at low water. For further information on the fossils the reader is referred to the list on p. 258.

As we proceed from Brook either westwards to Compton or eastwards to Atherfield, an ascending section in the same beds is provided in the cliff, the distance to be traversed in the former case before reaching the top of the Wealden beds being less on account of the greater steepness of the dip. We will first examine the cliffs westwards, as far as the great slip which marks the position of the Atherfield Clay (Plate II.).

On rounding the Point we find the cliff composed principally of red and purple marls for a distance of about 700 yards, the thickness of strata amounting to 439 feet. In the marls there occur beds of sandstone often conspicuous from their whiteness, and a few green bands containing lignite. Passing over some thin and impersistent sandstones near the Point, we meet the first noteworthy bed 170 yards further west, where there is seen in the upper part of the cliff a grey clay packed with lignite, resting on a white sandstone 5 feet thick, but thinning away westwards. This is overlain by purple and variegated clays, and 100 yards westwards a second bed of white sand-rock, 7 feet thick, succeeds. A third bed, 16 feet thick, is seen on the east side, and a fourth,

---

* *Quart. Journ. Geol. Soc.*, vol. xxxviii. p. 135. 1882.
† *Ibid.*, vol. xviii. p. 443. 1862.

9 feet thick, on the west side of Shippard's* or Compton Grange Chine, the last-mentioned rock being of a pinkish hue from the abundance of grains of pink quartz in it. At 190 yards distance from this chine we see the purple strata pass up into characteristic blue Wealden Shales with abundant *Cyrena, Paludina,* Cyprids, fish-remains, and fragments of ferns. These blue shales, which, like the Cowleaze beds, are interstratified with sands in the lower part, are about 222 feet thick, and are fully exposed up to and in a small chine 350 yards west of Compton Grange Chine, but beyond this they have been disturbed by slipping. They seem, however, to be succeeded by red marls at a point in the top of the cliff 50 yards west of the small chine, whether by a fault or natural superposition will be discussed subsequently.

Continuing along the top of the cliff, where the strata are in place, we see a thickness of 193 feet of purple marls with irregular white sand-beds and with three beds of grey or white clay and sand with lignite, the highest and lowest containing large tree trunks in addition to a great abundance of small fragments of wood.

These variegated strata pass up into blue shales and sandstones with bands of ironstone, which in the exposed parts have weathered into a cinder-like rock. About 27 feet of these blue deposits are seen in place, and they are followed by blue paper-shales with *Cypris* and slabs of *Cyrena* limestone with fish-bones, seen only in slips, but estimated to have a thickness of 65 feet. These are overlain by the Lower Greensand.

The question now arises whether the blue shales last described are the same beds as those near Compton Grange, the strata being repeated by a strike fault with a downthrow to the south; or whether there are two horizons at which this type of the Wealden series makes its appearance in the Isle of Wight, as on the mainland.

It is in favour of the theory of a fault, that neither at Atherfield 5 miles distant, nor at Sandown 15 miles distant, nor at Punfield 20 miles to the west, can more than one group of shales of this type be seen, and that only at the top of the Wealden series. The thickness also of the beds visible between Brook Point and the top of the lower blue shales is much the same as that between Brook Point and the top of the Wealden Shales of Shepherd's Chine, namely, at the former locality 676 feet, of which 454 are variegated, and in the latter 754 feet, of which 562 are variegated. The blue shales, moreover, strongly resemble the beds of Cowleaze and Shepherd's Chines.

But on the other hand, the differences in the two sections of Compton Bay are so great, though only a quarter of a mile apart, that even allowing for the variability of Wealden strata, it is difficult to suppose that the same set of strata appears in each. The variegated beds of the upper part are characterised by an abundance of lignite associated with white clays; in those below lignite is scarce, but several bands of sand-rock stand out

---

* Not to be confounded with Shepherd's Chine, near Atherfield.

conspicuously. In the uppermost blue beds the sandstones, except close to the base, are not prominent; in the lower they form a marked feature. The thickness, moreover, of the lower set reaches 222 feet, that of the upper only about 92 feet, while, lastly, fossils occur abundantly close to the base of the lower set of blue shales, but have not been found in the 27 feet of the upper set which are clearly exposed. The evidence is therefore rather more in favour of there being two horizons in the Wealden series of Compton Bay, at which fossiliferous shales occur. Which of the two sets of shales should be compared with the Wealden Shales of Shepherd's Chine remains doubtful. If we correlate the lower set with the shales of Shepherd's Chine, we have nothing to represent the upper 285 feet of Wealden Beds of Compton Bay. But no evidence can be found of so great an erosion of the Wealden Beds as the absence of the strata in question would seem to imply. We may more probably view the lower shales of Compton Bay as a local intercalation of this Wealden type among the variegated beds.

Before leaving Compton Bay we will refer briefly to the section of the Wealden Beds at Punfield, on the coast of Dorsetshire, already referred to. The Wealden Shales at that locality form a well-marked subdivision at the top of the Wealden group. They have a total thickness of $34\frac{1}{2}$ feet, cypridiferous paper-shales, hard limestone with *Cyrena* and *Paludina*, and some thin bands of sandstone being interstratified with them. Downwards they pass into white sandstone, grey clays with white sands or brown sandstone, and so into red marls. About 200 feet below them lie white clays and sands, with much lignite and concretionary lumps containing *Unio valdensis*. The total thickness of the variegated beds of the Wealden, near Punfield, has been estimated by various observers at 1,500 to 2,000 feet.

Descending Section of the Wealden Beds from Compton Bay to Brook Point. (See Plates II. and III.)

Perna Bed (Compton Bay).

		Ft.	In.
Beds seen only in land-slips, consisting of Cyprid shales with a hard band, containing numerous fish-remains in the upper part, bands of limestone and ironstone; estimated at		65	0
Blue and grey clay and sand		2	0
Sand		1	0
Blue shale		3	0
White sand and grit		3	0
Ochry band (cinder-bed) passing into a solid ironstone where less weathered		0	6
Blue shale		17	0
Cinder-bed, as above		0	6
Grey clay, with large trunks of trees		9	0
Purple marls		41	0
White sandstone and clay with lignite		9	0
Purple marls with sand-beds, about		55	0
Fine white sand		3	0
Pale purple clay		12	0

Wealden Shales.

	Ft.	In.
White clay, crammed with great masses of lignite and trunks of trees	5	0
Yellow and white clay, passing down	6	0
Purple marls, about	35	0
White sandy clay, with bones	6	0
Deep red marls, about	12	0

Here there is possibly a large fault, repeating the Wealden Shales of Compton Bay (see p. 8).

Wealden Shales.

	Ft.	In.
Blue shales, not well seen, about	20	0
Shales, seen in the west bank of a small chine	21	0
Paper shale, with Cyprids	0	8
Cyrena limestone	0	2
Shales with lines of sand, Cyprids here and there	12	0
Paper shales with Cyprids	2	6
Cyrena limestone	0	1–2
Paper shales with Cyprids (in the east bank of the small chine mentioned above)	14	0
Shales, not well seen	25	0
Shale, with lines of sand and grit containing ferns (rises from below the beach on the east side of the small chine)	51	6
Yellow and white sand-rock, with large grains of pinkish quartz	5	0
Blue shale	3	0
Sand-rock, as above	12	6
Blue shale, with thin ironstone in the lower part	5	4
Coarse grit, with grains of pink quartz ⎫		
Shale parting ⎬	2	10
Sandstone ⎭		
Blue shale	0	7
Ironstone, with *Unio, Cyrena, Paludina,* Cyprids, and "Beef"	0	6
Blue shale	6	6
Fine ochry and dusky sand	1	0
Fine white sand-rock	2	6
Shales, with *Paludina* and Cyprids	5	6
Lenticular ironstone	0	0–4
Sandy shales, with ferns	5	0
Shale	6	0
Shales, full of *Cyrena* and *Paludina*	3	0
Sandstone, with lignite	0	6
White sandy clay	1	6
Blue marly clay, with large concretions and obscure fossils	3	6
White and blue marly clay	2	0
Pale variegated marl	5	6
White sandstone, with irregular top	3	0
Purple marls, estimated at	78	0
White sandstone, containing an abundance of grains of pink quartz (crops out west of Compton Grange Chine)	9	0
Red, purple, and green marls of Compton Grange Chine	78	0
White sandstone (east of Compton Grange Chine)	16	0
Variegated marl	30	0
Red sandstone	2 to 4	0
Greyish blue marl	10	0
White sandstone	7	0
Purple marls	64	0
White band	1	0
Purple marls	40	0
Grey clay packed with lignite	2	0
White sandstone, thickening eastwards	0 to 5	0
Purple marls	36	0
Red sandstone and marl, thinning out east at Brook Point	6	0
Purple marls	41	0
Red marls	12	0
Grey sandstone	0 to 2	0

	FT.	IN.
Grey sandy clay - - - - - -	7	0
Ditto with much lignite (seen in Brook Point) -	6	0
Current-bedded white sandstone, with much pyrites in the upper part (forms the foot of Brook Point) - -	5	0+
The " Pine-raft "; numerous trunks embedded in sandstone.		
Variegated marls, seen in the fore-shore.		

BROOK TO ATHERFIELD.

We will now return to Sedmore Point, where we commenced the description of the series, and follow the coast eastwards. It will be remembered that the above beds described again come into view, but with a gentle and nearly uniform dip, at first a little north of east, subsequently a little south of east.

The sandstone with $1\frac{1}{2}$ feet of conglomerate at its base, which first appears half way up the cliff at Sedmore Point, thickens eastwards and runs for a distance of nearly a mile, before it finally descends to the beach 500 yards west of Chilton Chine. There also it presents at its base $1\frac{1}{2}$–2 feet of a gravel, composed of pebbles of sandstone with many small bones, though this conglomeratic band does not continue through the whole distance. Below this sandstone lie deep-red marls, and above it come red and green marls as at Sedmore Point. The latter may be well seen in Chilton Chine. They contain lenticular harder bands with potato-shaped calcareous concretions, and a little lignite. Another bed of sandstone comes in at the top of the cliff 250 yards east of the Chine, and descends to the beach about midway between Chilton and Grange Chines. This bed likewise has a gravelly conglomerate, about 6 inches thick, at its base. It contains quartz pebbles, small bones, and rounded pieces of wood similar to those composing the "pine-raft." It is much current-bedded, and of variable thickness, reaching sometimes as much as seven feet.

Near Chilton Chine the vertebral centrum of *Eucamerotus*, Hulke (*Ornithopsis*, Seeley), which has been described by Mr. Hulke,[*] was found, but from which bed is not known. Mantell records that bones were collected in 1829 near Bull-face Ledge also.[†]

Grange Chine has been excavated in deep-red and green marls, the green beds containing much lignite. On the east side and near the top of the chine a conspicuous black band two feet thick contains abundance of lignite, many fragments of bones, and *Unio valdensis* in some brown irony concretions. The bed descends to the beach 200 yards west of Ship Ledge, and the cliffs above it consist of red and green marls with several bands of hard sandstone, liable to rapid variations in thickness.

It may be observed here that the whole of the Wealden strata of the Isle of Wight are extremely irregular, and that the thick beds of sandstone which form conspicuous objects in the cliffs

---

[*] *Quart. Journ. Geol. Soc.*, vol. xxxv. p. 752. 1879.
[†] Geological Excursions round the Isle of Wight. 3rd Ed., p. 226.

occasionally thin out rapidly, even within short distances. This
may be observed in the case of the bed of greenish sandstone
seen in the cliff near Barnes Chine, which is reduced to a
thickness of 3 feet where it reaches the shore.

FIG. 5.

Sketch of Wealden Beds between Brixton Chine and Barnes Chine.

A. Variegated Marls.
B.B. Sandstones.
C.C. Red and purple Marls.

The skull of *Vectisaurus valdensis,* a Wealden Dinosaur, was
found by Mr. Hulke lying on the cliff-foot, 300 yards east of the
flagstaff near Brixton (Grange) Chine.*

Barnes Chine presents a section of red and mottled blue marls.
At the top of the eastern bank of the chine, a bone-bed, containing
also much lignite, was observed by J. Rhodes, the fossil
collector to the Survey.

We have now passed over a thickness of 562 feet of strata, and
at a point on the beach about 30 yards west of Cowleaze Chine
we reach the junction of the variegated beds and the Wealden
Shales. The nature of the junction may be gathered from the
following section which commences with the thick sandstone so
conspicuous in the chine and the long dip-slope of Barnes High.
The details were obtained from various points in the cliff below
this hill. The section of the same beds as seen at Cowleaze Chine
is given on p. 15.

---

* *Quart. Journ. Geol. Soc.,* vol. xxxv. p. 421.   1879.

Descending Section between Cowleaze and Barnes Chines

	Ft.	In.
Sandstone, very hard where washed by the waves, with nodules and veins of iron-pyrites and pebbles of clay. *Cyrena* abundant	2	0
Yellow sand and soft bright-yellow sandstone, current-bedded, and ripple-marked; carbonaceous in places	19	0
Grey and black shales, the upper part interlaminated with much sand in Cowleaze Chine; a band, crowded with *Paludina* and *Unio* near the top, and another with *Cyrena* and *Paludina* near the bottom	19	0
White sand and clay, with lignite	2	6
Current-bedded white rock	2	6
Reddish-blue sand and clay, with bone-fragments (*Hypsilophodon* Bed)	3	0
Red and variegated marls	44	0
White and yellow sand with tree-trunks, passing westwards into a sandstone 15 feet thick, and then splitting up and fingering out among red marls near the top of the cliff	9	0
Blue and purple marls, &c. (*see* p. 15).		

This locality has long been celebrated for its reptilian bones. In 1849, according to Mr. Hulke,[*] a block, containing a considerable portion of a reptilian skeleton, was found on the shore about 100 yards west of Cowleaze Chine. The skeleton was described as a young *Iguanodon Mantelli* by Professor Owen.[†] Another specimen was discovered and described under the same name by the Rev. W. Fox in 1868.[‡] These fossils were afterwards proved by Professor Huxley to be the bones of a new Dinosaurian, to which he gave the name *Hypsilophodon Foxii*.[§] Subsequently a great part of the skeleton of the reptile was exhumed by Mr. Hulke from the same stratum.[*] The bed, which rests directly on the variegated marls, forms the floor of Cowleaze Chine, and rises to the top of the cliff near Barnes High.

In 1874 the tibia and humerus of a reptile (probably *Hylæosaurus*) from the same locality were described by Mr. Hulke. The bones occurred " somewhere in the mottled purple and grey clays, therefore in the beds west of Cowleaze Chine, below the *Hypsilophodon*-bed."[‖]

The beds above the thick sandstone of Cowleaze Chine consist almost entirely of shales. Cypridiferous paper-shales, bands of ironstone and limestone, with layers of calcite, or " beef," and containing *Cyrena*, *Paludina*, and small oysters, occur at various horizons. (Plate III.) *Vicarya strombiformis* also, associated with *Cyrena*, is found in crowds at 1, 12, and 30 feet from the top, the appearance of hand-specimens with the two shells being precisely similar to that of specimens in the Museum of Practical Geology from the lowest beds of the Wealden at Burwash Wheel, near Hastings. The total thickness of the Wealden Shales of Atherfield is 192 feet.

* *Quart. Journ. Geol. Soc.*, vol. xxix. p. 532.   1873.
† Palæontographical Society's Publications.
‡ *Rep. Brit. Assoc.* for 1868 (Sections), p. 64.
§ *Quart. Journ. Geol. Soc.*, vol. xxvi. p. 3.   1870.
‖ *Ibid.*, vol. xxx. p. 516.   1874.

The uppermost bed of *Vicarya*, at one foot below the base of the Lower Greensand, contains a Cyprid, not previously noticed, and now described by Prof. Rupert Jones as *Cypris cornigera* (Fig. 6, 1). In the same pieces of shale with it there occur *Metacypris Fittoni*, and fish-bones.* Another new ostracod, described by Prof. Rupert Jones (*op. cit.*) under the name *Candona Mantelli* (Fig. 6, 2), occurs at 80–84 feet below the Perna Bed in the cliff between Shepherd's Chine and Atherfield Point. It is associated with *Metacypris Fittoni* (Mantell), small ; *Cypridea spinigera* (Sow.), young individuals ; *Cypridea tuberculata* (Sow.) ; *Cypridea valdensis* (Fitton) ; *Cyrena* ; and *Paludina*.

FIG. 6.

Cypris cornigera, Jones, and *Candona Mantelli*, Jones.

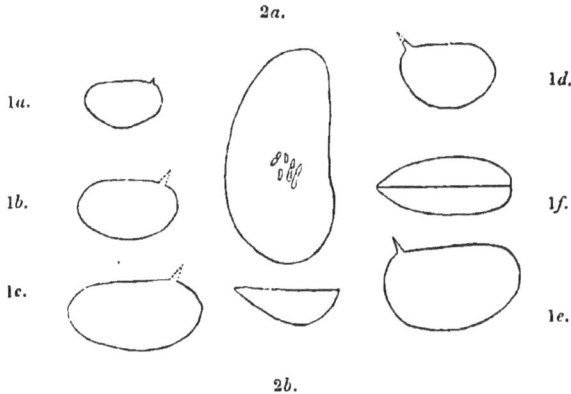

Fig. 1a. A right valve of a young individual.
„ 1b. A right valve of medium growth.
Cypris cornigera, Jones. „ 1c. A long and low (narrow) variety ; right valve.
„ 1d. A short and high (broad) variety ; left valve.
„ 1e. The largest specimen ; outline of left valve.
„ 1f. The largest specimen ; outline of edge-view of carapace.

Candona Mantelli, Jones. Fig. 2a. Outline of a right valve. (Anterior end placed upwards.)
„ 2b. Outline of the end view of the valve.

Magnified 20 diam. ; drawn by Mr. C. D. Sherborn, F.G.S.

Descending Section of the Wealden Beds from Atherfield to near Brook. (*See* Plates II. and III.)

	Ft.	In.
Perna Bed (Atherfield Point).		
Shales, with bands of *Vicarya*, 1 foot and 12 feet from the top, and Cyprids in the lower part	15	0
"Beef-bed"	0	2
Shales with Cyprids	8	10
Shales	6	0

* *Geol. Mag.* for 1888, p. 535.

		Ft.	In.
Wealden Shales.	Pale blue ironstone, with *Vicarya*, *Ostrea*, &c. abundant	0	2–3
	Shales	4	3
	Band, with Cyprids and fish-remains	0	0½
	Shales, with impersistent ironstone	9	6
	Lenticular sand and sandy shale	0 to 1	0
	Shale, with fish remains at the base	1	6
	Shale, with impersistent bands of ironstone, and bands of sand with ferns; Cyprids abundant in lower part	35	0
	Shales	14	0
	Dark limestone weathering red	0	0½
	Shales, with *Candona Mantelli*, Jones	9	0
	Shales, with a band containing *Unio*, *Paludina*, and Cyprids near the middle, and sandy beds, containing ferns, in the lower part	40	0
	Sandstone of Cowleaze Chine and Barnes High, massive, with bands of *Cyrena*	8	0
	Sandstone of Cowleaze Chine and Barnes High, thin-bedded, with shale	13	0
	Blue shales,* with *Unio* and *Paludina* in the top, and *Cyrena* and *Paludina* near the bottom	19	0
	White sand and clay*	2	6
	White rock	2	6
	Red sand, with bones (*Hypsilophodon Bed*)	3	0
	Red and mottled marls, rocky and ripple-marked at the top	44	0
	White and yellow sand, with fragments and large trunks of lignite, passing westwards into sandstone, and splitting up and dying away before reaching the top of the cliff	9	0
	Pale blue clay, becoming purple downwards	29	0
	Hard green bed, containing lignite and bones (seen in the top of Barnes Chine)	2	0
	Deep-red marls	6	0
	Purple and mottled marls	35	0
	Sandstone, with clayey beds (crosses Barnes Chine)	13	0
	Deep-red marls, purple below	28	6
	Conglomeratic grit, with an occasional pebble of quartzite, or of sandstone	3	0
	Pale mottled clay	14	6
	Green and white clays, with lignite	3	0
	Purple mottled marls	9	0
	Deep-red marls	13	0
	White sandy bed	3	0
	Pale purple and mottled marls	21	6
	Fine white sandstone (crosses the bottom of Ship Chine)	4	0
	Mottled marls	25	6
	Black bed of Brixton Chine; lignite, bones, *Unio valdensis*	2	6
	White sandy marl	3	0
	Mottled red marls of Brixton Chine, with a lignite bed near the middle	94	0
	Green sandy bed, with bones	2	0
	Red and white sandstone in beds of 1 to 3 feet, with partings of marl, and pockets containing shale and sandstone fragments; a band of gravel of sandstone fragments, 3 inches thick, at the base, with fragments of bones	17	0
	Mottled marls	49	0
	Pebbly band, lignite and pebbles of sandstone (top of east bank of Chilton Chine)	2	0
	Red and mottled marls	23	0
	Current-bedded sandstone (near the bottom of Chilton Chine) about	12	0

* These beds give a slightly differents ection in passing from Cowleaze Chine Barnes High. See p. 13.

	Ft.	In.
Mottled marls - - - - - -	28	0
Purple marls, with white concretions - - - -	4	0
Red marls passing down into white sandstone, with partings of marl, current-bedded in large sweeping curves - -	9	0
Massive sandstone, bands of bone and sandstone breccia running irregularly through; 6 to 18 inches of gravel at base, with bones. This bed thins away westwards, and is last seen at Sedmore Point - - - -	18	0
Deep-red and purple marls (at Sedmore Point) - -	20	0
Current-bedded sandstone of Sedmore Point - - -	8	0+

SANDOWN BAY.

The Wealden formation occupies a mile and a half of coast in this bay, and extends inland for a little over a mile. The axis of the anticline, which has already been described, lies nearly abreast of the stone fort, and trends a little north of west, in a direction parallel to the range of Brading and Bembridge Downs. The southern side of the anticline is entirely concealed by buildings on the cliff, and by sand on the fore-shore. The first exposure on the northern side is met with at the east end of the groins, where mottled clays with bands of sandstone form gentle undulations, with a general tendency to dip to the north-east. A short distance further on the dip increases to 11°, and finally to about 20° to the north-east, before the Wealden beds are lost to sight below the Lower Greensand of Redcliff.

The Wealden series is divisible here as in Brixton Bay into a lower group of variegated clays, and an upper group of fossiliferous shales. The lower group forms the low cliff or bank which extends as far as Yaverland Fort. It consists of mottled red, purple, and white marls, but is much obscured by slipping.

The fort stands on a low escarpment formed by a bed of sandstone about 8 feet thick; possibly the same bed that forms the corresponding feature of Barnes High in Brixton Bay, for the base of the blue fossiliferous shales is found at about the same distance below it in the two localities. This sandstone is seen again in the road-side south of Yaverland, and in a sand-pit 300 yards south-west of Sandown Farm. There it exceeds 18 feet in thickness, and dips to the south-west at 9°.

The details of the beds above and below the sandstone in the cliff are as follows :—

	Ft.
Fine black shale, Cyprids very abundant.	
Blue sandy shale, with lines of brown grit - -	10
Sandstone, about - - - - -	8
Blue shale, base not seen - - -	10
Blue fossiliferous shales, not well seen, about - -	30
Purple and mottled marls.	
	58

The beds above these are much obscured by slips, but can be seen to consist of shales of the usual type of the upper group, without any of the purple variegated marls. The junction with the Lower Greensand can be exposed by digging, as will be described, but the top beds of the Wealden are not clearly seen. The details in the following section are therefore quoted from Professor Judd's paper on the Punfield Formation.[*]

Lower Greensand.

		FT.	IN.
Blue paper-shales - - - - - -		0	9
,, ,, light-coloured and pyritic -		1	0
Dark-coloured paper-shales (with *Cypridea valdensis*), and several layers of nodular ironstone - - -		4	0 ?
" Beef " - - - - - - - -		0	1
Limestone, crowded with *Cyrena* and a few oysters -		0	6
" Beef " - - - - - - - -		0	2–3
Finely laminated pyritic clay - - - -		0	9
Ferruginous band, almost entirely made up of shells (oysters and small univalves) - - - -		0	3
Other beds of dark blue laminated shales, with occasional beds of limestone, imperfectly exposed; seen to 30 or 40		0	0

(bracketed as "Wealden Shales")

The total thickness of the Wealden Shales, as estimated from the breadth of outcrop and the dip, is about 170 feet.

The same assemblage of fossils occurs here as in Brixton Bay. Fragments of the thin bands of limestone containing *Paludina* and *Cyrena* may be found in abundance upon the beach, together with pieces of lignite, while the Cyprids occur in profusion in certain bands of finely laminated paper-shales. A pelvis and the external metacarpal bone of the right foot of Iguanodon have been discovered in the sandstone below Yaverland Fort.[†]

Vertebræ, a femur, and ribs of the same animal are stated by Mantell to have been found near the same spot.[‡]

A femur was found also in the low cliff of Weald Clay to the west of Sandown Fort, a part that is now obscured. The beds are stated to have dipped slightly to the west.[§]

It will be observed that, if the sandstone under Yaverland Fort is the same bed that forms Barnes High, the horizon of the Hypsilophodon band is clearly fixed in Sandown Bay; but no remains of this reptile have yet been discovered. Mantell notes that some " grey sandstone, interspersed with clay," near Yaverland Fort, " several cones of a plant allied to the *Zamiæ*, mixed with fragments of lignite, have been discovered." [||]

For further particulars concerning the fossils the reader is referred to the fossil lists on p. 258.

---

[*] *Quart. Journ. Geol. Soc.*, vol. xxvii. p. 220. 1871.
[†] Rev. Dr. Buckland. *Proc. Geol. Soc.*, vol. i. p. 159. 1826-33.
[‡] Geological Excursions round the Isle of Wight, 3rd Ed., p. 98.
[§] T. F. Gibson. *Quart. Journ. Geol. Soc.*, vol. xiv. p. 175. 1858.
[||] Geological Excursions round the Isle of Wight, 3rd Ed., p. 99.

CHAPTER III.

LOWER GREENSAND OR UPPER NEOCOMIAN.

INTRODUCTION.

THIS formation occupies the greater part of the southern or Cretaceous area of the Isle of Wight, and forms important escarpments, such as that which runs from Compton Bay by Brook, Mottistone, and Brixton, or the succession of bold shoulders which dominate the upper parts of the Medina and Yar valleys, and on one of which Godshill is situated. But the most complete sections are to be obtained in the four coast-sections of Compton Bay, Atherfield, Shanklin, and Redcliff at the east end of Sandown Bay.

At Redcliff the thickness of the Lower Greensand is about 600 feet; at Atherfield it has increased to over 800, but at Compton Bay, about 16 miles west of Redcliff, the thickness is reduced to 399 feet. Lastly, at Punfield, 20 miles west of Compton Bay, it is no more than 198 feet. It would seem then that the direction in which the strata thicken most rapidly lies a little east of south.

The Lower Greensand of Atherfield was made the subject of the most exhaustive examination by Dr. Fitton in the years 1824–47. The results of his work were embodied in a large number of papers, but chiefly in a paper read before the Geological Society in 1845.* Not only was the thickness at Atherfield found to be greater than elsewhere in the Island, but fossils were very much more abundant. The rich collection made by Dr. Fitton showed that the fauna of the Lower Greensand was both distinct from that of the Upper Cretaceous Rocks above, and possessed nothing in common with the Wealden Shales below, there being in fact a complete palæontological break at the base of the formation. This is the more noticeable in that the lower beds of the Lower Greensand are, like the Wealden Shales, of a clayey character.

Later observations have shown that this complete contrast in the fauna was caused by an abrupt change in the physical geography of the area in which the Lower Greensand was distributed, and was preceded by some erosion of the Wealden Beds. The abruptness of the change is indicated by the following evidence:—

(1.) The division of the Lower Greensand from the Wealden Shales is everywhere absolutely sharp, so much so that the two can be cleanly separated by a knife-blade.

---

* *Quart. Journ. Geol. Soc.*, vol. iii. p. 289. 1847. References to his other papers will be found in the Bibliography at the end of this book.

(2.) The base of the Lower Greensand is a thin line of coarse grit, containing rolled fragments of fossils (Ammonites and other marine forms) which must have been derived from some marine beds, exposed outside the limits of the freshwater Wealden Beds, together with an occasional pebble of sandstone larger in size, and resembling the sandstones which are interstratified in the Wealden Beds. There are also in this grit numerous broken bones, teeth, and scales of fish, and at Atherfield it contains fragments of *Vicarya strombiformis*, the gasteropod which is so abundant in the top of the Wealden Shales at this spot. The fragments occur only in the grit, which is about three quarters of an inch thick, and have doubtless been washed out of the surface of the Wealden Shales. At Punfield this grit has yielded a well-rounded pebble of white silicified wood, precisely similar to the wood in the Lower Purbeck Beds.

(3.) The Wealden Shales, where the junction is exposed, often present the appearance of having been disturbed and broken up for a distance of a foot or two below the base of the Lower Greensand.

(4.) In Wiltshire the Lower Greensand overlaps the Wealden Beds so rapidly as to indicate an actual unconformity.[*] As a result of this overlap it passes westwards on the Upper Oolites, a fact which provides a clue to the source of the rolled fossils of marine species, which occur as pebbles at the base of the Perna Bed in the Isle of Wight.

As far as the Isle of Wight is concerned, however, there is not sufficient evidence to establish an unconformity between the Lower Greensand and Wealden Beds. That the bedding of the two is strictly parallel is proved by the persistence of the Wealden Shales at the top of this formation, not only in the Isle of Wight, but both to the east and west on the main-land. The change of sediment is such as might have been produced by the sudden conversion of a partially land-locked estuary or lake into a bay open to the sea, whether by subsidence or by the washing away of a barrier.

On this theory we must suppose that a Lower Greensand sea with its proper fauna was in existence at the time when the Wealden Shales were still being deposited in the land-locked area. This supposition is in accordance with the sequence observed in the north of England. For the Upper Neocomian deposits of Yorkshire, as shown by Professor Judd, contain the same fauna as the lowest of the Isle of Wight Neocomian beds, namely, the Atherfield Clay. We are thus compelled to suppose the Middle and Lower

---

[*] Geology of England and Wales, by H. B. Woodward, 2nd Ed. 1887, pp. 352, 354, 375.

Neocomian strata of the north to have been contemporaneous with a part of the Wealden Beds of the south, the one having been deposited in an area open to the sea, the other in a basin that remained land-locked until a later part of the Neocomian period. The history of the great freshwater deposits, of which in the Isle of Wight we have only the upper part, is beyond the scope of this Memoir, and will be treated more fully in the General Memoir on the Cretaceous Rocks.

The Lower Greensand of the Isle of Wight is divisible into four groups, capable of being traced throughout. But at Atherfield, where they are most fully developed, Fitton made six principal divisions and sixteen minor groups. In the following table Fitton's groups are compared with those adopted in this Memoir, and with those in use in the Weald of Kent and Sussex.

The only point in which a material difference between the two classifications exists, occurs in Fitton's Division F. A portion of this has now been separated under the name of Carstone, while the lower part of it is grouped with E., to which it is lithologically allied, under the name of Sand-rock Series. The lowest member of Fitton's Group XV., a thick bed of clay, is taken as the top of the Ferruginous Sands, in consequence of the similarity of the deposit to a band of shale which forms the top of the Sandgate Beds at Pulborough.* The Perna Bed, though palæontologically of the greatest interest, is too thin to be separately mapped. The names used have been adopted as far as possible from those who first investigated the beds.

The term Shanklin Sands was proposed by Fitton† for the whole of the Lower Greensand to avoid confusion between this formation and the Upper Greensand, and was used in this sense by Martin. But subsequently the name became restricted to the upper beds of the Lower Greensand, and having been made to include a varying proportion of the deposit by various authors, and its original meaning, as intended by its author, having been lost, it has been thought better to abandon it. The name Vectine, also proposed by Fitton, and subsequently modified into Vectian by Mr. Jukes-Browne,‡ has never come into general use. (*See also* p. 2 on the use of Vectian for the Fluvio-Marine Series.)

Fitton, 1845. (Atherfield.)	Geological Survey, 1887. (Isle of Wight.)	Geological Survey. (S.E. England.)
XVI. Various sands and clays F	Carstone.	Folkestone Beds.
XV. Upper clays and sandrock E	Sand-rock Series.	

---

* Geology of the Weald (Geological Survey Memoir), p. 136.
† *Ann. Phil.*, 2, viii. p. 461.
‡ *Geol. Mag.* for 1885, p. 298.

Fitton, 1845. (Atherfield.)		Geological Survey, 1887. (Isle of Wight.)	Geological Survey. (S.E. England.)
XIV. Ferruginous beds of Blackgang Chine -			
XIII. Sands of Walpen Undercliff - -			
XII. Foliated clay and sand			
XI. Cliff-end sands -			
X. Second Gryphæa bed -	D	Ferruginous Sands.	Sandgate Beds.
IX. Walpen and Ladder sands - -			Hythe Beds.
VIII. Upper Crioceras group			
VII. Walpen clay and sands			
VI. Lower Crioceras group			
V. Scaphites group -			
IV. Lower Gryphæa bed -			
III. The Crackers - -	C		
II. The Atherfield Clay -	B	Atherfield Clay.	Atherfield Clay.
I. Perna Mulleti bed -	A		

These divisions pass one up into the other, without any sharp line of demarcation, except in the case of the Sand-rock Series and the Carstone. Here the boundary is rather more sharply defined, and can be followed with little difficulty through the central parts of the Island. The Carstone everywhere passes up into the Gault.

In describing the Lower Greensand it will be convenient to take the localities in order from west to east as before, commencing with Compton Bay.

COMPTON BAY.

The base of the Lower Greensand in Compton Bay is not seen *in situ* in consequence of a great slip of Atherfield Clay and of the upper Wealden beds described on p. 8. It is not difficult, however, to find among the ruins masses which show the junction as clearly as if it were in place. The following details were noted in a fallen mass :—

		FT.	IN.
Atherfield Clay -	Clay, mottled red and grey.		
Perna Bed -	Calcareous and ferruginous grit, with *Modiola*, &c. -	1	0
	Green sandy clay - -	0	9
Wealden Shales -	Blue paper-shale, broken up into a breccia for a distance of about 1 foot below the base of the Lower Greensand - - -	3	0 +

In every case where the junction was exposed, the same brecciated appearance in the surface of the Wealden Beds was observable, sometimes extending to a depth of 2 feet into the Wealden. There can be little doubt that it indicates that a certain amount of erosion of these beds took place before the Lower Greensand was deposited. In addition to the particles of quartz which give to the Perna Bed its gritty character, there are in it small rolled phosphatic nodules.

The Atherfield Clay, excepting the top beds, can be seen only as a flowing mass of pale-blue clay, with phosphatic concretions. Its thickness consequently is difficult to determine, but so far as can be judged it is like the other beds considerably thinner than at Atherfield, and may be estimated at about 60 feet.

The succeeding beds are clearly exposed, and are shown in descending order in the following detailed section :—

Compton Bay.

		Ft.	In.
Carstone, 6 ft.	Brown sand, with 3-inch pebble-band at the base, containing rounded quartzite pebbles up to ¾ inch in diameter, some phosphatic pebbles, and many pieces of wood. Cylindrical phosphatic nodules also occur - - - - -	6	0
	Blue clay - - - - - -	2	6
	Pebble-band with quartzites, &c.. 0-3 inch ⎫		
	Grey and greenish sand, with a layer of pyritised wood 8½ feet from the top, and scattered frag- ⎬13 ments near the top, about 12½ feet - - ⎭	13	0
Sand-rock Series, 81 ft. 6 ins.	Pebble-band, as above, 6 inches - - -		
	Bright-yellow sand, with an irony seam at the base - - - - - -	10	0
	Clean white sand and blue clay, interbedded in wavy laminæ, and giving out copious chaly- beate springs ("foliated series") - -	56	0
	Clayey grit, weathering green, with a band of quartzite pebbles, 5 inches thick, at the base -	26	0
	White sand like gannister - - - -	2	0
	Dark sand and clay intermixed, with much vege- table matter in the upper part, and looking like a rootlet-bed* - - - - -	3	0
	Band of small quartzite pebbles - - -	0	3
	Sand like gannister - - - - -	5	0
	Very black and sooty-looking sand or silt -	7	8
	Lighter do. striped - - -	10	0
	Band of soft yellow rolled phosphatic nodules, with some quartzites - - - -	0	1½
	Lighter coloured and striped "sooty" sand, with many small soft yellow phosphate pebbles near the base - - - - - -	4	0
	"Foliated" sand and clay as above, passing down into paper-shale - - - -	5	8
Ferruginous Sands, 25] ft. 6½ ins.	†Very green gritty sand, with hard pale-yellow phosphates, some cylindrical, some rounded -	3	6
	Brown sandstone - - - - -	1	2
	Green grit as above - - - - -	1	6

* This and the other dark sands were tested by Mr. C. Tookey for manganese, but found to contain none. The colouring matter appeared to be carbonaceous.

† This and the seven beds following it crop out in the west side of Compton Chine. Its green colour is due to an abundance of grains of glauconite. See p. 255 for an analysis of a specimen from this bed.

	Ft.	In.
Brown sandstone · - - - -	1	2
Green and grey silty sand, with fucoidal markings	1	0
Brown sandstone - - - -	1	0
Green and grey silty sand, with fucoidal markings	2	0
Brown sandstone, with small pebbles and pieces of lignite scattered throughout; an impersistent band of silty sand in the middle -	42	0
Green silty sand, passing down - - -	11	6
Clay - - - - - -	3	0
Brown and red grit, made up largely of rounded grains coated by iron oxide; forms the cliff east of Compton Chine - - -	54	0
Yellow sand, much fretted by the weather in the upper part - - - - -	20	0
Pale-green sandy clay, with light-grey nodules containing fossils, and passing down into -	10	0
Yellow sand, clayey in parts - - -	15	0
Grey silty sand, with bands of soft yellow sandstone below - - - - -	21	0
Atherfield Clay. 60 ft. Pale-blue clay with Perna Bed at the base; estimated at - - - - -	60	0
	399	0½

The precise correlation of the beds in this section with those
of Atherfield is impossible. As will be seen subsequently, the
beds are not only very much thinner, but have changed their

FIG. 7.

The Sand-rock Series in Compton Bay.

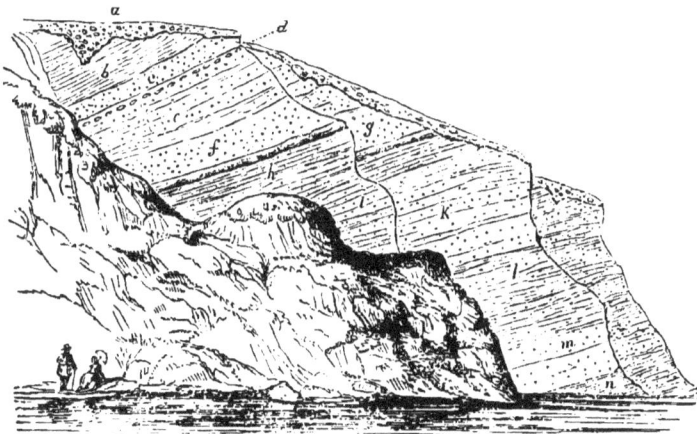

a. Soil and gravel.	g. Ochry band.
b. Gault.	h, i, k, l. Blue clay and white sand
c. Carstone, or ferruginous grit.	interlaminated in varying propor-
d. Pebble sand.	tions.
e. Blue clay and sand, with small pebbles and lignite.	m. Chiefly sand, throwing out much chalybeate water.
f. Bright yellow sand.	n. Very green and gritty clay.

characters. Fossils are also comparatively scarce in Compton Bay. Dr. Fitton identified a "mass of brownish clay and sand" which lies next above the Atherfield Clay, as the Lower Lobster Bed, or the lowest part of the Crackers sub-division of Atherfield, and a prominent portion in the lower part of the brown and red grit as the Lower Gryphæa bed of Atherfield.

The upper beds present a general resemblance to those which form the upper part of Blackgang Chine, though they are very much thinner, and contain none of the bands of sand-rock which form so distinctive a feature in that chine. The abundance of water strongly impregnated with sulphate of iron, which issues from them, is a noticeable feature. As will be seen, the chalybeate spring near Blackgang issues from the same beds. The annexed wood-cut (Fig. 7) represents the general arrangement and appearance of these upper beds in the cliff.

ATHERFIELD.

The Lower Greensand here attains a greater development than in any other part of the Isle of Wight, and has yielded a rich suite of fossils. Its thickness has been variously estimated at 808 feet by Dr. Fitton, at 833 feet by Ibbetson and Forbes,* and at 752 feet 11 inches by Mr. Simms. The description of it will be taken from west to east, that is in ascending order of the strata.

The Atherfield Clay and Perna Bed.

After leaving Compton Bay the Perna Bed is not seen again till we reach Cowleaze Chine. It is here well exposed under the

Fig. 8.

Perna Mulleti, Desh.

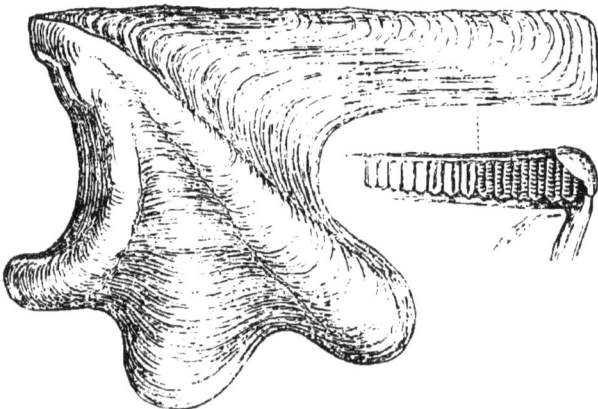

---

* The thickness given by these authors is 843, but the total obtained by adding up the figures given in their table is only 833.

bridge by which the military road crosses the chine. It re-appears in the top of the cliff 300 yards south of the chine, and slants down thence to the beach 150 yards east of Atherfield Point, the dip, as calculated from the heights and distances on the Ordnance Map, being 1 in 24, or about 2½°.

The section of this bed in the cliff is frequently obscured by the slipping of the Atherfield Clay, but is now (1887) admirably exposed 250 yards north-west of the point.

Section of the Perna Bed near Atherfield Point.

		FT. IN.
Perna Bed {	Calcareous and ferruginous stone, with many fossils - - - - -	2 6
	Blue fossiliferous clay, based by a gritty seam with phosphatic nodules and fish-remains. Panopœa occurs in the clay in the position of growth - - - - -	2 7
Wealden Shales (see p. 14).		
		5 1

The brecciation of the top bed of the Wealden, which has been described at Compton Bay, is not observable here, but the line of demarcation between the blue purely argillaceous shale, with its numerous bands of fresh or brackish water shells, to the rather sandy clay with numerous marine forms, is sufficiently striking. The gritty base of the clay, moreover, points to some erosion having taken place. The grit varies in thickness rapidly, and is sometimes absent. Dr. Fitton, in allusion to it, remarked that " the remains of fishes, chiefly teeth and small fragments of

FIG. 9

Exogyra sinuata, Sow.

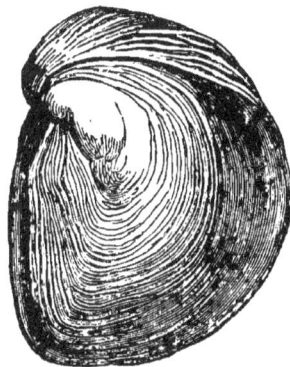

bones, are mixed with coarse quartzose gravel at the bottom of this bed [the Lower Perna Bed] ; and occurring thus immediately

over the Wealden, or even in contact with it, it is not unreason-
able to suppose that the fish were killed by the change from fresh
water to salt."* Remains of fishes were identified by Sir Philip
Egerton, and a small Saurian phalanx by Professor Owen.

The Perna Bed was so named by Dr. Fitton in consequence of
its containing great numbers of *Perna Mulleti*, Desh. (Fig. 8),
which has not been found in any of the other beds. *Exogyra*
(*Gryphæa*) *sinuata* also occurs in abundance and of a large size.
The rest of the fossils will be found distinguished in the fossil
list on p. 261.

The Atherfield Clay, which was also named by Dr. Fitton, is
of a pale-blue colour, and, unlike the Wealden Shales, is devoid
of lamination; it contains numerous flat concretionary nodules.
" Among the fossils the most common in the lower portion is
Pinna robinaldina, d'Orb. Ammonites are not unfrequent; and
the remains of a turtle . . . were obtained here." (Fitton,
op. cit., p. 296.) The thickness of the Atherfield Clay is about
60 feet, according to Fitton, but 99 feet according to Ibbetson
and Forbes, who include the Lower Lobster Bed in the subdivision.

The Lower Lobster Bed is an impure fuller's earth, abounding
in remains of *Meyeria* (*Astacus*), from which fossil it takes its
name. It is now grouped with the Atherfield Clay on purely
lithological grounds, the natural base of the ferruginous sands
which constitute the overlying group occurring above and not
below the Lower Lobster Bed. The thickness of the bed is 25
feet 6 inches according to Fitton, 29 feet according to Ibbetson
and Forbes.

The Ferruginous Sands.

This division of the Lower Greensand attains a thickness at
Atherfield of about 520 feet by Fitton's measurements, or 508 by
those of Ibbetson and Forbes.

The lowest bed of the group, bed No. 5 of Fitton, and named
by him the Crackers, from the noise made by the waves in the
slight rocky prominence formed by the rock, consists of coarse
grey or brown sand, about 20 feet in total thickness. It contains
two layers of ferruginous and calcareous concretionary masses,
abounding in fossils. Some of the masses in the lower layer
" are 6 or 7 feet long, and a foot to 18 inches in thickness, and
almost composed of *Gervillia anceps* (*ariculoides*), with *Trigonia
dædalea, Ammonites Deshayesii*, &c." (Fitton, *op. cit.*, p. 298.)
In the upper layer Dr. Fitton noted coniferous wood bored by
Teredo, and in the upper part of the sand, *Thetis*, a large *Astacus*,
and *Ammonites Deshayesii*. " In the lower part, great numbers
of *Panopæa* (*Myacites*) *plicata*, Sow., are found in it standing

---

* *Quart. Journ. Geol. Soc.*, vol. iii. p. 294 (1847).

obliquely upwards." *Pinna* occurs also in clusters. The prominence formed by this rock will be found at the foot of the cliff,
600 yards east of the Coastguard Station.

Fig. 10.

Panopæa plicata, Sow.

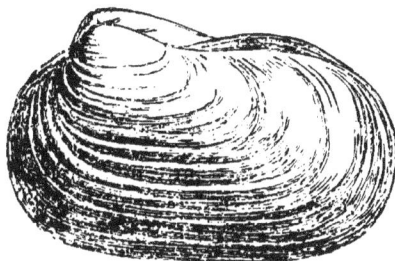

The overlying set of beds (forming the upper part of Fitton's
Crackers Group, Nos. 6–10) embraces a thickness of about 40
feet. It consists of brown clay, 16 or 17 feet thick, in the lower
part, and of clay mixed with sand in the upper part. The beds
are fossiliferous throughout, and are known as the Upper Lobster
Beds, from the occurrence in them of remains of *Meyeria* (*Astacus*)
vectensis.

Fig. 11.

Gervillia anceps, Desh.

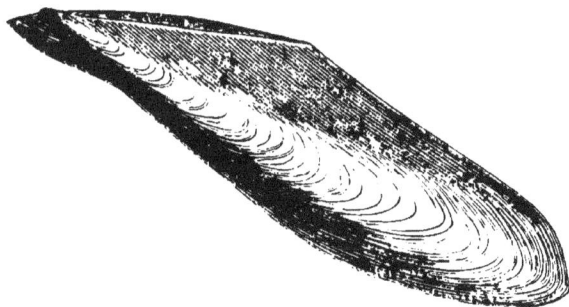

Group IV., or the Lower Gryphæa [Exogyra] Group of Fitton,
has at its base a bed of rust-coloured sand about 21 feet thick.
This is overlain by two feet of sand containing *Gerrillia* (*Perna*)
alæformis, but chiefly remarkable for the great abundance of
Terebratula sella, Sow., which, though ranging from the base of
the Lower Greensand to the top of the Ferruginous Sands
(Group XIV. of Fitton), is nowhere so numerous as here.

The bed with *Exogyra sinuata*, which comes to the shore under Atherfield High Cliff, and forms a reef running out through the beach about 350 yards west of Whale Chine, is next in succession. It is about 10 feet thick, the lower part consisting of brown and reddish sand with spherical grains of oolitic iron-ore, and containing *Pinna rubinaldina*, D'Orb., while the upper part forms the reef in which numerous large *Exogyræ* are conspicuous.

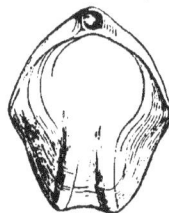

FIG. 12.

Terebratula sella, Sow

Group V., or the Scaphites group of Fitton, has a thickness of 50 feet 4 inches, and may be divided into three beds, the lowest of which is brown and rust-coloured sand about 20 feet thick, and containing large *Exogyra sinuata*, *Ostrea carinata*, &c., and at the bottom layers of *Serpulæ*, *Terebratulæ*, &c. Nodules in layers containing *Ancyloceras* (*Scaphites*) *gigas* and *A. Hillsii* lie next above this sand, and are succeeded in ascending order by about 27 feet of dark-grey sandy clay, with the large *Exogyra sinuata*, in the upper part. A reef containing conspicuous clusters of *Serpulæ* runs out from the cliff at this point.

Group VI., or the Lower Crioceras beds, contains several ranges of this fossil, imbedded in sand. The lowest range rises from the beach on the west of Whale Chine; the highest crosses the bottom of the chine. The group is 16 feet 3 inches in thickness.

Group VII., the Walpen clay and sands, consists of a dark-green mud at the bottom, about 27 feet thick, with nodules including *Exogyra* and *Ammonites Martini*, and of an upper division, clayey above and sandy below, about 33 feet thick, containing *Panopæa* (*Myacites*) *mandibula*, *Pinna robinaldina*, and a *Dentalium*. The clay-beds of this group form the undercliff, on to which Ladder Chine opens. They rise from the beach about 200 yards east of the chine, cross Whale Chine, and reach the top of the cliff 700 yards west of Whale Chine. Their position is always marked by the springs they throw out, except close to the east side of Whale Chine.

Group VIII., the Upper Crioceras beds, is 46 feet 2 inches thick, and contains four or more ranges of *Crioceras Bowerbankii*, with *Ammonites Martini*, *Gervillia solenoides*, *Terebratula sella*, and *Trigonia alæformis* (*T. vectiana*, Lyc.). The top bed of the group rises on the east of Walpen Chine, crosses Ladder Chine, and may be seen in the chasm beneath it. The whole group crosses Whale Chine also.

Group IX., the Walpen and Ladder sands, consists of greenish and grey sand, about 42 feet thick, with a layer of lenticular masses of dark olive-green stone at the base, containing numerous fossils. About 6 feet higher up is a thin band, consisting for the greater part of *Serpulæ*, apparently twisted together, associated with *Terebratula sella* and other fossils.

Group X., or the Upper Gryphæa group, includes about 16 feet* of sand, with some clay. In the lower 12 feet there are several ranges of *Exogyra sinuata*, and nodules with *Enallaster* (*Brissus*) and *Ammonites Martini*. The ferruginous matter of this bed is in some places distinctly oolitic, like that of Group IV. The upper part of the group is a greenish sand with *Exogyra sinuata*, this being the highest point in the Atherfield section which has yielded that species. Small fragments of vegetable remains (*Lonchopteris Mantellii*) occur not only in these beds, but nearly throughout the entire formation. In the lower part of this group they are associated with *Inoceramus*.

Group XI., the Cliff End sands, about 20 feet in thickness, consists of sands with a thin bed of clay with *Trigonia dædalea* in the lower part, and in the upper part of dark bluish and green sand, with many cylindrical stem-like and branching concretions, containing pyrites.

Group XII., the Foliated Clay and Sand, consists of thin alternations of clean greenish sand, with dark-blue clay, and much pyrites. The bed includes also lenticular masses of coarse current-bedded sand-rock. It is about 25 feet thick, and from its yielding nature forms an extensive undercliff on the west side of Blackgang Chine. But it is most clearly exposed to view on the buttress of rock which forms the south side of Walpen Chine, where, however, it can be reached from above only. The dip in this part of the section may be calculated by tracing this bed down to the beach. It amounts to 1 in 26, or a trifle over 2°.

In general character this group is closely allied to the Sand-rock series, and it was correlated by Dr. Fitton with a bed which has been taken as the base of that series at Shanklin. Traced inland this bed passes by Pyle, Corve, and Kingston, cropping out at the foot of a marked feature all the way (*postea*, p. 30), and thence striking westwards seems to die way in beds distinctly of the ferruginous type.

Group XIII., the sands of Walpen Undercliff, is about 97 feet thick. It has at its base a bed of loose white sand or sand-rock, about 10 feet thick, which rises to the top of the cliff on the south side of Walpen Chine. Above this bed, which he calls the First Sand-rock, Dr. Fitton noted the following in descending order :—

	Ft.	In.
Light green and yellowish sand, giving a bright-green streak under the pick	25	9
Brown sand with *Astarte Beaumontii, Pinna, Pecten,* and *Terebratula*	1	6
Moist greensand	12	6
Sand, based by a coarse gravel with pebbles of quartz and Lydian stone	29	8

Above these are brown sands with polished particles of iron-ore, and sands with beds of dark-green or black coherent mud.

---

* There are some slight discrepancies in this and other cases between the thicknesses given in the text and in the table of Dr. Fitton's paper.

Group XIV., the Ferruginous beds of Blackgang Chine, forms
the upward limit of the fossiliferous beds of the Lower Greensand. The beds appear above the shore at a point 600 yards
north-west of Rocken End, and form a vertical foot to the cliff
as far as Blackgang Chine. Here the undercliff formed by
Groups XII. and XIII. commences, and the harder beds of Group
XIV., rising slowly in the cliff, form a step between this undercliff
and a similar feature formed above by Group XV. The cascade
in the lower part of Blackgang Chine, which was ascertained by
Fitton to be 91 feet in height, is caused by the comparative hardness of the ferruginous bands of Group XIV. This group crops
out in the top of the cliff on the south side of Walpen Chine, and
strikes thence in a bold escarpment through Pyle, Corve, and
Kingston.

The details of the group are given by Dr. Fitton as below :—

Ferruginous Bands of Blackgang Chine.

	Ft.	In.
Ferruginous concretions, immediately above the cascade	1	0--6
Brown and yellow sand	5	0
Ferruginous concretions, with many vacant moulds of fossils, most abundant near Walpen High-Cliff	1	0
Sand, with fossils	7	0
Ferruginous sand-rock, with fossils	5	0
	19	6

The species found in this group can be identified in several
cases with those of the Perna Bed, at the very bottom of the
Lower Greensand. Among them may be named *Panopæa
plicata*, Sow., *P. neocomiensis*, D'Orb., *Corbula striatula*, Sow.,
Thetis minor, Sow., *Trigonia caudata*, Ag., *Pinna robinaldina*,
D'Orb., &c.

The next overlying bed, forming the lower member of Fitton's
Group XV., is a great mass of clay, between 35 and 40 feet thick.
It occupies the shore for a distance of 350 yards, first rising into
sight near a waterfall 200 yards north of Rocken End. It forms
a step in the cliff as far as Blackgang Chine, where it widens
out into an undercliff. The most convenient place for examining
it will be found from 500 to 600 yards west of Cliff Terrace, near
the top of the cliff, where the shale of which the bed largely consists
has been cut back by wind and rain into a broad shelf, entirely
bare of vegetation. This bed forms the top of the great division
of the Lower Greensand, which we have named the Ferruginous
Sands.

The Sand-rock Series.

This series, like the other beds of the Lower Greensand, attains
its greatest development in the southern part of the Island, its
thickness being 186 feet by Fitton's measurements, or 208 by
those of Ibbetson and Forbes, while at Compton Bay it amounted
to 81½ feet only. Here also it contains in their typical form
those beds of slightly coherent white or yellow quartz sand,

which form so conspicuous a feature in the upper part of Blackgang Chine, and to which the name sand-rock is singularly applicable. Three distinct bands of this deposit occur, namely, the beds referred to by Fitton as the fourth, third, and second sand-rock. The second or lowest occupies the beach from Rocken End for a distance of 200 yards northwards; but is partly concealed by slips of Chalk and Greensand. Thence it may be traced continuously to the top of the cliff 500 yards west of Cliff Terrace, where it is seen overlying the great clay-bed previously described. The third or middle bed, and the fourth at the top of the series, may be traced from the chalybeate spring to a point on the east side of Cliff Terrace, where they reach the top of the cliff.

The following descending section of the series was made in the neighbourhood of the chalybeate spring, 600 yards south-east of Southland House :—

Section of the Sand-rock Series near the Chalybeate Spring.

	FT.
Carstone (for details, see p. 57).	
Grey sand with wood, large concretions, and seams of clay; a line of quartz pebbles at the base - - - -	20
Grey and yellow sand interlaminated with clay -	7
Current-bedded yellow sand-rock, with wood; thins away southwards (4th sand-rock of Fitton) - - -	14
Laminated sand and clay, with wood; throws out the chalybeate spring - - - - - - -	22
A variable bed; contains clay with partings of sand, sometimes nearly all sand, and passes down into - -	16
White sand-rock (3rd sand-rock of Fitton) about - -	25
Variable sand and clay, with a line of nodules about the middle - - - - - - - - -	60
White sand-rock (2nd sand-rock of Fitton) - - -	20
	184

The interlaminated sands and clays in this section are identical in character with the " foliated bed " 56 feet thick of the Compton Bay section (pp. 22, 23), and like it throw out chalybeate water, derived doubtless from the decomposition of iron pyrites.

The Chalybeate or Sand-rock Spring was first noticed about the year 1800. It was found to flow at the rate of 100 to 150 gallons a day, and gave the following analysis[*] :—

16 ounces yielded :—
Carbonic acid gas, 3 cubic inches.
Solid ingredients, dried at 180°, 80·5 grains.

	GRAINS.
Sulphate of iron - - - - -	41·4
Sulphate of alumina - - - -	31·6
Sulphate of lime, dried at 160° - - -	10·1
Sulphate of magnesia - - - -	3·6
Sulphate of soda - - - - -	16·0
Chloride of sodium - - - -	4·0
Silica - - - - - -	·7
	107·4

Temperature, 51°. Specific gravity, 1·0075.

[*] Dr. Marcet, *Trans. Geol. Soc.*, Ser. 1, vol. i. p. 213. 1811.

From the chalybeate spring eastwards the Sand-rock series is almost entirely concealed by the slipped Greensand and Chalk of the Undercliff. The upper beds of the series are seen in a bold bluff between Rocken End and Knowles, and again in the lower part of the cliff below Niton. Here a white sandstone also is exposed above the beach, about 100 feet below, which seems to be the third sand-rock of Fitton. The last exposure occurs in Binnel Bay, where interlaminated sands and clays are exposed at the base of the cliff. From this point eastwards there is no rock seen in place till we reach Monk's Bay at Bonchurch. The description of the Carstone or uppermost sub-division of the Lower Greensand of this neighbourhood will be found on pp. 57, 58.

Sandown to Bonchurch.

The Atherfield Clay and Ferruginous Sands.

Though nearly the whole of the Lower Greensand is exposed in this coast section, the beds are not so conveniently situated for examination as at Atherfield, and have yielded far fewer fossils.

The Perna Bed and Atherfield Clay rise from the beach near Sandown Pier in a low cliff, but are concealed by buildings; nor is the former exposed now at low water, as seems formerly to have been the case. The overlying beds consist of green grey and brown sands, so far decomposed as to render the identification of the groups of Atherfield impossible. But specimens of *Crioceras* were found by Captain Ibbetson in a quarry, not now identifiable, near the shore between Small Hope Chine (the north end of Shanklin sea-wall) and the Barrack Hill, Sandown. The horizon would seem to correspond approximately with that of the Crioceras ranges of Whale Chine. Some of the sands north of Little Stairs Point are very dark-coloured, and contain small fragments of wood impregnated with pyrites.

At Little Stairs Point a fault is clearly exposed, a rare circumstance in the Isle of Wight. The fault ranges about west-north-west, and throws the beds down to the south. Soon after passing this fault the beds assume a horizontal position, or nearly so, and we meet with the first marked bed in the section. It consists of ferruginous sandstone, studded with clusters of *Exogyra sinuata* and *Ostrea frons* (= *O. prionota*,) and identified by Fitton (*op. cit.* p. 317), with part of his Second Gryphæa Group X. Above it occurs a bed composed of alternations of dark slaty clay with greenish sand, which Fitton recognised as his Group XII. At the top of the cliff is an iron sand.

Chalybeate water issues from these strata. The spring known as Shanklin Chalybeate Spa was first noticed by Dr. Fraser, physician to Charles II. It has been analysed by Dr. A. H. Hassall with the following result :—

Chalybeate Spa, Shanklin Esplanade.

Chemical Composition.		Combined as follows :—	
	GRAINS.		
Total residue	23·46 per gallon.	Carbonate of lime	7·66
Lime	5·64 ,,	,, magnesia	2·35
Magnesia	1·90 ,,	,, protoxide of	
Potash	0·25 ,,	iron	2·13
Soda	2·01 ,,	Sulphate of lime	3·28
Sulphuric acid	2·81 ,,	,, magnesia	1·32
Chlorine	3·23 ,,	Chloride of potassium	0·40
Iron	1·03 ,,	,, sodium	3·04
Silica	1·40 ,,	,, magnesium	0·85
Nitrogen as nitrates		Silica	1·40
and nitrites	—	Volatile and combustible	
Free ammonia	—	matter	0·14
Organic nitrogen	0·01 ,,		
Hardness, 9·30.			

The horizontality of the beds (excepting in a very gentle anticline south of Little Stairs fault) is maintained as far as Shanklin Chine. Here a south-south-westerly dip sets in, which gradually brings the upper strata down to the beach in succession, the angle of dip, as calculated from the heights on the Ordnance Map, amounting to 1 in 30, or a trifle less than 2°.

The strata last described contain oolitic iron ore, and are identified by Fitton with a part of his Group XIII. They sink below the beach on the south of Shanklin Chine, and are succeeded upwards at a few feet distance by a richly fossiliferous bed, in which Fitton obtained *Vermicularia, Serpula, Waldheimia (Terebratula) pseudojurensis,* Leym., *T. sella,* Sow., *Rhynchonella sulcata,* Park. (*T. multiformis,* Fitton), *Rhynchonella gibbsiana,* Sow. (*T. gibbsiana,* Fitton), and *Anomia, Exogyra, Pecten, Lima,* &c. Ten feet and eighteen and a half feet higher up respectively are two ranges of *Exogyra sinuata,* first discovered by Captain Ibbetson.[*]

Next above these lies the sandstone which forms a reef called Horseledge by Fitton,[†] and which yields ferruginous nodules with *Panopœa plicata,* Sow., *Trigonia alæformis,* Park., *Thetis minor,* Sow., *Gervillia anceps,* Desh., *Terebratula sella,* Sow., *Rostellaria robinaldina,* D'Orb. This was said by Fitton to resemble his Group XIV.

A clay-band, 8 feet thick, which rises from the beach about 300 yards north of Luccomb Chine, corresponds to the thick clay which lies next above the cascade in Blackgang Chine (the lower part of Group XV. of Fitton). It makes a small undercliff or ledge in the cliff, and crops out 300 yards south of Shanklin Chine, whence it may be traced through the brick pit at Lower Hide, by Apse Farm, to the brick pit, now disused, at Sandford. This band forms the top of the Ferruginous Sands.

[*] *Rep. Brit. Assoc.* for 1844 (Sections), p. 43.

[†] This seems to be the reef marked Yellow Ledge on the Six-Inch Ordnance Map, and is about 350 yards south of the reef marked as Horse Ledge.

It will be noticed that the fossiliferous group described above corresponds to beds at Blackgang, in which only a few fossils occur. On the other hand, the strata between Little Stairs and Sandown, though corresponding to richly fossiliferous beds at Blackgang, have yielded no fossils. These differences are principally due to the condition of the rock. Fossils are seldom preserved in any part of the series near the surface of the ground, but only in the deep-seated strata that are exposed at the foot of the cliffs, and the weathering of the beds, which has reached a depth varying according to local circumstances, has extended below the level exposed in the Sandown cliffs. This weathering consists chiefly in the replacement of carbonate of lime by carbonate of iron, and the conversion of the latter into peroxide of iron, the effect being to destroy the coherence of the rock and to impart to it a brown colour. The original condition of the rock was probably that of the hard greyish and calcareous concretions, in which alone fossils are found in perfection, even at Atherfield.

The Sand-rock Series.

This division is finely exposed in the cliffs from Bonchurch to Knock Cliff. Its base is very clearly marked by the ledge or undercliff formed by the clay last described. A second, but smaller ledge, is formed by a bed of very green clayey grit, at times more clay than grit, which lies about 20 feet higher up. A descending section is as follows :—

Sand-rock Series at Luccomb and Knock Cliff.

		Ft.
Carstone (p. 59).		
Sand-rock Series	Bright yellow and white sand with laminæ of blue clay in planes of current-bedding. A few bands of very green sand throwing out chalybeate water - - -	35
	White and grey sand - - -	50
	Very green clayey grit, forming a ledge in the cliff, and throwing out water -	8
	White and ashy grey sand and sand-rock -	20
Ferruginous Sand, &c.		

113

The lower part of the series may be most conveniently studied at the top of Knock Cliff, and in Luccomb Chine. The upper beds are accessible in the cliff between Luccomb and Bonchurch, the last exposure being in Monk's Bay. The inland sections of these beds in the neighbourhood of Shanklin are unusually good, and will be described subsequently (p. 46).

SANDOWN TO CULVER CLIFF.

The position of the base of the Lower Greensand is marked here as in Compton Bay by a great founder of the cliff, and at the present time (1887) the junction is easily accessible throughout the greater part of the hollow from which the slip has taken place. The section of the Perna Bed is similar to those which have been described before. The base line of the Lower Greensand is sharp and definite, the lower beds are conglomeratic, and the surface of the Wealden Shales shows signs of disturbance and slight erosion. Lastly, the fossils characteristic of each formation are found close up to, but never transgressing the boundary. The Perna Bed is not only visible in the cliff, but reappears in the foreshore below Redcliff Foot, and forms a long straight reef running out to sea a little south of east.

Southwards from the slip caused by the Atherfield Clay, the cliff consists of ferruginous sands and becomes mural, continuing so until the softer beds of the Sand-rock series are reached. On the yellow and white sands and blue clays of this series there rests a great thickness of Carstone, which passes up into the Gault. A small fault crosses the cliff at an oblique angle at this point, running W. 30° N., and throwing the beds down to the north. It is best seen in the base of the Carstone, which it crosses about half way up the cliff.

The Gault forms a small gully descending the cliff obliquely, and occupied by a footpath. This formed a convenient starting point for the following section :—

Section of the Lower Greensand at Redcliff.

		Ft.	In.
Gault, blue micaceous clay passing down into			
Carstone, 72 ft. 9 ins.	Brown clayey grit, becoming more sandy below ; small scattered pebbles, and a line of pale phosphatic concretions made up of grit and grains of iron oxide 9 feet from the top	10	3
	Pebbly band, with small quartzites -	0	6
	Brown sand with many scattered quartzite pebbles, and phosphatic concretions as above at several horizons. Wavy lines of iron oxide, and some beds with many grains of oxide	60	0
	Loose brown sand and grit -	2	0
Sand-rock Series, base uncertain, about 93 ft. 6 ins.	White sand and blue clay interlaminated -	12	0
	Do. with occasional lines of blue clay	32	0
	Striped sand and clay -	9	0
	Do. chiefly clay and very sulphury -	4	0
	Seam of iron oxide -	0	6
	Bright-yellow and white sand, with ferruginous band at base -	31	0
	Grey striped sand and clay -	2	0
	White sand -	3	0
	Blue and striped sandy clay (?=40 feet clay of Blackgang) -	21	0
	Hard brown sandstone -	3	6
	Grey sand, " soot-coloured " -	6	0
	Pebbly bands, containing small quartzites, phosphates, and iron oxide -	2	0

		Ft.	In.
	Dark-green or bluish clay and sand - -	1	0
	Ferruginous pebbly band with small phosphates and pebbles of iron oxide - -	1	6
	Soft yellow sand - - - -	6	6
	Dark clayey sand - - - -	6	0
	Pebbly band, containing many rolled phosphatic casts of ammonites and bivalves -	0	4
	Pale-brown ferruginous sand - - -	3	0
	Pebbly band, with small quartzites and numerous flakes of iron oxide - -	0	2
	Pale-brown sand with flakes of iron oxide -	11	0
	Brown pebbly grit with small quartzites and grains and flakes of iron oxide - -	4	0
	Loose pale-green sand - - -	17	0
Ferruginous Sands, about 367 ft. 6 ins.	Greenish grit with many wavy seams of iron oxide - - - - -	3	0
	Brown and green gritty sand - -	3	0
	Dark-green or nearly black clayey sand -	6	0
	Brown sand with flakes and grains of iron oxide - - - - -	68	0
	Greensand, with a vivid green streak ; lines of clay occasionally ; a layer of broken oysters 9 ft. from the base. Forms a smooth vertical wall - - - -	60	0
	Brown and reddish brown sandstone with grains of iron oxide very abundant about 20 feet from the top ; forms the cliff on which Redcliff Fort stands - -	114	0
	Green sandy clay with wood and a line of large nodules - - - -	2	0
	Fine and very clayey sand with wood ; lines of nodules in the upper part, and veins of iron oxide - - - - -	14	0
	Seam of brown iron oxide - - -	0	5
	Fine grey clayey sand - - -	2	0
	Band of blood-red iron oxide - -	0	1
	Fine grey clayey sand - - -	10	0
	Fine white clayey sand - - -	2	0
Atherfield Clay, 83 ft. 4 ins.	Pale-blue clay with pale-blue nodules, weathering brown - - -	77	0

Perna Bed, 6 ft. 4 ins.

		Ft.	In.
Calcareous and ferruginous grit with many fossils, 1 ft. 6 ins. to - -		2	0
Passing down into pale-blue sandy clay with fossils - - -		3	6
Impersistent grit, with scales and bones of fish and phosphatic pebbles, some of which are rolled ammonites and bivalves ; about - -		0	3
Pale-blue sandy clay with fossils -		0	6
Grit, as above - - -		0	0½ —1

	Ft.	In.
	617	1

It will be observed from this section that the thickening of the Carstone, which was noted between Compton Bay and Blackgang, and still more between Blackgang and Shanklin, is still progressing in an easterly direction. The Sand-rock Series and Ferruginous Sands on the contrary, as previously noted, thicken in a southerly direction. In the series of comparative sections forming Plate III. these differences are clearly presented.

The occurrence of a band of rolled phosphatic nodules in the upper part of the Ferruginous Sands has attracted the attention of several observers.* The nodules seem to be on the same horizon as those noted at Compton Bay, but in the "coprolite bed" 4 inches thick at Redcliff, are larger, harder, and better preserved. Among the specimens Mr. Keeping identified *Ammonites biplex*, Sow., *A. cordatus*, Sow., *Pleurotomaria* sp., *Cardium striatulum?* *Lucina* sp., *Myacites* sp., *Cytherea rugosa?* *Arca contracta*, Phill., all being fragmentary and much rolled. There occurred also quartzite and other pebbles, as large as walnuts.

Up to the present this bed has not been discovered near Shanklin or at Blackgang, nor is its horizon marked by any break in the sequence of the strata. It was probably a near-shore deposit, and did not extend southwards in the direction in which presumably the deeper portions of the Lower Greensand sea lay. Near Godalming, on the contrary, it is largely developed according to Mr. Meyer, who describes it as resting on an apparently eroded surface of the sands beneath, and as constituting a well-marked basement-bed to an upper division of the Lower Greensand (*op. cit.*, p. 10).

PUNFIELD COVE.

Before quitting the description of these fine cliff sections of the Lower Greensand, we will briefly notice the sequence of beds in Punfield Cove. Lying 20 miles west of the Isle of Wight, this locality gives further opportunity of observing the changes in the strata which we have already seen in progress within the limits of the Island.

The section of the Lower Greensand in Punfield Cove is as follows. (*See also* Plate III.) :—

	Ft.	In.
Gault.		
Carstone, seen only in lumps lying about ; apparently about	0	4
Yellow sand, not well seen, about - · -	10	0
Very sandy dark clay with selenite (perhaps the thick clay of Blackgang) - · - -	15	0
White sandstone with white quartz pebbles -	20	0
Brown sandstone, and yellow sandstone with shales	15	0
Interlaminated sands and clays, the latter traversed by numbers of small tubes filled with sand (? worm-burrows) - · -	15	0
Ferruginous sand and hard sandstone with *Leda* -	12	0
Interlaminated sands and clays with some thicker bands of yellow and white sand - · -	61	0
Limestone with wavy seams of lignite and many fossils (the "Marine Bed" of Professor Judd), variable, but about - · - -	0	10

(Left margin, bracketed: Ferruginous Sands and Sand-rock Series, 148 feet 10 inches.)

* Meyer, On the Lower Greensand of Godalming. (*Geologist's Assoc.*), 1869. Woods, *Geol. Mag.* for 1887, p. 46.

			FT.	IN.
Atherfield Clay, 49 feet 3 inches.		Reddish clay, becoming pale-blue below, very fossiliferous in the lower part	28	0
		Soft yellow sandstone, with a few fossils	1	0
		Pale-red clay, bluish in parts, a few fossils	8	6
		Four bands of very hard grey sandstone; no fossils	2	9
		Red clay, a few fossils in the lower part	6	0
	Perna Bed	Dark-green sand, with small pebbles and grit, many fossils	1	0
		Pale-blue sandy clay with many small pebbles (rolled bivalves, Ammonites, &c.), and larger pebbles of sandstone, wood, &c., at base; many fossils	2	6

Wealden Shales (*see* p. 9).

198 5

The lumps of Carstone contain many pebbles, up to half an inch in length. Its thinness is in accordance with what has been indicated in the Isle of Wight, where it thins from about 70 feet at Sandown to 30 feet near Bonchurch, to 12 feet near Blackgang, and to 6 feet in Compton Bay.

The Sand-rock Series is not easily distinguished unless the dark clay with selenite, 15 feet thick, be taken as the representative of the thick clay of Blackgang Chine (35–40 feet thick). A large part of the Ferruginous Sands has assumed a character which in the eastern part of the Isle of Wight is seen only in the Sand-rock Series, namely, that of interlaminated white sand and blue clay (the "foliated sands and clays" of Fitton). In Compton Bay this change is foreshadowed by the appearance of thin beds of this type, interstratified with ferruginous sands, considerably below the base of the Sand-rock Series.

The very fossiliferous limestone, 10 inches thick, corresponds in position with the Crackers, the most fossiliferous zone in the Atherfield section.

The Atherfield Clay presents no unusual features, except that there are beds of sandstone at two horizons in it. The recognition of the Perna Bed, and of the usual sharply defined line dividing it from the Wealden Shales, was a satisfactory point. The rolled phosphatic pebbles in the Perna Bed are slightly larger and more abundant at Punfield than in the Isle of Wight, and more frequently recognisable as the casts of bivalves and Ammonites. This, as well as the changes in the overlying beds, indicates that in working westwards we approach the old shore line of the Lower Greensand sea.

The fossils in the following list, except where otherwise noted, were collected for the Survey by John Rhodes, and have been identified by Messrs. G. Sharman and E. T. Newton. The specimens marked thus * are inserted on the strength of their having been recorded from the "Marine Bands of Punfield" by Prof. Judd in the *Quart. Journ. Geol. Soc.*, vol. xxvii. p. 215. Those marked † are added on the authority of Mr. Meyer, *ibid.*, vol. xxviii. p. 252 and vol. xxix. p. 73.

Fossils from the Lower Greensand of Punfield.

The Atherfield Clay and the limestone above it.

Wood.
Crustacean, fragment.
*Serpula.
†Terebratula sella, *Sow.*
†Anomia lævigata, *Sow.* (collected by the Survey also).
†Arca cornueliana, *D'Orb.*
*   „   cymodyce, *H. Coquand* (young).
†   „   Raulini, *Leym.*
     „   sp.
*†Astarte, sp.
†Cardita neocomiensis, *D'Orb.*
†Cardium (Arca) Austeni, *Forbes.*
*†   „   impressum, *Desh.*
†   „   subhillanum?, *Leym.* (collected by the Survey also).
Corbula striatula?, *Sow.*
†   „   sp.
†Cyprina, sp.
†Cytherea parva, *Sow.*
*†Exogyra Boussingaultii, *D'Orb.*
*†   „   sinuata, *Sow.*
†   „   tombeckiana, *D'Orb.*
*Isocardia nasuta, *H. Coq.*
*   „   sp.
*Leda scaphoides, *P. and C.*
Lima, sp.
†Lucina, sp.
*Modiola giffreana, *P. and R.*
†   „   simplex, *Leym.*
*Orthostoma Verneuili, *Vil.*
†Ostrea Leymerii, *D'Orb.*
†Panopæa neocomiensis, *Leym.*
†   „   Prevosti, *Leym.*
†   „   sp. (= P. plicata, var. of Atherfield).

†Pecten (Neithia) neocomiensis, *D'Orb.*
†   „   „   robinaldinus, *D'Orb.*
†   „   „   sp.
*   „   „   „ (very small).
*Perna rauliniana, *P. and R.*
*†Pholadomya semicostata, *Ag.*
     „   sp.
*†Plicatula asperrima, *D'Orb.*
†   „   carteroniana, *D'Orb.*
†Solecurtus Warburtoni, *Forbes.*
*Tellina? gibba, *H. Coq.*
†   „   vectiana, *Forbes.*
†Thetis lævigata, *D'Orb.*
†Trigonia (Atherfield sp.).
†Venus, sp.
*Actæonella oliviformis, *H. Coq.*
*Actæon Esqueræ, *De Verneuil and De Lorière.*
*   „   pradoana, *De V. and De L.*
*Cerithium Pailleti, *De V. and De L.*
*   „   Vilanovæ, *De V. and De L.*
*Fusus? neocomiensis, *D'Orb.*
*†Natica lævigata, *Desh.*
*   „   pradoana, *Vil.*
*Neritopsis minima, *De V. and De L.*
*Pleurotoma Utrillasi, *De V. and De L.*
*Trochus Esqueræ, *De V. and De L.*
*Turritella Tournali, *H. Coq.*
*Vicarya Lujani, *De V. and Coll.*
*   „   pizquetana, *Vil.* (collected by the Survey also).
*   „   Pradoi, *De V. and De L.*
Ammonites Deshaysii, *Leym.*
*Lamna (teeth).
*Pycnodus (teeth).

A band of soft sandstone in the Atherfield Clay.

Arca Raulini, *Leym.*
Exogyra, sp.

Panopæa plicata, *Sow.*
Solecurtus (cast of).

The Perna Bed.

Multizonopora rimosa, *D'Orb.*
Arca corneueliana?, *D'Orb.*
     „   Raulini, *D'Orb.*
Astarte, sp.
Avicula depressa, *Forbes.*
Cardita fenestrata, *Forbes.*
Cardium subhillanum, *Leym.*
Cypricardia undulata?, *D'Orb.*

Exogyra subplicata, *Röm.*
Lima lingua?, *Forbes.*
     „   sp.
Lucina, sp.
Panopæa plicata, *Sow.*
Pecten interstriatus?, *Leym.*
P. quinquecostatus, *Sow.*
Tellina, sp.

CHAPTER IV.

LOWER GREENSAND—*continued.*

INLAND SECTIONS.

(1.) ALONG THE CENTRAL DOWNS.

The Atherfield Clay.

No section of any importance occurs in this division away from the coast, and the tracing of a base-line has consequently been a matter of some difficulty The clue to the position of the boundary is provided by the topographical feature and change of soil produced by the Ferruginous Sands above.

The Ferruginous Sands and Sand-rock Series.

These two groups will be conveniently taken together in description. As previously remarked, they pass one into the other. Commencing our description on the west, we find the Ferruginous Sands rising into a characteristic escarpment, slightly lower than the Chalk Downs, which runs eastward from Compton Bay on the north side of Brook, Mottistone, and Brixton. The higher part of the ridge is formed by the iron-sand which comes down to the beach on the west side of Compton Chine. The more massive iron-sand which forms the cliff on the east side of Compton Chine crops out in the southern slope of the hill, and gives rise to the terrace of deep-red sand on which Brook Church stands. The position of the Sand-rock Series is marked by the abundance of white sand in the soil.

At Mottistone a ravine has been cut through the Ferruginous Sands. The top of the Atherfield Clay seems to occur at the Church. The clay is overlain by a great thickness of ferruginous clayey sands with a marked bed of brown iron-sand, which seems to be the same as that on the east side of Compton Chine. At the top of the ravine the following descending section may be traced in beds which form the passage between the Sand-rock Series and the Ferruginous Sands :—

Near the Long Stone, Mottistone.

	FEET.
White sand, about	20
Ironstone	$\frac{1}{2}$
Grey and " sooty " silt and sand	15
Grey silt	6
Red clay, grit, and sand	10
Ferruginous grit	2
Dark " sooty " silt	12
Ferruginous grits, &c.	—
	65½

These beds are seen again, but less clearly, in the lane to Calbourne by Black Barrow, this hill itself being composed of very fine white and grey sand of the Sand-rock Series. But the best section occurs by the road-side at Rock. There the Sand-rock Series consists of current-bedded crimson, pink, brown, buff, yellow, and whitish sand; a beautiful combination of colours, the crimson being very rich. Above this sand lies a band of pebbly iron-stone constituting the base of the Carstone.

The Lower Greensand escarpment is breached at Rock by the stream from Bottlehole Spring, but rises again on the east of this valley into a bold hill, many of the lanes up which provide good sections. The upper boundary of the Atherfield Clay seems to run along the upper road in Brixton, and the strata next above it consist of yellow sandstone, brown or reddish in places, and with a few thin clayey bands. At the foot of the steeper and unculti-vated part of the hill there runs a bed of deep-red iron-sand with abundant spherical grains of iron-oxide as well as rounded quartz grains, which seems to be the same bed that extends from the east of Compton Chine under Brook Church. Immediately over it lies a bed of yellow and white sand, with wavy laminæ of clay, closely resembling the Sand-rock Series. This series, however, comes on nearer the top of the hill, where bright-pink, pale-red, yellow and white sand-rock is repeatedly exposed.

The escarpment becomes insignificant south of Shorwell, where it is crossed by the stream from which this village takes its name. Yafford stands on the Atherfield Clay, but a slight rise in the ground, and the brown sandy soil indicate the base of the Ferru-ginous Sands, and show that the strike has changed to nearly south. Near Yafford Mill, a pit shows buff sand and loam overlain by a little gravel, and at Wolverton iron-sand rests on greensand, the dip being north-north-east at 10°. The Shorwell and Atherfield road-cutting near this farm is made through brown and green current-bedded sand at a slightly higher horizon; while at Haslett brown sand appears with bands of ferruginous grit, and in the upper part a band of white sand. It is difficult to detect here the horizon of the iron-sand which we traced as far as Brixton. It might be expected to run near Wolverton, and through Smallmoor, connecting itself there with a well-defined bed which we shall subsequently follow up from near Blackgang.

The sections in the Sand-rock Series are more numerous. The beds of rock, which become a noticeable feature above Brixton, increase in number and thickness eastwards, and form small features along the strike near West Court and Presford. They are generally white, though tinged here and there with red or yellow. So abundant is the white sand soil on these strata that some of the fields on the east side of Bucks had the appearance of being partly covered with snow in the dry summer of 1887.

The dip of the rocks in this neighbourhood has diminished to 8°, and grows less as we proceed eastwards. The various sub-divisions accordingly each occupy a wider belt, and at the same time display more fully their characteristic features in the form of the ground. The Ferruginous Sands stretch away in a broken table-land to the cliffs of Atherfield and to the southern hills of the Island. The Sand-rock Beds form a series of rounded hills, capped by the Carstone, and fringing the more continuous escarpments formed by the Chert Beds of the Upper Greensand and by the Chalk, while a belt of ground, characterised by its gentle slopes and generally by its comparative lowness, marks the position of the Gault. These features are all well displayed in the valley followed by the Chillerton road near Billingham. The best section in the Sand-rock Series occurs by that road-side; the Ferruginous Sands are well exposed in the road-cuttings at Kingston.

Near Cridmore the upper part of the Ferruginous Sands contains beds of bright-yellow and white sand, much like the Sand-rock Series, and making it difficult to decide on a boundary line.

After passing the Medina, however, the base of the Sand-rock Series is marked by a bed of coarse white quartz-grit. The bed is seen south of the Star Inn and near Upper Yard, but more clearly in a small pit, 300 yards north-west of Birchmore. There, and in the road-cutting close by, it may be described as a fine gravel, so large are the grains of quartz. The sands above this bed are seen in a pit south of Pagham; they are white and current-bedded with lenticular ferruginous beds. The few sections in the beds below show brown and yellow ferruginous sands.

The next section occurs in the Sand-rock Series in the lane running east from Blackwater Station. Here white sand and sand-rock were formerly dug. The base of the series is marked by springs and other indications of clay-beds. The same beds are repeatedly exposed in the lanes about Marvel, and are now being dug in a large sand-pit in Marvel Wood, where the following section is exposed :—

Marvel Wood Sand-pit.

		FEET.
Carstone; a ferruginous grit, cemented irregularly in bands by iron-oxide; some of the lower beds contain small pebbles. Top not seen - - - - - - -		12
Sand-rock Series.	Grey sand with fragments of clay, with the appearance of being a reconstructed bed (*see also* p. 56), resting on the edges of the current-bedding planes of - - -	3
	White sand with lines of blue clay - -	30+
		45

The strata dip, so far as can be judged, to the south-west at a gentle angle; but a few yards further on rapidly roll over and plunge down to the north. From this point eastwards the series runs in a narrow belt near and parallel to the central Downs of the Island. The centre of the anticlinal axis described above seems to strike nearly east from Little Whitcombe to the north side of Marvel Farm, and thence towards Horringford, where further evidence of its position may be seen.

A large sand-pit at Standen provides the following section of the Sand-rock Series :—

Standen Sand-pit.

	Ft.	In.
Green and grey sand, current-bedded - -	12	0
Yellow sand-rock - - - - -	2	0
Ironstone with a few small pebbles - -	0	6
Yellow and grey loamy sand and clay - -	10	0
Dark-blue clay - - - - -	15	0
Ironstone, about - - - - -	0	6
Grey pebbly sand, passing down - - -	6	0
Loose yellow and white grit - - -	12	0
Fine sand - - - - - -	8	0
Clay-bed - - - - - -	0	6
Fine white sand-rock - - - - -	9	0+
	75	6

The bottom of the pit is probably about 15 or 20 feet above the base of the Sand-rock Series, but a considerable thickness of beds, consisting in part of fine-grained buff and brown sand, occurs in the hill-side above, before we reach the base of the Carstone. The dark-blue clay may be the upper of the two clays seen near Shanklin, but correlation in so variable a series is mere guess-work.

Almost the only section of the Ferruginous Sands in the Black-water valley occurs in the road-side near Stone, where green and ferruginous sand and deep-brown sand with many grains of iron oxide, are exposed. Similar sands extend along the southern slopes of St. George's Down. On the north side of the Down, 300 yards south of Garrett's, a sand-pit has been opened near the top of the Ferruginous Sands ; the beds exposed are dull-green sands with lines of soft concretions, and are traversed by several small faults, which run nearly east and west, and throw the beds down a foot or two to the south. The dip is northwards at 23°.

The next sections occur near Arreton and Merston. A road-cutting south-west of the former place exposes red sand containing many grains of iron oxide, the dip being north-east at 13°, while 300 yards north of Merston Cross pale sand is seen, dipping south-south-west at 7°. Here then we have the continuation of the anticlinal axis, which we noticed at Marvel. Obscure casts of fossils occur in a band of ironstone on the road to Merston, 600 yards south-west of Arreton Church.

At Redway and near Horringford Station red and brown irony sand may be seen, the latter locality yielding specimens of *Venus* and other fossils according to Mr. Norman.* Apparently the same beds are exposed in the road in Newchurch. Here and

---

* A Popular Guide to the Geology of the Isle of Wight, p. 56. (1887.)

wherever elsewhere visible, namely, east of Wackland, and on Skinner's Hill, they are nearly horizontal, but the Sand-rock Series, on the other hand, near Heasley Lodge dips north at 20°. The anticlinal axis therefore must run nearly along (or a little north of) the River Yar at Newchurch.

At Knighton a little irregularity occurs in the trend of the great central axis of the Island, in consequence of which the Lower Greensand dips at a more gentle angle, and the characteristic features of its subdivisions are better shown. The Sand-rock Series is seen in a deep lane and pit, 400 yards east of Knighton Mill, and in many spots around Kern, as a brown, red and white sand, while above it the Carstone makes a fairly pronounced feature. Good exposures of the Ferruginous Sands occur about Alverstone Farm and on the road to Brading. At the former place, grey and green sand passes under red and brown sand, with many grains of iron oxide. The dip is westerly at 5°–10°, but sweeps round to north at 21° at Adgestone. Here then we fix another point on the line of the Marvel Anticline, and join it on to the fold which brings up the Wealden Beds of Sandown Bay.

The dip of all the strata increases, and their outcrops become proportionately narrow near Yarbridge. A pit in the lowest of the Ferruginous Sands, near Morton Farm, shows brown sandstone dipping north-north-east at 35°, while the Sand-rock Series appears in a pit and road-cutting 400 yards west of Morton as a white sand with traces of blue clay.

(2.) AROUND THE SOUTHERN DOWNS.

In describing the Atherfield section we spoke of a bold escarpment or terrace formed by the ferruginous beds of Blackgang Chine (Group XIV. of Fitton), which runs through Pyle, Corve, and Kingston. There are many sections in the roads descending the hill at these places. On the top and extending nearly to the brow of the terrace, soft, brown, buff, and white sand appears similar to the sand at Cridmore (p. 42), and approaching the type of the Sand-rock Series. Lower in the hill-side, greyish-green sand follows, weathering brown, and of considerable thickness. On descending to the foot of the escarpment, we find a line of springs and a belt of peaty ground marking the outcrop of a soft and clayey bed, doubtless the "foliated sand and clay" of Walpen Chine (Group XII. of Fitton). The escarpment spoken of runs through Kingston, and, sweeping thence to the south-west round Gun Hill, points for Haslett and Wolverton, but becomes obscure in that neighbourhood.

A second terrace is formed locally by a thick bed of red and brown sand with numerous grains of iron-oxide. This feature includes the bold brow known as Warren Hill, three quarters of a mile west of Corve, and stretches thence by Dungewood towards Small Moor. There, like the other terrace, it also becomes obscure, so that whether it is a continuation of the bed which we traced by Brook Church must be left in doubt.

It will be noticed that the source of the Medina at Chale Green is situated on the upper of these two terraces. The valley of the river gains in depth northwards, while the strata, except for some very gentle undulations, remain horizontal. It is probable that the depth thus gained is sufficient to let the stream reach the "foliated sand and clay," and that this may account for the width of the alluvial flat; but there is no section to prove it. The hills are capped by buff and white sand, while their sides are formed of brown and grey sands with an occasional seam of iron-oxide.

The Sand-rock Series is exposed at Chale Farm, Gotten, and at the north end of St. Catherine's Down, with its usual character of fine soft white sand. But its outcrop, though broad, is partly overspread by Gault, which, owing to the influence of percolating water, has flowed down over the intervening Carstone.

We now enter the drainage area of the (East) Yar. Blake Down, here forming the watershed between this river and the Medina, is a long spur of the uppermost beds of the Ferruginous Sands, capped with flint-gravel. As the river is about 100 feet below the highest strata of this spur, the "foliated sand and clay" might be expected to be reached. There can be little doubt that this is the case, for a terrace, closely resembling that of Pyle, Corve, and Kingston, runs through Godshill, north of Sandford, towards Lessland, and perhaps to Branston. From the foot of the bold brow which terminates this terrace at Godshill springs wander through wide peaty marshes, as at Corve, while the brow itself is composed of a ferruginous sand and greyish green sand, exposed to considerable depth in the road-cuttings.

The lower beds of the Sand-rock Series are seen in a pit near Sibbecks, which gives the following section :—

	FEET.
Soft sand with seams of clay - - - -	20
Soft yellow and white sand-rock (perhaps the third sand-rock of Fitton) - - - - -	18
Thin-bedded yellow and white sand with brown loamy partings - - - - - -	6+

Similar beds are seen in the grounds of Wydcombe, Redhill, Fairfields, and under the gravel at Ford Farm. Near Itchall a pit exposes the top of the series, namely, white sandstone, more than fifteen feet thick, overlain by eight feet of Carstone. The base of the series is difficult to fix throughout the neighbourhood of Chale Green, but a blue clay seen in the brook south of Roud, in the lane at Russell's Farm, and in the high-road north-east of this farm, is presumably the same bed which we have already noticed at the top of the Ferruginous Sands at Shanklin.

The characteristic scenery produced by the Sand-rock Series and the overlying Carstone is admirably shown around Sainham and Godshill Park. The base line of the Carstone, the beds being nearly horizontal, meanders round a number of short but deep valleys, the sides of which are composed of bright-white sand and sand-rock.

A remarkably coarse grit has been already described as occurring at the base of this series near Blackwater; a somewhat similar bed may be noticed in a lane south of Sandford, but not elsewhere. The clay-bed of Roud, however, referred to above, seems to have been well developed at Sandford, where it was formerly worked for bricks, and where it is still exposed to a depth of 8 feet. An outlier of the Sand-rock Series occurs here, its top capped with gravel, its sides showing the usual white sand soil, while a line of springs around its base marks the position of the clay-bed.

Crossing the Wroxall stream, we find a sand-pit near Winstone, showing 10 feet of white sand, and another by the side of the railway half a mile east of Winstone, presenting more than 18 feet of white sand with thin lines of clay. The neighbouring railway cutting is much overgrown, but reveals some white sand in the upper part. The base of the series is marked near Rill by a fall in the ground and the issue of springs.

In Apsecastle Wood and the adjoining valleys, the features of the Sand-rock Series are finely shown, a remarkably good section having been opened out in the railway cuttings. We may conveniently take up the description at the east end of the cuttings, where we left it in speaking of Shanklin. It will be remembered that two clay-beds occur in Knock Cliff. The upper appears to be the one worked in a brick-pit west of Gatten, where, however, it seems to be impersistent. The lower bed is worked by the side of the railway at Lower Hide, where it is a stiff dark-blue clay. The sand between the two beds is dug in a pit on the opposite of the line, which exposes :—

		FEET.
Brown irony sand - - - - -		4 to 6
Coarse grit or fine gravel - - -		1 to 3
White sand - - - - - -		14+

The railway cutting commencing 500 yards east of Lower Hide gives a more complete section of these sands and of the upper clay, which has here again developed itself. A descending section runs as follows :—

Railway Cutting three-quarters of a mile west of Shanklin.

	FEET.
Dark clayey sand - - - - -	4
Dark-green sandy clay with scattered grit and pyritised wood - - - - - -	15
Brown pebbly and ferruginous grit with wood, about -	½
White sand with black grains - - -	2
Hard brown pebbly rock - - -	2
Coarse brown grit with numerous concretions - -	5
Grey sand or white sand with black grains - -	5
White sand-rock with bright-yellow and brown staining	14
Dark sands - - - - - -	3+
	50½

The strata dip gently (at about 2° to 3°) a little to the south of west, and the green clay slopes down to the level of the rails in the next cutting. The sands lying upon this clay are dark and ferruginous, but are not well seen.

The upper clay-bed, seen near Upper Hide, runs along the valley in Apsecastle Wood, where it has caused a good deal of slipping; the lower clay-bed occurs at Apse Farm, but elsewhere is overspread by a downwash of sand.

The Ferruginous Sands between these localities and the River Yar form an undulating tract, in part overspread with river-gravel, but in part rising into flat-topped hills, capped with gravel. The dip, if any exists, is too gentle to be detected in the small sections that occur, except on Blackpan Common.

The features of this tract suggest that the same beds which form the escarpments of Pyle and Kingston, and of Godshill, extend here across the valley of the Yar in a neck of about a mile in breadth. The base line of the beds on the east side of the neck seems to run from the cliff near Little Stairs Point, by the west of Lake, past Borthwood, across the river near Alverstone, and thence eastwards. The western boundary which we have already traced through Godshill to near Branston, seems to be continued in the hill on which Newchurch stands, and to trend thence eastwards, but all evidence of its position is lost in the valley.

INDICATIONS OF CONDITIONS UNDER WHICH THE LOWER GREENSAND WAS DEPOSITED.*

" At the close of the deposition of the Wealden, there appears to have been a sudden depression of the bed of the great fresh-water estuary, and an influx of the sea. The first effect of such an influx would be the destruction of the animals in the estuary not adapted for living in salt water; hence we find a total destruction of the Wealden animals, the remains of which accumulate towards the point of the junction of that formation with the Lower Greensand,—a fact which indicates the nature of the change. Even the *Cerithium* [*Vicarya*], although belonging to a genus many species of which are capable of living in the depths of the sea, was destroyed, notwithstanding that its appearance, only in the uppermost beds of the Wealden, indicates that its presence there was due to the commencement of the very state of things which eventually destroyed it. That the depression was of some extent, though not, perhaps, of very many fathoms, is indicated by the nature of the animals which lived in the first-formed sea-bed, and which, when they died, were often embedded in the fine and probably fast-depositing mud, in the vertical position which it

* On the Section between Blackgang Chine and Atherfield Point, by Capt. L. L. B. Ibbetson and Prof. Edw. Forbes. *Proc. Geol. Soc.*, vol. iv. p. 409 (1844).

is the habit of animals of such genera as *Pinna* and *Panopæa* to assume when alive.*

"After this a temporary change followed, when an influx of sand, mingling with the calcareous mud, caused a state of sea-bottom peculiarly favourable to the presence of animal life. In this way were called into existence a multitude of species which were added to those which had appeared before them. This was, in fact, such a state of sea-bottom as is now presented by great shell-banks; but it does not seem to have lasted long, and new depositions of mud appear to have extinguished some forms, whilst others suffered by the change only in the diminution of their numbers. In the midst of this muddy epoch, a temporary and peculiar condition of sea-bottom, forming what are now called the Crackers, called forth the presence of numerous mollusca, at first of various species of the genus *Gervillia*, and afterwards of *Auricula* [*Arellana*], *Cerithium*, *Dentalium*, and other univalves, which appear to have enjoyed but a brief existence (as species) in this locality, since similar conditions were never afterwards repeated. The greater number of the Gasteropodous mollusca of the English Lower Greensand are found within this very limited range. At the close of the deposition of this great mass of clay there was for a time a great multiplication of the individuals of certain *Brachiopoda*, which had commenced their existence in the lowest beds. Thus *Terebratula Gibbsii* [*Rhynchonella gibbsiana*] suddenly appears in immense abundance, covering the bottom of the sea, and predominating over the animals among which it had previously been but thinly scattered.

"This lowest zone of *Terebratulæ* marks the commencement of a new state of sea-bottom where sands predominated over the clays, each interval of deposition being usually marked by the presence of a layer of *Gryphæa* [*Exogyra*] *sinuata*, the period of rest being almost always sufficient to enable the *Gryphæa* to attain its full growth. Other bivalves are found with it, but in comparatively small numbers, and not such as are of gregarious habits. During the whole of this period enormous *Cephalopoda*, including species of *Crioceras* and *Scaphites* [*Ancyloceras*], frequented these seas, and when dead formed the nuclei round which calcareous and sandy matter collected and formed nodules. The death of these animals seems to have been connected with the periodical charging of the sea with sediment; hence we find them usually alternating with the zones of *Gryphæa*, and forming irregular bands in the intervening sedimentary deposits.

* " The same decided change from dark-coloured fresh water marls containing *Melanopsis* (or *Melania*) [*Vicarya*] and *Cypris* to marine beds, occurs round the edge of the Weald, and was very well exposed at Haslemere during the cutting of the London and Portsmouth Railway, a few years back. In company with Professor Ramsay and Mr. F. Drew, I examined the passage beds, and found in the brown clay abundant tracks of marine worms, and the *Panopæa*, vertical in their old burrows, within an inch or two of the dark marls. A great *Perna*, a coral (*Holocystis elegans*), and numerous other fossils, occur in plenty just above these."— J. W. SALTER.   *See* Geology of the Weald, p. 114 (*Mem. Geol. Survey*).

" In the midst of this epoch of *Gryphœa* there is a sudden reappearance of the muddy deposits, during the predominance of which those animals adapted for such a sea-bottom, and which had survived the deposition of the fullers' earth, again multiplied, but the species which had become extinguished were not replaced by representative forms. This, however, did not last long, the sand again predominating with its zones of *Gryphœa* and lines of *Crioceras* nodules.

" A temporary multiplication of *Terebratula sella* suddenly marks a change in the zoological conditions,—for the *Cephalopoda* disappear, although the zones of *Gryphœa*, which animal does not appear to have been affected by the change, (probably a change in the depth of the sea,) go on as before, there being, however, no alternating lines of nodules. It would seem that the sea began to shallow, probably from elevation of the sea-bottom, until at last the *Gryphœa* itself disappears, the bands exhibit traces of the influence of currents, and become more gravelly ; lignites, indicating a shallow sea, become common, form belts in the ferruginous sand, and in one place a bed in the wavy blue sand, at a time when much iron was deposited. The deposition of the peroxide of iron appears to have been connected with the disappearance of the majority of mollusca, though *Trigonia, Thetis,* and *Venus* occasionally occur in considerable numbers. In the uppermost strata scarcely any animal remains are found, and everything appears to indicate a barren and shallow sea, previous to a new state of things, when a fresh series of clays (forming the Gault) being deposited, the majority of the animal forms which characterise the clays of the Lower Greensand disappear, and are replaced by distinct species, representative in time."

CORRELATION WITH THE MAINLAND AND THE CONTINENT.

Dr. Fitton first pointed out the identity of the fossils in the Atherfield Clay of the Isle of Wight with those of a clay in Sussex and Kent,* which corresponded to the Atherfield Clay, except in the absence of the fossiliferous stone known in the Isle of Wight as the Perna Bed. The calcareous nodules of the " Crackers Rock " were considered by him to represent the thick limestone (Kentish Rag) of Hythe, Maidstone, &c. The Carstone and Sand-rock Beds of the present Memoir were identified by him as the upper division of the Lower Greensand which he had described at Folkestone, that is to say, the Folkestone Beds of the Geological Survey ; while the great mass of beds intervening between the Sand-rock Series and the Crackers group were correlated with his middle division at Folkestone, now known as the Sandgate Beds. Lastly, he noticed that the Ferruginous Beds of Blackgang Chine (Group XIV.) and the corresponding bed of Horsledge, near Shanklin, contain the same species as are found in the Sandgate Beds at Parham Park

---

* *Proc. Geol. Soc.,* vol. iv. pp. 198, 208, and 396 (1843).

and other places in Sussex, and near Sandgate,[*] thus obtaining
further evidence of the correctness of the correlations given above.
According to Mr. Meyer[†] the coprolite bed at Redcliff (described on p. 37) corresponds to a pebble-bed at Godalming which
he considered to represent "a break in the hitherto continuous
deposition of the Greensand," and which he traced by Dorking,
Nutfield, and Maidstone towards Folkestone. This bed he took
as the base of his Folkestone Beds or upper division of the Lower
Greensand. It cannot, however, be followed through the Isle of
Wight, nor, when present, is it accompanied by any appearance of
a break.

But while this line fails us, we find that the base of the Folkestone Beds, as drawn by the Geological Survey,[‡] corresponds well
with the line at the base of the Sand-rock Series, which was inde-
pendently selected as a boundary capable of being traced through
the Isle of Wight. During the present year a brief visit was paid
to that part of the Lower Greensand outcrop in the Weald, which
lies nearest the Isle of Wight, for the purpose of comparing the
strata in the two areas, the result being to confirm in every particular the conclusions arrived at by Fitton. Lithologically, the
brightly coloured clean quartz-sands of the Folkestone Beds at
Pulborough, Midhurst, and Petersfield closely resemble the Sand-
rock Beds of the Isle of Wight. In both Sussex and the Isle of
Wight, moreover, these sands pass down into a group in which
beds of shale are conspicuous, and which is more evenly bedded
and more mixed with loam than the Folkestone Beds.[§] At
Pulborough a band of shale, 30 feet thick, and taken by
Mr. Gould as forming the top of the Sandgate Beds,[‖] corresponds closely in character and position to the thick clay-band of
Blackgang Chine, and of the railway cutting near Shanklin,
described on p. 46. The identification on the mainland, however,
of the rock now mapped in the Isle of Wight under the name of
Carstone is attended with some difficulty. The description of
this rock and its probable relations will form the subject of the
succeeding chapter.

The great development of beds of corresponding age on the
Continent has been pointed out by Professor Judd,[¶] of whose
conclusions the following is an abstract. The *Rhodanien* of
Switzerland, which forms a complete link between Upper Neocomian (*Aptien*) and Middle Neocomian (*Urgonien*), has been shewn
by M. Renevier[**] to be the equivalent of the Perna Bed, Atherfield
Clay, and Crackers of the Atherfield section. Among the fossils

---

[*] *Quart. Jour. Geol. Soc.*, vol. iii. p. 311. 1847. *See also* Geology of the
Weald (Geological Survey Memoir), pp. 136, 137.
  [†] On the Lower Greensand of Godalming (*Proc. Geol. Assoc.*), 1869, p. 10.
  [‡] Geology of the Weald (Geological Survey Memoir), 1875, pp. 138–144.
  [§] The difference is greater than appears at the first view of sand-pits in the
two subdivisions. The Folkestone Beds are used commercially for building sand,
the Sandgate Beds for moulding purposes.
  [‖] Geology of the Weald, p. 136.
  [¶] *Quart. Journ. Geol. Soc.*, vol. xxvii. pp. 223–5. 1871.
  [**] *Bull. de la Soc. Géol. de France*, 2me sér. tome xii. p. 89.

Ammonites Deshaysii, which occurs in the "marine band" at Punfield [the top of the Atherfield Clay] abounds in the higher beds of the Neocomian, but is not known in the *Urgonien* or any lower bed. *Vicarya Lujani* and several other of the Punfield shells are well-known and characteristic *Rhodanien* forms.

In the east of Spain* the upper and middle Neocomian rocks are greatly developed, and contain beds of coal and jet which are extensively worked. They are divisible into three series, namely :—

An upper series of variegated clays and brightly coloured sands (crimson, grey, violet, and white), 600 feet in thickness, probably in great part freshwater, but containing a few marine shells of Upper Neocomian affinities.

A middle series, consisting of ferruginous sandstones and limestone, alternating with sandy clays, and containing ten beds of coal, lignite, or jet at Utrillas, where they are 530 feet thick. These beds contain the same fossils as the "marine band" of Punfield. They are characterised by six species of the gasteropod *Vicarya*, three of which occur at Punfield, and one in the *Rhodanien* of Switzerland. Hardly a fossil is found in the "marine band" of Punfield [the top of the Atherfield Clay] which does not also occur in these Spanish beds.

A third and lowest series, consisting of about 500 feet of alternations of limestones, sandstones, and marls, with jet and coal, and containing *Urgonien* fossils.

* See also H. Coquand. Description géologique de la formation crétacée de la Province de Teruel. *Bull. Soc. Géol. de France*, sér. 2, tome xxiv. p. 144 (1868).

CHAPTER V.
LOWER GREENSAND—*continued*.
THE CARSTONE.
INTRODUCTION.

THIS name has been given to a coarse and highly ferruginous grit, which may be traced continuously at the base of the Gault through the Isle of Wight. Wherever fully exposed the Carstone is seen to pass up into the Gault; on the other hand a fairly sharp line at its base separates it from the Sand-rock Beds, with an appearance even of slight erosion at times, though we have no evidence of an actual unconformity. The feature produced by this comparatively hard grit, capping the soft sands of the Sand-rock Beds, is especially prominent where the beds are nearly horizontal. It is most marked at Marvel Wood, near Shide, and in the neighbourhood of Godshill.

The Carstone varies considerably in thickness within the Island. From 6 feet at Compton Bay it expands to 12 feet near Blackgang, to 30 feet near Bonchurch, and to no less than 72 feet at Red Cliff. At Punfield, on the other hand, it seems to be represented by a few inches only of pebbly grit, but is not seen there in place. The Carstone, therefore, thickens towards the north-east, while the other subdivisions of the Lower Greensand increase towards the south.

The Carstone corresponds to the upper part of Fitton's Group XVI. The present name* has been adopted on account of the resemblance the rock bears to the Carstone of Lincolnshire and Norfolk, of which there is reason to suppose it to be the stratigraphical equivalent. For the Carstone of those Counties passes up into the Red Chalk, which there occupies the position of, and partly represents the Gault. Moreover, further south we find that the Gault when it makes its appearance passes down into a grey clay with phosphatic nodules, which in its turn shades into a lower light brown sand with phosphatic concretions and numerous fossils.†

These fossils, as pointed out by Mr. Teall, are found in the south of England to occur in the Gault, and in the *Ammonites mammillaris* zone, which lies next below the Gault. He infers, therefore, that "the Norfolk Neocomians [Carstone] are found to resemble both stratigraphically and palæontologically the Folkestone Beds of the South" (*op. cit.*, p. 22). But we have already pointed out that the Folkestone Beds as a whole are comparable to the Sand-rock Series. It remains to be seen whether any sub-

---

* The name is applied locally in the Weald to the portions of the Folkestone Beds, which have been cemented by brown iron oxide into a hard rock.
† The Potton and Wicken Phosphatic Deposits (Sedgwick Prize Essay for 1873) by Mr. J. J. H. Teall. Cambridge, 8vo., 1875, p. 20.

division of the Folkestone Beds corresponding to the Carstone of
the Isle of Wight can be recognised on the Mainland. The
Carstone thickens in the Isle of Wight towards the north-cast, yet
in the part of the Weald which is nearest to the Island, the
Folkestone Beds preserve their character of fine-grained quartz
sand up to within a foot or two of the base of the Gault. But on
the other hand the base of the Gault invariably consists of a more
or less pebbly grit, or of a sand with phosphatic nodules. At
Steep Common, near Petersfield, the Gault is green and sandy
towards the base, contains phosphatic nodules, and rests on a
"brown and green sand, with large pebbles, and at one place
phosphatic nodules at base." * Further east, near Midhurst and
Pulborough, the base is formed by a pebbly grit, varying from
3 to 10 inches only in thickness, but conspicuous from its extreme
hardness and from its deep-brown or blood-red colour. The
pebbles in this band range up to half an inch in length, and
their presence, together with the gritty character of the rock, distinguish it, even apart from its hard ferruginous cement, from the
fine-grained sand of the Folkestone Beds. Elsewhere in the
Weald the base of the Gault is marked by nodules of phosphate
of lime or of iron pyrites, the hard pebbly grit described above
being confined to the neighbourhood of Midhurst and Pulborough.
Associated with the nodules, and likewise in a phosphatic state,
there are fossils of Gault affinities, viz., *Ammonites Beudantii,
A. mammillaris, Exogyra conica, Inoceramus Salamoni, Natica
gaultina,* and others, which have led to the remark that the
Folkestone Beds are more closely connected with the Gault than
with the underlying Sandgate Beds. In 1859 Professor A. Gaudry
remarked that the sands at the top of the Lower Greensand at
Folkestone and Wissant in the Bas-Boulonnais contain *Ammonites
mammillaris,* and proposed to group these sands with the Gault on
that account.† In 1868 Mr. Topley noticed that at Folkestone
the Folkestone Beds both pass lithologically up into the Gault,
and also contain in their upper part "nodules with Gault-like
fossils," ‡ and the same view of their relationship was taken by
M. Barrois, who mentions that not only are several fossils of the
Ammonites mammillaris zone, which in France is included in the
Gault, found in the upper part of the Folkestone Beds, but that
the brachiopods which occur in this zone are especially abundant
in the lower part of the same strata. He concludes that unless
the Folkestone Beds, like the *A. mammillaris* zone, are classed
with the Gault, there is no satisfactory upper limit to the Aptian
in England.§ Mr. Price, on the other hand, would retain the
zone of *A. mammillaris* in the Upper Neocomian.‖

---

* Geology of the Weald, p. 142.
† *Bull. Soc. Géol. de France,* sér. 2, vol. xvii. p. 32. 1860.
‡ On the Lower Cretaceous Rocks of the Bas-Boulonnais, &c. *Quart. Journ.
Geol. Soc.,* vol. xxiv. p. 474. 1868.
§ L'Age des "Folkestone Beds" du Lower Greensand. *Ann. Soc. Géologique
du Nord,* t. iii. p. 23. 1875.
‖ Monograph of the Gault, 1880, p. 35.

The nodules and fossils referred to above occur in three to four feet of sand, which form the top only of the Folkestone Beds. This sand, which both passes up into, and possesses this palæontological affinity with the Gault, seems to be an expanded representative of the grit-band of Midhurst and Pulborough, which also passes up into the Gault. The grit-band, as before explained, is sharply marked off from the underlying mass of the Folkestone Beds ; if the sand with Gault fossils could also be separated from the Folkestone Beds, we should no longer have to face the anomaly of the upper member of the Neocomian group being characterised by a Gault fauna, and should also be able to point in the Wealden area to a basement-bed to the Gault corresponding to the Carstone of the Isle of Wight. At present, however, it must remain uncertain whether an upper portion of the Folkestone Beds can be separated off, as an equivalent to the Carstone of the Isle of Wight, or whether the Carstone changes horizontally into a sand of the usual Folkestone Beds type during its passage northeastwards below the Hampshire Basin.

The fossils of the Carstone of the Isle of Wight, so far as they go, indicate as close a relationship with the Upper Neocomian as with the Gault. Two forms, however, occur which are not known below the Folkestone Beds, viz., *Lima parallela*, Sow., which ranges through the Gault, and *Ammonites Beudantii*, Brong., which occurs in the *A. mammillaris* zone both in England and France, as well as in the zone between the Upper and Lower Gault, to which it gives its name. The following is the complete list :—

Fossils of the Carstone.

Wood (Bonchurch and Dunnose).
Echinoderm, fragment (Bonchurch).
Enallaster (Hemipneustes) Fittoni, *Forbes;* as a pebble (Bonchurch).
Crustacean fragment (Bonchurch).
Hoploparia longimana, *Sow.* (Sandown and Dunnose).
Avicula (Bonchurch and Blackgang).
Astarte (Sandown).
Cardium (Bonchurch and Sandown).
Exogyra (Sandown and Blackgang).
Leda scapha?, *D'Orb.* (Sandown).
Lima (Blackgang).
Lima parallela, *Sow.* (Blackgang).
Nucula (Blackgang).
Panopæa? (Fitton, Blackgang).
Pecten orbicularis, *Sow.* (Bonchurch, Dunnose, Sandown, Blackgang).
Pecten quinquecostatus, *Sow.* (Sandown).
Plicatula carteroniana, *D'Orb.* (Sandown).
Tellina (Sandown).
Venus? (Fitton, Blackgang).
Actæon (Sandown).
Pleurotomaria (Blackgang).
Solarium (Forbes, Blackgang).
Trochus (Bonchurch).
Ammonites, fragment (Blackgang).
    ,,   Beudantii, *Brong.* (Blackgang).
Lamna, tooth of (Dunnose).

COMPTON BAY TO REDCLIFF.

At Compton Bay the Carstone is a brown sandstone, having as its basement layer a band, three inches thick, of quartzite pebbles, ranging up to three-quarters of an inch in length, with rolled phosphatic pebbles, many bits of wood, and cylindrical concretions which seem to have been formed in place. Though the beds below also contain pebbly bands, they appear to be more of the type of the Sand-rock Series, and to be divided from the Carstone by a hard and fast line. Upwards the Carstone passes gradually into the Gault, the nature of the junction being shown in Fig. 7 (p. 23) and in the accompanying sketch by Professor E. Forbes.

FIG. 13.

Junction of the Gault and Lower Greensand in Compton Bay.

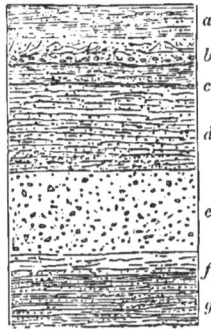

FT. IN.

a. Dark blue sandy clay (Gault).
b. Brown sand with a pebble-band, three inches thick, at the base, containing quartz-pebbles, many pieces of wood, and some phosphatic pebbles (Carstone) - - 6 0
c. Blue sandy clay - - - - - - 2 6
d. Grey and greenish sand with small quartzite pebbles at the top and the bottom, and with a layer of pyritised wood, 4 feet from the base - - - - - 13 0
e. Bright-yellow sand - - - - - 9 0
f. A ferruginous band, about - - - - 1 0
g. Irregularly interlaminated white sand and blue clay (for the continuation of this section, *see* p. 22).

Eastwards from Compton Bay there is no section of the Carstone, though its position can be determined with some accuracy by the nature of the soil. In the section of the Sand-rock Series at Rock (p. 41) the base of the Carstone is exposed, but no more.

There are indications, however, of the steady thickening of this subdivision eastwards. Not only does the outcrop widen, but south of Coombe Tower the rock begins to form a distinct escarpment, which gradually becomes the best marked feature in the Lower Greensand. Wherever exposed the rock consists of a brown and ferruginous grit.

By the side of the high road from Chale to Chillerton a pit shows the base of the Carstone, consisting there of a ferruginous grit with a few pebbles at the base, and resting on sand and clay with markings resembling fucoids, about 6 feet thick, under which lies white sand. The escarpment continues to grow in importance, but excepting in a lane near Roslin, presents no sections till we reach Rookley Green, the road-cutting south of which place shows yellow and white laminated sand and loam (Sand-rock Series) in the lower part, and ferruginous sand and loam with some clay nearer to Rookley Green. Thence the Carstone sweeps round to the east and north of Rookley, and crosses the same road south of Blackwater, in a cutting where it rests on white sand.

It is next seen in small pits near Park Cottage, but is better exposed in a road-cutting at Sandway, 300 yards east of White-croft, where it rests on the white sand previously alluded to (p. 42).

A short distance to the north, at Marvel Wood, the Carstone rises into one of the boldest escarpments in the Isle of Wight, of which the section was given on p. 42. It here rests on sands in which current-bedding is very conspicuous. The definiteness of its base, taken together with the manner in which it crosses the edges of the current-bedding planes of the strata below, gives a strong appearance of unconformity, which is heightened by the fact that the grey sand, 3 feet thick, on which the Carstone reposes, looks as if it had been "reconstructed" from the clays and white sands of the Sand-rock Series. The mapping of the Island as a whole did not, however, support the idea of an un-conformity at this horizon, though there may have been local erosion and redeposition. The base of this subdivision may be followed along Marvel Wood to the head of the valley on the west side, where two small pits give a similar section.

The Carstone is next seen in the lanes near Newclose House, but, owing to the rapidly increasing dip, the outcrop becomes narrow, and the escarpment insignificant. On the east side of the Medina it is seen in the lane leading up the hill past Standen. The upper beds of the Sand-rock Series are also brown here, but may be distinguished without difficulty from the coarse ferruginous grit of the Carstone.

From St. George's Down eastwards the position of the Carstone is marked by a slight rise in the ground, and the highly ferruginous soil. The rock is exposed in the road-side at Great East Standen, but does not appear again till we reach a small opening 300 yards south-east of Heasley Lodge, where it rests on buff sand.

At Knighton it forms a fairly well-marked feature, and is exposed in the wooded bank on the east side of the stream, and again in the valley a quarter of a mile west of Kern. East of Kern the dip increases and the outcrop narrows down to a mere line. There is a small exposure 250 yards north-west of the Roman Villa at Brading.

This brings us to the coast section at Redcliff, the section of which was given on p. 35. The Carstone here, as everywhere, passes up into the Gault, and shows at this locality a greater thickness than in any other part of the Isle of Wight, namely, 72 feet 9 inches. A small fault, previously alluded to, is clearly shown in the Carstone, and in some of the beds below it. Such phosphatic concretions as occur consist of cemented masses of grit, and seem to have been formed in place. The whole rock is markedly ferruginous.

From Niton and Blackgang to Shanklin and Bonchurch.

We will now trace the course of the Carstone around the southern hills of the Island, proceeding as before from west to east. The exposures about the Undercliff near Blackgang are numerous and easily accessible. The Carstone forms the brow of a shelf in the cliff, which is occupied by the Gault, or more usually by the débris of Upper Greensand and Chalk that has slid down over the Gault. This brow may be traced continuously from Chale to the Chalybeate Spring. It reappears above Knowles, and near the foot of the cliff below Niton presents its most eastern exposure. Still further east the southerly dip is believed to carry the Carstone down to the level of the beach, but no rock appears *in situ* to determine the point.

The following section was noted above the Chalybeate Spring :—

		Ft.	In.
Gault ; blue clay passing down.			
Carstone ⎰ Brown grit, interbedded with grey clay, and containing phosphatic nodules in the upper part		8	0
Blue clay		3	0
Reddish-brown grit, very red in parts		1	0
Line of small quartz pebbles with rolled phosphatic nodules up to 2 inches in diameter		0	2
Sand-rock Series (for details, *see* p. 31).			
		12	**2**

In the cliff below Niton we find the following details :—

		Ft.	In.
Gault ; blue clay passing down.			
Carstone ⎰ Brown grit		3	3
Clay-parting		0	1
Brown grit with phosphatic nodules		1	4
Brown sand and clay		6	0
Pebbly and ferruginous band		0	6
Sand-rock Series ; grey sand with seams of blue clay, seen to 4½ ft.			
		11	**2**

In Blackgang Chine, and on either side of it, the Carstone with
the base of the overlying Gault is repeatedly exposed, but a little
north of the Chine, reaching the top of the cliff, it strikes inland,
its base being exposed in the hill on the south side of the high
road near Cliff Terrace.

On proceeding inland along the outcrop of the Carstone, we
are soon struck with the fact that it is more often than not overspread with Gault clay. The appearance of the ground at once
supplies the explanation. Over large areas the clay from the
Gault outcrop has slid down and spread itself as a skin over a
more or less even slope of Carstone, but is still easily distinguished
by the hummocky appearance of the ground it occupies, as well as
by the character of the soil. In some places the clay has flowed
down in the form of mud-rivers, keeping usually to the lines of
hollow in its descent, but overspreading also many of the higher
parts of the Carstone feature. The course and limits of these
mud-rivers or gutters may be distinguished, for many years after
they have ceased to move, by the large sods of turf which have
been torn off and heaped in a little irregular bank along their
edges, and by the lines which still serve to indicate where the
mass of moving clay was traversed by long curving cracks, convex
in the direction of movement. The mud-rivers extend sometimes
to a distance of a quarter of a mile or more beyond the base of the
Gault.

The sections along the western slope of St. Catherine's Down
are few and poor, but at its extreme north end pebbly Carstone
rests on buff and white sand. On its east side the guttering
of the Gault, assisted by the slight easterly dip of the strata, has
been more than usually extensive, but the Carstone near Wydcombe forms a characteristic feature. It may be followed round
the south side of the house, and is seen at a small waterfall
350 yards south-east of it. Near here three outliers of Carstone
cap conspicuous hills, the lower portions of which consist of
white sand and sand-rock. The base of the Carstone appears in
two sand-pits 300 yards west, and the same distance north of
Itchall, which show clayey sand and ironstone resting on white
sandstone. A similar section occurs at Sheepwash, where the
Carstone forms a fine escarpment, corresponding to the feature at
Marvel Wood, which we have already described. The strata
being nearly horizontal, the Carstone runs for a long distance
along the tops of steep spurs of white sand and sand-rock that jut
out from the hill-side. Presenting everywhere the same ferruginous character, it may be readily distinguished from the series
below. The slipping down of the Gault is especially noticeable
south of Godshill Park. Redhill, where there is a good section
of the Carstone, has been named, like Redhill in Surrey, from the
ferruginous colour of the soil.

In Appuldurcombe Park and about Wroxall, a large area is
occupied by slipped Gault; but the Carstone appears by the side
of the road north of the village, and its base is well exposed at
Yard Farm, where it rests on white sand

At Winstone, a fine example of a mud-slide is crossed by the railway cutting, now grassed over. Another a little to the east has travelled down a hollow in the hill-side, and is now being dug for bricks. On the hill-side above the brick-pit a small opening has been made in the Carstone.

From here to Shanklin occasional small sections serve to fix the position of the Carstone, but call for no particular notice. In the great cliff-section, however, which extends from Knock Cliff to near Bonchurch, this subdivision is finely shown. It strikes the coast half a mile north of Luccomb Chine, and forms thence the brow of the cliff to Monk's Bay, where it comes nearly to the beach. West of this, through everywhere hidden by landslips, it probably descends to the level of the beach, as is believed to be the case near Niton. Everywhere it passes up into the Gault, and rests with a sharply-marked base on the brightly coloured sands of the Sand-rock Series. The following section was noted in Monk's Bay :—

		Ft.	In.
Gault.	Blue micaceous clay passing down.		
Carstone	Blue micaceous clay with lines of grit -	3	0
	Brown ferruginous rock with derived phosphatic concretions containing oolitic grains of iron oxide - - - -	1	0
	Sandy and gritty blue clay, passing down -	1	0
	Clayey brown grit with nodules as above -	3	0
	Brown grit - - - -	6	0
	Brown grit with many small pebbles -	20	0
	Pebbly band, with quartzites up to half-an-inch in length - - - -	0	3–6
Sand-rock Series.	Bright-yellow and white sand.		
		34	6

A well-rolled specimen of *Enallaster* (*Hemipneustes*) *Fittoni*, Forbes, was found as a phosphatic nodule in the clayey brown grit, 3 feet thick. This fossil is recorded as occurring at Horseledge (p. 261), and more abundantly in the same beds at Atherfield, and in the Hythe Beds at Hythe. Its occurrence therefore as a derived specimen in the Carstone is significant.

CHAPTER VI.

THE GAULT AND UPPER GREENSAND.

1. THE GAULT.

INTRODUCTION.

THE Gault, which rests quite conformably on the Carstone, may be described generally as a blue or bluish grey clay, more or less sandy, and with minute spangles of mica. It contains little or no calcareous matter, such proportion of this material as may have been originally present having been converted into sulphate of lime, which in the form of small crystals of selenite sometimes occurs in considerable quantity. The fossils are few, and distributed at rare intervals.

In thickness the Gault varies from 120 feet at Culver to 146 feet at Blackgang, and 139 feet in Compton Bay.   At Punfield, where, however, it is difficult of measurement, it is about 111 feet thick.   In its upper part it becomes sandy and lighter in colour than in the lower beds, so as to pass almost insensibly into the Upper Greensand.   The proportion of sand increases westwards in these passage beds, so that at Punfield the name of Gault, as indicating a clay, becomes quite inapplicable.   In the extreme west (Black Down) the whole formation seems to pass into a sand.

LANDSLIPS.

The Gault has received the name of the " blue slipper "* in the Isle of Wight, from its tendency to give rise to landslips, or of " Platmore," a name which was in former days applied to any close black earthy stone or clay.   The beautiful and romantic scenery of the Undercliff or " Back " of the Island has been mainly caused by the sliding of the Chalk and Upper Greensand over the unctuous surface of the Gault clay, the tendency to slide being principally due to a rather pronounced seaward southerly dip, and to the outburst of springs at the junction of the porous Upper Greensand and impervious Gault.

---

* The term "slipper" is applied in the Island to any bed which gives rise to landslips.

Through the greater part of the Undercliff the slipped materials assumed a position of rest before the commencement of the historic period. It seems likely that in the belt of ground occupied by the slip, the southerly dip was steeper than it is in the existing cliff, and that the strata now forming this cliff will never be in a position to slide so readily as those portions that have already gone. Still, as the sea, in the course of centuries, removes the fallen *débris* which forms the coast, the movements will doubtless be renewed from time to time. Indeed, at Blackgang and Bonchurch, the west and east ends respectively of the Undercliff, there have been great slips within the present century.

The following account of the East End Landslip, which took place in 1810 in Bonchurch and Luccomb, is taken from one of Mr. Webster's letters, dated May 27th, 1811, and published in Sir Henry Englefield's Isle of Wight (p. 131) :—

" I was surprised at the scene of devastation, which seemed to have been occasioned by some convulsion of nature. A considerable portion of the cliff had fallen down, strewing the whole of the ground between it and the sea with its ruins ; huge masses of solid rock started up amidst heaps of smaller fragments, whilst immense quantities of loose marl, mixed with stones, and even the soil above with the wheat still growing on it, filled up the spaces between, and formed hills of rubbish which are scarcely accessible."

" Nothing had resisted the force of the falling rocks. Trees were levelled with the ground ; and many lay half buried in the ruins. The streams were choked up, and pools of water were formed in many places. Whatever road or path formerly existed through this place had been effaced ; and with some difficulty I passed over this avalanche which extended many hundred yards."

" Proceeding eastward, the whole of the soil seemed to have been moved, and was filled with chasms and bushes lying in every direction . . . . I perceived, however, on my left hand, the lofty wall of rock which belonged to the same stratum as the Undercliff."

This description of the scene is equally applicable at the present day, except that the ruins are covered with vegetation. Huge pinnacles or slices of the Upper Greensand have moved down a few feet only and remain with their upper parts resting against the parent cliff, but separated from it below by a narrow cleft, along which it is possible to squeeze one's way. The top of the Gault is everywhere concealed by fallen rock.

At the west end of the Undercliff, under Gore Cliff, a great slip took place in 1799, and the movement has been renewed from time to time ever since. A letter, dated Niton, February 9th, 1799, and published in the Isle of Wight Magazine for the same year, is quoted by Mr. Norman as follows :*— " The whole of the ground from the cliff above was seen in motion . . . . The

---

* Geological Guide to the Isle of Wight, 8vo., Ventnor, 1887, pp. 187–189.

ground above, beginning with a great founder at the base of the cliff immediately under St. Catherine's Down, kept gliding forward, and at last rushed on with violence, totally changing the surface of all the ground to the west of the brook that runs into the sea, so that now the whole is convulsed and scattered about, as if it had been done by an earthquake. . . . . The cascade which you used to view from the house at first disappeared, but has now broken out and tumbled down into the withey-bed, of which it has made a lake."

Mr. Norman relates that an enormous mass of rock by the road beneath Gore Cliff "once formed part of a large pinnacle which had become loosened from the cliff and overhung in a manner extremely threatening to the safety of the public. The authorities decided upon its removal by means of gunpowder. In its fall it carried with it tons of adjacent rock and *débris*, entirely blocking and destroying the roadway made round the landslip of 1799" (*op. cit.,* p. 189). The roadway has again been threatened with destruction (1887) by the constant slipping of the Gault, some of the rain gullies having cut their way into the slope as far as the seaward fence of the highway.

The most striking feature in the central parts of the Under-cliff is the succession of short escarpments produced by the fall of slices of the Upper Greensand cliff. These portions range in size from mere blocks up to slices of half a mile in length. They have broken off along the vertical joints by which the sandstone is traversed, and as their bases slid forward over the Gault, have slowly acquired a steep landward (northerly) dip. The process has been repeated several times, thus producing at different levels in the Undercliff a series of Upper Greensand escarpments, separated by deep hollows, which have been not uncommonly occupied by natural lakes. The distance to which they have descended varies indefinitely. Above Bonchurch a very long but narrow slice has moved a few feet only, and still forms the principal face of the cliff. But many others, with a portion of Chalk above them, have descended to the beach some 300 feet below, and from a quarter to half a mile distant.

Such wholesale slipping is, generally speaking, confined to the coast, but some large masses of Greensand have slid down on all sides of St. Catherine's Down, and from the shoulder which separates Shanklin and Luccomb. The slipping down of the Gault in great mud-rivers all round the southern Downs has already been noticed (p. 58). It does not take place along the Central Downs of the Island, where the dip is generally at a steeper angle and into the hill-side.

DESCRIPTION OF SECTIONS.

The best section of the Gault is afforded in Compton Bay, where nearly the whole deposit may be examined, the section being as follows :—

Section of the Gault in Compton Bay.

FEET:

Upper Greensand (for details see p. 68).

Gault
- Passage Beds.
  - Hard blue clayey bands with fucoidal markings alternating with sandy bands, containing iron pyrites - - - 6
  - Pale blue silty sand or sandy micaceous clay with fucoidal markings, weathering yellow - - - - - 30
- Clay, as above, but of a deeper blue - - - 8
- Greenish clay - - - - - - 2
- Blue clay as above, with fish-scales, &c. in several bands - - - - - - 20
- Blue clay - - - - - - 73

Carstone (for details see p. 55).

139

The passage up from the Gault is illustrated in the accompanying sketch (Fig. 14), made in the cliffs at Compton during the progress of the geological survey of the Island in 1852.

Fig. 14.

Junction of the Upper Greensand and Gault in Compton Bay.

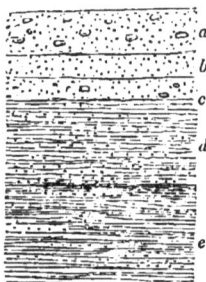

FT. IN.

a. Upper Greensand. Hard concretionary band, with phosphatic nodules - - - - - 1 0
b. Passage by a bluish sand with thin fucoidal markings, into - - - - - - 0 6
c. Green sandy band with a few nodules - - - 0 6
d. Dark blue sandy clay - - - - - 2 0
e. Paler and darker beds with small nodules: FOSSILS, *Gryphæa, Vermicularia, Arca* (rare).

The passage beds, in the former Edition of this Memoir, were included with the Upper Greensand. Lithologically, however, they are more nearly allied to the Gault, with which they have usually been grouped of later years.

Downwards the Gault passes into the Carstone as described on p. 55. In its lower part Mr. Norman observed *Inoceramus sulcatus, Natica gaultina,* and *Ammonites dentatus* (var. of *A. interruptus,* D'Orb.), the last-named occurring as a brittle coal-black material, the inner whorls permeated by a phosphatic substance.[*]

At Blackgang the numerous sections in the lower part of the Gault have been noticed in the description of the Carstone. *Inoceramus sulcatus* and *I. concentricus* have been found in a gulley west of the hotel. The top of the Gault appears in Gore Cliff, this being the only spot in the Undercliff where it is not concealed by fallen rubbish. The beds are similar to those at Compton Bay, and the thicknesses differ but little. According to Mr. Simms[†] there are here 43 feet of light-coloured Gault (passage beds), and 103 of blue Gault, giving a total of 146 feet.

The sections in the cliff from Bonchurch to Knock Cliff show the lower beds of the Gault only. The passage downwards into the Carstone may be conveniently examined in the brow of the cliff near Bonchurch (p. 59).

In Sandown Bay the position of the Gault is marked by a narrow hollow in the cliffs. The passage beds into the Upper Greensand above and the Carstone below are there exposed, but the rest of the deposit is concealed by vegetation. The top layers consist of alternations of blue sandy clays and sands with *Vermicularia*, about 15 feet thick, and the lower beds of darker blue micaceous clay. The total thickness of the Gault here is about 120 feet.

Through the central parts of the Island, the Gault occupies a narrow belt of low ground, separating the Upper and Lower Greensands. When not overspread by a downwash of sand, the soil of this belt is wet and rush-covered, and presents a characteristically different appearance from that of the strata above and below. But as a rule the Gault is entirely masked, and sections are exceedingly rare.

The passage beds into the Upper Greensand are seen in a lane 100 yards south-west of Rill, near Chillerton. At Gossard Hill, near Rookley, where a long shoulder of Gault, capped by an outlier of Upper Greensand, juts out across the Medina, a brick-pit has been opened; but only the weathered surface of the Gault is worked, a pale-blue or nearly white structureless clay. A better section is provided in the brick-pit at Bierley, near Niton, where the lower beds of the Gault are exposed.

The brick-pits by the side of the railway between Wroxall and Shanklin are worked in Gault that has slipped down the hill-side below the true outcrop (p. 59). One of the most noticeable features in connection with the outcrop of the Gault, is the copious supply of water which it throws out nearly all round the southern Downs of the Island. The greater part of the strata over-lying this clay being of a permeable nature, the rainfall is absorbed by them, and is thrown out in a line of springs along the top of the first impermeable bed it encounters. The springs are of course most copious along the hill-sides where the Gault is at the lowest level, the

---

[*] Geological Guide to the Isle of Wight, p. 70.
[†] *Quart. Journ. Geol. Soc.*, vol. i. p. 76 (1845).

underground water naturally moving down the dip-slope of the beds; but, the dip being very gentle, there are springs along nearly the whole Gault outcrop. The most copious occur at Wydcombe, Bierley (utilised for the Niton and Whitwell Water-works), Niton, Whitwell, south and south-east of Wroxall, and in Greatwood Copse near Shanklin. The natural spring which formerly issued at the last-mentioned locality was utilised for the Shanklin Water-works, the supply of water having been some-what increased by driving a heading into the hill along the junction of the Upper Greensand and Gault. Ventnor is sup-plied by a spring issuing from the same strata, and met with in driving the railway tunnel. Several springs take their rise in the same neighbourhood, and were formerly used to drive a mill in Ventnor Cove.

Along the central chain of hills the springs are less frequent, owing to the steep inward dip of the strata. But a fine spring issues at Bottlehole Well near Brixton, and another, issuing, however, in the Upper Greensand, gives its name to the village of Shorwell. About Chillerton and Gatcombe, where the dip is very gentle, numerous springs rise along the sides, and particularly at the heads of, the valleys.

At Knighton there are good springs, which, supplemented by a well, are utilised for the supply of Ryde.

CORRELATION WITH THE MAINLAND.

The zones into which the Gault of Folkestone has been divided by Messrs. De Rance* and Price† have not been recognised in the Isle of Wight, and it is the opinion of the latter that the Gault of the Island is of Upper Gault age (Monograph of the Gault, p. 27). This opinion was founded on the occurrence of *Inoceramus sulcatus, Ammonites rostratus, Solarium ornatum, Belemnites ulti-mus,* &c. Of these *Ammonites rostratus,* and *Inoceramus sulcatus* are confined to the Upper Gault, but *Belemnites ultimus* ranges throughout the deposit, while *Solarium ornatum* occurs in the Lower, as well as in the Upper Gault. On the other hand *Ammonites dentatus* is a variety of *Ammonites interruptus* which gives its name to the lowest zone of the Gault at Folkestone, from which it would seem that the Lower Gault also is represented in the Isle of Wight. This might be likewise inferred from the absence of any break in or below the Gault of the Island. A complete list of the fossils will be found in Table III. of Appendix II.

UPPER GREENSAND.

INTRODUCTION.

This rock forms one of the most conspicuous features in the Island, namely the cliff which overhangs the Underclliff from

---

* *Geol. Mag.* for 1868, p. 163.

† *Quart. Journ. Geol. Soc.,* vol. xxx. p. 342, 1874, and a Monograph of the Gault, 8vo. London. 1880.

Bonchurch to Blackgang, and which reappears inland in the bold brows of St. Catherine's Down, Head Down, Gat Cliff, and St. Martin's Down (Cook's Castle Crag). In the central range the same rock forms the bold ridge of Rams Down, which is scarcely less conspicuous than the Chalk Downs themselves.

The existence of these striking features is due to the hardness of a bed composed of alternations of chert and sand, and underlain throughout the central parts of the Island by a band of freestone. The position of the base of the Chert Beds has been indicated on the map by a broken line in the central and southern parts of the Island, principally on account of their topographical importance.

Above the Chert Beds a variable thickness of glauconitic sands passing up into the Chalk Marl is known as the Chloritic Marl. Below the Chert Beds there lie from 70 to 90 feet of sands, called "malm," with bands or lenticular masses of chert and cherty limestone or "rag." Other local names of less common occurrence are "hassock" for the sands, "whills" for sandstone, "shotter-wick" for chert, "firestone" for a stone formerly employed for lining hearths, and "rubstone" for a stone once used for whitening hearths or doorsteps.

The thicknesses of the Malm Rock and Chert Beds are given for different localities in the Isle of Wight, and for Punfield, in the following table, the thickness of Gault at the same spots being appended to show that the Upper Greensand and Gault thicken and thin together, and not one at the expense of the other.

	Punfield. Feet.		Compton Bay. Feet.		Gore Cliff. Feet.		Culver.
Chert Beds -	6 ⎫		13 ⎫		27 ⎫		
	⎬ 45		⎬ 86		⎬ 121½ - -	80
Malm Rock -	39 ⎭		73 ⎭		94½ ⎭		
Gault -	- 111		- 139		- 146		- 120

The Malm Rock passes downwards into the strata which have been above referred to as "passage beds" into the Gault. A convenient base for this subdivision has been selected near Ventnor by Mr. Parkinson[*] in a band of chert nodules from which the carapace and rib-bones of a fresh-water tortoise (*Plastremys lata,* Owen) were obtained by Mr. Norman, and the remains of *Hoploparia Saxbyi,* M'Coy, by Mr. Saxby.[†] In other parts of the Island the base has been drawn where the clayey bands begin to predominate over sandy beds.

The zone of *Ammonites inflatus* occurs, according to Mr. Parkinson, rather more than 20 feet from the base, while *Ammonites rostratus* attains its greatest development about 11 feet from the top of the Malm Rock. By Dr. Barrois, however, the Malm

---

[*] *Quart. Journ. Geol. Soc.*, vol. xxxvii. p. 370 (1881).
[†] *Ann. Mag. Nat. Hist ,* vol. xiv. p. 116 (1854).

Rock (excluding a few feet at the top) was grouped with the passage-beds into the Gault as the zone of *Ammonites inflatus.**

The most important bed commercially is the band of freestone, from 3 to 5 feet thick, above alluded to as occurring a short distance below the base of the Chert Beds. This freestone is not recognisable in the east or west ends of the Island, but has been largely worked as a building-stone in the southern hills, being especially conspicuous in the cliff between Blackgang and Bonchurch. Between it and the Chert Beds lie one or two bands of "firestone" and "rubstone."

The Chert Beds attain their fullest development near Ventnor. In Sandown Bay they can scarcely be recognised. The chert, though used for road-metal, is not much worked, except in gaining access to the freestone below. Some of the beds of chert are crowded with the spicules of sponges.

Dr. Hinde† remarked of the Chert Beds of the quarry at Ventnor Station that they "so abound with spicules that they may be considered as a continuous sponge-bed. . . . . The chert is usually of a light brown tint, and in thin sections under the microscope it is seen to be filled with spicules and spicular casts imbedded in a translucent matrix of chalcedonic silica. The spicules are likewise of chalcedony, and their canals are infilled with glauconite. Another variety of chert, also very abundant, is of a grayish or greenish-white tint; it differs from the former in that the matrix is of amorphous silica, while the inclosed spicules are of chalcedony. The chert bands . . . . are enveloped in an outer crust, of varying thickness, of white or yellow siliceous porous rock, which is interspersed with the empty moulds of spicules.

"In some of the thicker masses of chert there are cavities or pockets filled with spicules, loosely mingled in a grayish silicocalcarous powder, in which there are also numerous well-preserved foraminifera, chiefly of the genus *Textularia*. The spicules in these cavities have undergone a remarkable alteration in structure; they appear to have lost their original silica, which has been replaced by glauconite and some other silicate of a greenish-white aspect. The replacing material has only partially filled the form of the original spicules, and thus they look like mere shadowy casts of complete spicules. These in many cases are peculiarly distorted and contracted." Spicules occurred in the lower beds in the quarry also, but not so abundantly.

By Dr. Barrois the Chert Beds and the freestone below them were correlated with the Warminster Beds. A specimen of *Clathraria Lyellii*, a cycadeous plant, which it will be remembered occurs in the Wealden Beds, has been obtained from the Upper Greensand by Capt. Ibbetson in bastard freestone at the

---

* Recherches sur le Terrain Crétacé Supérieur de l'Angleterre et de l'Irlande, p. 107.
† *Phil. Trans.*, vol. 176, p. 418. 1886.

base of the Chert Beds.* Another specimen has been recorded
by Mr. Parkinson from the Chert Beds at Steephill, about 10 feet
below the Chloritic Marl.† A femur of a reptile is stated by
Mantell to have been found at Bonchurch three or four feet above
the firestone.‡

For the other fossils the reader is referred to the tabulated lists
at the end of the volume.

COAST SECTIONS.

1. *Compton Bay.*

The following details were observed in the cliff forming the west
side of Compton Bay:—

			FEET.
Chalk Marl (see p. 83).			
Chloritic Marl (see p. 81).			
Upper Greensand	Chert Beds.	Green sand with 10 or 12 bands of chert, light-brown outside, blue inside - - -	13
	Malm Rock.	Darker green sand, light-green when dry, with small scattered phosphatic nodules and lenticular masses of chert or rag - - -	32
		Sandstone, jointed and weathering into caves at the foot of the cliff. Many black nodules scattered throughout - - - -	41
Gault	-	- Passage Beds (see p. 63).	
			86

The Chert Beds are not so well developed here as in the central
parts of the Island, and the chert itself is more calcareous. The
freestone bed also, so marked a feature in the Undercliff, cannot
be recognised.

2. *Blackgang to Shanklin.*

Gore Cliff shows the Upper Greensand in a form that is typical
of the central and southern parts of the Island. The Chert Beds
form a vertical face, deeply scarred by the weather, each band of
chert forming a ledge, while the soft sands between have been
scooped out by the wind. At the foot of this vertical part of the
cliff the 5-foot bed of freestone runs for some miles and can
generally be recognised at a glance. The Malm Rock below
forms a steep, often precipitous slope.

---

* Notes on the Geology and Chemical Constitution of the various Strata in the
Isle of Wight, p. 25. See also *Quart. Journ. Geol. Soc.*, vol. xxxvii. p. 372. The
specimen is incorrectly stated by Mantell (Geol. Excursions in the Isle of Wight,
pp. 215, 217) to have been found in the Chalk Marl.
† *Quart. Journ. Geol. Soc.*, vol. xxxvii. p. 372. 1881.
‡ Geological Excursions, pp. 179, 180.

Gore Cliff.

		FT.	IN.
Chloritic Marl (see p. 81).			
Chert Beds. Alternations of chert and sand -	-	27	0
Firestone and rag - - -	-	2	0
Freestone { Bastard freestone 1 ft. } Freestone - 4 „ } -	-	5	0
Malm Rock { Sand with rag - - - -	-	,58	6
„ with many ledges of rag - -	-	27	0
Blue clayey sand - - -	-	1	6
Blue micaceous sandstone - -	-	0	6
Gault - Passage Beds.			
		121	6

A still more convenient spot for examining the upper part of the Greensand, known as the Cripple's Path, slants up the cliff, south-east of the village of Niton. The Chloritic Marl, however, is not seen.

At Ventnor the section of the beds is given by Mr. Norman as follows :—

Section above Ventnor.

		FT.	IN.
Chert Beds, { Alternations of chert and sandstone beds, 21			
24 feet. { to 24 in number - - -	-	24	0
Firestone - - - -	-	2	0
Rag - - - - -	-	1	0
Freestone Bed [bastard in upper part]	-	5	0
Rag - - - - -	-	1	0
Sandstone - - - -	-	3	0
Rag - - - - -	-	0	8
Soft sandstone - - -	-	4	0
Black band - - -	-	1	0
Malm Rock, { Soft yellow sandstone ("Whills") -	-	7	0
81 feet. { Rag - - - - -	-	1	0
Compact reddish sandstone - -	-	10	0
Rag - - - - -	-	1	0
Compact reddish sandstone - -	-	10	0
Mammillated rag - - -	-	0	8
Soft yellow micaceous sands with concretions	30	0
Dark coloured rag - - -	.	1	0
Dark clayey bed - - -	-	2	0
Hard blue chert, with crushed *Inoceramus* -	-	0	8
Gault - Light-grey sandy micaceous clay.			
		105	0

Near Shanklin in some quarries where the "free-stone bed" is worked for building, and the beds above and below it for road-making, the following sections were noted.

Quarry on the south side of the Luccomb Valley.

	FT.	IN.
Alternations of chert ("shotterwick") and sand (top not seen)	15	0
Rag in lenticular masses - - - - -	0	0–8
Firestone - - - - - -	2	6
Rag - - - - - - -	0	6–12
Firestone - - - - - -	3	0
Rag - - - - - - -	0	6–12
Firestone or Rub-stone (a stone formerly used for whitening hearths, &c.) - - - - -	0	8
Freestone - - - - - -	4	0

Quarry on the north Side of Greatwood Copse.

	Ft.	In.
Chert, rag, and sand (top not seen) - - - -	15	0
Rag - - - - - - - -	0	0–6
Firestone - - - - - - -	2	0
Rag - - - - - - - -	0	0–8
Firestone - - - - - - -	2	0
Rag - - - - - - - -	0	0–12
Rubstone - - - - - - -	0	8
Freestone - - - - - - -	4	0
Rag - - - - - - - -	1	0
Inferior stone or malm - - - - -	5	0
Rag - - - - - - - -	1	0
Inferior stone - - - - - -	2	0

3. *Culver Cliff.*

In this section the layers of chert, so conspicuous near Ventnor, are represented by a few lenticular masses only, or by layers of a hard flinty stone. The freestone also can no longer be distinguished, and the whole group shows a loss in thickness of 18 feet.

Culver Cliff.

FT. IN.

Chloritic Marl (see p. 81).			
Chert Beds and Malm Rock.	{ Green sand with lenticular masses of black chert at 9–11 feet from the top, and some bands of hard grey stone - - -	80	0
Gault	- Passage Beds (see p. 64).		

INLAND SECTIONS.

1. *Along the Central Downs.*

Although numerous inland sections lay open the Upper Greensand, the whole subdivision is rarely exposed at one spot. An exception occurs in the road-cutting north of Brook, where the following beds are seen :—

Road-cutting three-quarters of a mile north of Brook Church.

FT. IN.

			FT.	IN.
Chalk Marl	{ Alternations of chalk and marl (top not seen), passing down - - -		120	0
	{ Rocky chalk, very impure, and with glauconite, passing down - - -		5	8
Chloritic Marl, 11 feet 6 inches.	{ Green sand with phosphatised Ammonites, &c. irregularly hardened into stone in the upper part - - - - -		11	6
Upper Greensand, 107 feet.	{ Chert Beds, 10 feet 6 inches. } Cherty lumps in sand		10	6
	{ Malm Rock, 85 feet.	{ Greenish sand with great lenticular and oval masses of rock - - -	85	0
Gault	- Passage Beds, not clearly seen.			

The Chert Beds are seen in a by-road above Dunsbury, and make a small but well-marked escarpment for about 600 yards westwards. The next exposure occurs in the road from Brixton to Calbourne where the Chloritic Marl, 14 feet 2 inches thick, abounds with phosphatised Ammonites. The Chert Beds appear also, but the greater part of the Malm Rock is concealed by a thick stratified talus of chalk.

Proceeding eastwards we find the Chert Beds at Coombe Tower beginning to form the feature, which becomes so conspicuous in the central and southern parts of the Island. In this neighbourhood the chert, white in colour and accompanied with much chalcedony, is exposed repeatedly all along the crest of the escarpment to Shorwell, where it is quarried, or rather dug, for building.

East of Shorwell the escarpment becomes steadily bolder, and we find blue chert associated with the white along the crest of the hill. At the east end of this hill, over the Chillerton road, freestone is worked in a quarry below the Chert Beds, this being the most westerly appearance of the bed so prominent about Ventnor.

Between the bold escarpment of Rams Down and the Chalk Downs runs the long winding valley of Chillerton Street, a slight prolongation of which would convert the Chert Beds of Rams Down into an outlier. This valley owes its existence to some springs issuing at the junction of the Greensand and Gault, along the line of a gentle syncline, which is indicated by the relative dips in Rams Down (from 4° to 5°), and in the nearly horizontal Chalk. The trough becomes more marked near Sheat, and in Gossard Hill.* Near the former place the Malm Rock dips north-east at 10°, and the Gault, striking right across the valley of the Medina, runs for nearly a mile eastwards around Rookley, while on the top of the shoulder thus formed, an outlier of Upper Greensand makes a narrow ridge, capped with chert and striking nearly due east and west, with a dip to the north of 8° to 10°. The north side of the syncline is not well defined, as the beds gradually assume a horizontal position. It might perhaps be more correctly described as a monocline, like that of the central axis of the Island, but on a small scale. (*See* also Horizontal Sections, Sheet 43, No. 2.)

Numerous old quarries in the Chert Beds and underlying freestone roughen the brow of the hill above Gatcombe and Whitcombe. On mounting this eminence, we find a long dip-slope stretching away westwards to the boundary of the Chalk Downs, which is generally marked by a rise in the ground.

In the valley of the Medina near Shide the Upper Greensand disappears from sight till we reach Great East Standen. In two large pits, however, long since completely overgrown, between West Standen and Great East Standen, "malm" is reported to have been dug. So far as can be judged the pits have been opened in the lowest beds of the Chalk Marl.

At Arreton, while the topographical feature of the Upper Greensand is well marked, the Chert Beds no longer form as definite a subdivision as heretofore. The stony bands, to which this feature is

---

* A bold hill near Rookley, so named in the old edition of the Ordnance Map.

due, seem to come in at a rather lower horizon, while the chert itself is impersistent. The escarpment becomes conspicuous at Knighton, where the dip is gentle, and is separated from the Downs by a deep valley. Some old quarries on either side of the Knighton valley have exposed friable green sand with cherty lumps. The springs previously alluded to (p. 65), issue at the base of the Chalk Marl. The sand is well exposed in a lane at the east end of Knighton East Wood.

This brings us to Yarbridge, where there is a fine section in the Chalk Marl, ending, however, at its junction with the Chloritic Marl. The latter can be seen in the sides of the lane which runs along the foot of the Down westwards, while the sand below it is shewn in the lane leading to Morton, 100 yards west of the High Road. The Chert Beds are not distinguishable.

East of the Yar the scarped ridge of the Greensand stands out prominently, and excepting a break at Yaverland, continues to do so till it presents on the coast the section which has already been described.

2. *Around the Southern Downs.*

On the west side of St. Catherine's Down several small pits occur along the scarped brow formed by the chert and freestone, the former material being used for road-metal The outcrop of the Upper Greensand is narrow, but steep, and on the broader slope of Gault lie many huge masses of Greensand that have slipped bodily down. The long flat-topped spur of St. Catherine's Down which juts out to the north, and marks the line of strike, is capped with a strip of Chert Beds, about 1,300 yards in length, but only from 50 to 80 yards in breadth, and terminates northwards in a remarkable semicircular hollow, which seems to have been formed by a landslip. The chert is worked for road-metal in small pits here, and on Head Down. West of Niton some old quarries range along the outcrop of the chert and freestone.

Another fine brow, known as Gat Cliff, is formed by these beds in Appuldurcombe Park. The dip being southerly, the boldest front is presented to the north. Here also a long line of old quarries marks the outcrop of the freestone.

In the valley south-east of Wroxall, along which the railway passes, several sections may be observed. The cutting by which the tunnel is approached has been made in the Malm Rock, the Gault, so far as can be seen, lying about the level of the rails. At the south end of the tunnel the rails are about eight feet below the freestone; the tunnel descends southwards at the rate of 1 in 173, and is about 1,300 yards in length. From these data it may be calculated that the dip of the strata to the south amounts to 1 in 38 or an angle of rather less than 2°.

St. Martin's Down which terminates northwards in nearly as bold a brow as that of Gat Cliff contains chert bands of exceptional thickness.

CHAPTER VII.

THE CHALK.

INTRODUCTION.

THIS formation extends completely across the Island in an east and west direction from the Needles to Culver Cliff. It may be examined both in the sea-cliffs and in the numerous pits with which its surface is covered throughout the entire distance between those points. It forms a range of elevated undulating hills, conspicuous from afar on account of their altitude, and the bold rounded outline they present to the eye, as well as from their bare and uncultivated surface, which is covered with a short grass, and is rarely used for any other purpose than the pasturage of sheep. In consequence of the high angle at which the Chalk dips throughout the greater part of its range from west to east, the breadth of surface occupied by it is inconsiderable compared with that of most of the strata above and below it, while, on the other hand, its horizontal extension increases in proportion as the inclination of the strata diminishes. For this reason, from Alum Bay to Mottistone Down, and from Carisbrook to Culver Cliff, between which localities the Chalk is nearly vertical, it constitutes a mere ridge of high land, scarcely a quarter of a mile broad in Afton Down. Between Mottistone Down and Carisbrook, where the strata become less inclined, the width of the band of Chalk exceeds three miles. For the same reason, the outliers of Chalk on the south side of the Island between St. Catherine's Down and Shanklin Down, although of inconsiderable thickness compared with the depth of the entire formation, yet in consequence of being nearly horizontal extend over a comparatively wide surface. Throughout the central range of the Island the dip of the Chalk gradually increases in amount towards its higher strata, becoming nearly vertical at its junction with the overlying Tertiary formations.

The well-known rocks called the Needles are large wedge-shaped masses of Chalk standing out in the sea, isolated from the main body of Chalk by the wasting action of the waves upon the coast. A lofty spire of chalk, which once rose as the most conspicuous of the group and chiefly suggested the name to these rocks, fell down in 1764. Conspicuous as they look from the land, the Needles appear of much larger dimensions when viewed from the sea. A base of 60 feet in diameter has been levelled on one of them for the foundations of the lighthouse, which was removed to it in 1858 from High Down, where it originally stood.

Other masses of chalk, consisting of the lowest beds of the flinty Chalk, form similar but smaller isolated rocks in the sea near the base of the cliffs on the east side of Freshwater Bay. These are shewn in the accompanying sketch, Fig. 15, by Prof. Edward Forbes, made in the year 1852, for his Memoir on the Tertiary Fluvio-Marine Formation of the Isle of Wight (Fig. II., p. 4).*

---

* These sea-stacks of Chalk seem to have undergone considerable diminution since this sketch was made.

Fig. 15.

Freshwater Bay from the West. From a sketch by Prof. E. Forbes.

The cliffs of flinty Chalk at the two ends of the Isle of Wight are among the finest to which this formation gives rise in the British Isles. The brow of the part known as the Main Bench, near the Needles, which is vertical and descends sheer into the water, was determined by the Ordnance Survey to be 416 feet above the datum-level, while the Grand Arch, which forms the east side of Scratchell's Bay and overhangs considerably, is 300 feet in height. It will be noticed that the flinty chalk alone is capable of forming these vertical or overhanging cliffs. Both here and at Culver, wherever chalk without flints rises next the sea, there is a beach of chalk blocks, and a more or less accessible slope at the foot of the cliff.* It is in the Chalk-with-flints also that the numerous caves of the neighbourhood of Freshwater, the Needles, and the extreme point of Culver Cliff, have been excavated.

In the Chalk of the Isle of Wight the following sub-divisions are recognisable. The thicknesses of the Middle and Lower Chalk have been obtained by direct measurement in Culver Cliff, that of the Chalk-with-flints by estimation.

		FEET.
Upper Chalk	Chalk-with-flints, about - -	1,350
	Chalk, nodular, but without flints -	15–25
	Chalk Rock, a line of green-coated nodules.	
Middle Chalk	Thick-bedded chalk, with thin partings of marl - - - -	166
	Nodular chalk (? Melbourn Rock) and marl (? Belemnitella Marl) - -	14
Lower Chalk	Massive chalk - - -	86
	Thin-bedded chalk and marl in numerous beds - - -	120
	Chloritic Marl - - -	7–15

The principal sub-divisions were first recognised by Mr. Webster in 1812 (Sir H. Englefield's Isle of Wight, p. 236). He used the names Chalk with Flints, Chalk without Flints, and Chalk Marl. The Chalk without Flints, he remarks, "differs from the former only in the absence of flints, in the beds being thicker, and the chalk being sometimes a little harder." The Chalk Marl is described by him as consisting "of chalk and an intimate mixture of clay . . . . It may be readily distinguished from chalk by its falling to pieces on being wetted and dried again."

In 1865 Mr. Whitaker† identified a line of green-coated nodules, occurring some 8 or 10 feet below the lowest course of flints, as the representative of the bed which he had previously in Berkshire named "Chalk Rock," and had taken as the topmost bed of the Lower Chalk, i.e., of the Middle Chalk of the above table.

---

* The coast from the Needles to Freshwater can be examined by boat only. The point on the east of Freshwater Bay and that of Culver Cliff can rarely, if ever, be passed on foot.

† Quart. Journ. Geol. Soc., vol. xxi. p. 400.

In 1875 M. Barrois published his *Description Géologique de la Craie de l'Ile de Wight*, in which he gives an exhaustive account of the literature, physical features, zones, and fossil contents of the Chalk. The sub-divisions which he adopted are as follows in descending order:—

Craie Blanche (Chalk with flints).	Zone à *Belemnitelles* 80 mètres (= 262¼ feet). Zone à *Micraster coranguinum*, Ag. 160 mètres (= 524½ feet). Zone à *Micraster cor-testudinarium*, Gold. 50 mètres (= 164 feet). Zone à *Holaster planus*, Ag. 20 mètres (= 65½ feet).
Craie Marneuse (Chalk without flints).	Zone à *Terebratulina gracilis*, D'Orb. 20 mètres (= 65½ feet). Zone à *Inoceramus labiatus* Schloth. 40 mètres (= 131 feet).
Grey Chalk, Chalk Marl	Zone à *Scaphites æqualis*, Sow. 35 mètres (= 115 feet).

He stated that his zone of *Inoceramus labiatus* has as its base a bed of very hard yellowish nodules imbedded in a greenish grey marl ; this he correlated in 1876 with the Melbourn Rock.[*] He gave additional particulars concerning the Chalk Rock of Mr. Whitaker, which occurs near the top of his zone of *Terebratulina gracilis*, and noticed a third nodular horizon in the lower part of the zone of *Holaster planus*.

Thus it will be seen that there is a general agreement as to the main divisions of the Chalk. The names Upper, Middle, and Lower Chalk are here used in place of those formerly employed, so as to bring the nomenclature into accordance with that of the mainland.

The Lower Chalk, which passes insensibly down into the Chloritic Marl, consists of alternations of chalk with shaly and pale-blue marl, in beds of six inches to two feet in thickness. Towards the lower part it is impure, and contains glauconite, or even rolled phosphatic nodules, but upwards the proportion of chalk increases at the expense of the marly bands, the more massive rock thus produced constituting the "grey chalk" of some authors. This sub-division forms generally the first rising ground at the foot of the Downs. It has been extensively dug for agricultural purposes, and the old pits have yielded a great number of fossils, among which *Ammonites rhotomagensis, A. varians*, and *Scaphites æqualis* are the most persistent.

The Middle Chalk, of which the Melbourn Rock constitutes the base, consists of massive beds of chalk from 3 to 6 feet in thickness, with partings of marl 2 or 3 inches thick. It forms the steeper part of the slope of the Downs, and is exposed in the upper part of many of the pits in which the Chalk Marl has been

---

[*] *Ann. Soc. Geol. Nord*, t. iii., juin 1876.

dug. *Inoceramus mytiloides* occurs in great profusion towards the upper part of the Middle Chalk.

The Upper Chalk occupies the whole area of the Downs except the steep slope in which the lower sub-divisions crop out as just described. It consists of a great thickness of white chalk with numerous lines of flints. Towards the base the flints become more sparse and grey, and gradually disappear, but below the lowest flint there occur nodules of hard siliceous chalk, having the form of flints, but the texture of chalk. This flintless portion of the Upper Chalk varies from 15 to 20 feet in thickness and has the Chalk Rock for its base.

The line engraved on the map shows the position of the Chalk Rock, but on so small a scale as the one inch scale, especially where the dip is high, represents pretty closely the base of the flinty chalk.

The flint in the Chalk occurs for the most part as irregularly shaped nodules, but sometimes as tabular layers either coincident with the stratification or filling cracks and joints. Those flints which occur parallel with the bedding, are of a different age from those filling the cracks and joints. The former have been derived from siliceous matter, frequently and perhaps in most instances deposited contemporaneously with the calcareous sediment of which the Chalk is composed, around sponges and other organised bodies, the forms and internal structure of which are still preserved. The latter, on the contrary, are of more recent origin, having been carried by percolating water, holding silica in solution, into cracks and joints, where they occur as thin plates of black flint, from ¼ to 1 inch in thickness, frequently separated by a central hollow, or porous grey layer. These subsequently introduced flints are, as might be expected, unfossiliferous, instead of abounding in fossils, as is the case with those of contemporaneous formation.

The cracks and joints filled with this secondary flint were not improbably due to the movements which upheaved the rocks of this region. These movements will be shewn in a later Chapter to have taken place at a late Tertiary date. The redistribution of the silica was thus probably in progress after the Chalk and the flints in it had been buried beneath a great thickness of Tertiary Beds, and had assumed their present consistency. There is no reason to doubt that in certain situations the transposition of the silica is still in progress.

In the parts of the Island where the strata are most highly inclined, the fossils in the more plastic strata, such as the Chalk Marl, are greatly distorted by pressure. The flints also which appear to be whole when viewed *in situ*, are found on closer examination to be nearly all broken, so that when extracted from the quarry they fall to pieces. The cracks are mostly filled with chalky matter, and the flints themselves appear to have been squeezed into the body of the Chalk, under the influence of the elevatory force by which it has been made to assume its present highly inclined position. These appearances are not observable

where the Chalk is in a comparatively undisturbed state.
Shattered flints may be observed in the large Chalk pits south
of Newport and on Arreton Down; also on Ashey Down, where
the Chalk is rather hard, as is most frequently the case where
it is inclined at a high angle.   The distortion of the fossils is
noticeable in the pit in the Chalk Marl at Yarbridge, described
on p. 88.

At Sun Corner, near the Needles, as noticed by Mr. Whitaker,[*]
" there is a bed of some thickness, in which the layers of flint are
so close together that they form nearly as much of the rock as
the chalk itself."   This intensely flinty zone occurs towards the
base of the flinty chalk.   It is not recognisable in Culver Cliff,
but on the other hand the flints are very large immediately below
the base of the Tertiary Beds at Brading (p. 96).

In Culver Cliff a marked flintless zone in the Upper Chalk,
about 350–400 feet above its base, was first noticed and described
by Mr. Whitaker as follows :—" Here, in the midst of the Chalk
with layers of flint at every three or four feet, is a space some
forty or fifty feet thick, with only one seam of tabular flint, but
with four lines of green-coated nodules, like those of the Chalk-
rock . . . . . but perhaps of a deeper colour."   The following
fossils have been obtained from one of the bands of green nodules :—
a sponge, a coral, *Cardiaster pillula*, Lam., *Serpula plexus*, Sow.
(adhering to one of the nodules), and *Rhynchonella plicatilis*,
Sow.   These nodules were submitted, for examination under the
microscope, to Mr. W. Hill, who kindly furnished the following
information concerning them.   He found them to consist mainly
of the fine amorphous material of the chalk, with a somewhat
unusual number of large and perfect foraminifera, and with many
sponge spicules, the silica of which had been replaced by calcite.
The colouring appeared to be sometimes due to a green material,
much of which had accumulated in the interior of foraminiferal
cells, but the whole of the amorphous material was sometimes tinted
green with no apparent change in its constituent particles.   There
were no isolated grains of glauconite, such as appear in the somewhat similar nodules of the Chalk Rock.   After treatment of the
nodules with hydrochloric acid, the residue was a dull-greenish soft
and earthy-looking material, a large part of which occurred as the
casts of foraminifera and the canals of sponge spicules.   In some
of the nodules Mr. Hill noticed ramifying cylindrical perforations,
filled with a white material, sharply defined from, and shewing in
strong relief against the remainder of the nodule.   In the larger
perforations there was a greater proportion of foraminifera and
shell fragments than in the material of the nodule.   In the smaller
ramifications the infilling material was like that of the nodule, yet
always shewed a clearly defined edge.

Nodules showing somewhat similar peculiarities occur at several
horizons in the Chalk of the Mainland, and have been remarked
by Mr. Hill to be usually accompanied by an exceptional abun-

* *Quart. Journ. Geol. Soc.*, vol. xxi. p. 401.   1865.

dance of organic remains. He considers that the fact that young shell-fish are frequently attached to their exterior, and that the material now filling the so-called perforations seems to have been introduced after the formation of the nodule, leads to the conclusion that the nodules were formed on the sea-bottom contemporaneously with the deposition of the Chalk, and formed at one time a suitable home for some kind of boring animal. There is no structure in the nodules that points especially to sponges as having been their origin. On the other hand Mr. Hill remarks that their occurrence in strata exceptionally rich in fossils is suggestive of their having resulted from the decay of organic matter.

The finest sections of the Chalk-with-flints form the precipices of Scratchells Bay and Culver Cliff. In each case the lines of flint enable the eye to follow the bedding from a distance, and to take in at a glance the regularity of the great curve in which the Chalk rises from beneath the Tertiary, and arches over the Secondary formations (see Section, Plate I.).

The thickness of the Upper Chalk can be arrived at by calculation only. Lines of section have been plotted across four different parts of the central line of Downs, giving a mean thickness of 1,350 feet, or rather more than the 1,017 feet assigned to this subdivision by M. Barrois.

In describing the sections, it will be convenient, after dealing with the Chloritic Marl, to take the three sub-divisions of the Chalk together, for it generally happens that the same pit, or group of pits, provides sections of parts of all of them.

THE CHLORITIC MARL.

The Chloritic Marl received its name from the abundance of grains of green colouring matter in it, formerly regarded as chlorite, but now recognised as glauconite. Although a calcareous deposit, it is remarkable for the number of phosphatised casts of Ammonites, Turrilites, and other fossils it contains. These were at one time worked for phosphoric acid on St. Catherine's Down, but the attempt was soon abandoned. The Chloritic Marl varies in thickness, being 13 feet at Compton Bay, 11½ feet at Brook, 7 feet at St. Catherine's, 8½ feet above Ventnor, and 15 feet at Culver Cliff. The variation is perhaps accounted for by the fact that no definite line can be traced between it and the Chalk Marl above. The lower beds of the Chalk Marl not only contain an abundance of sand and glauconite, but sometimes also rolled phosphatic nodules, not distinguishable from those in the Chloritic Marl (see section at Compton Bay, p. 83). It is generally difficult to decide at what exact horizon the proportion of sand in the rock falls so far below the proportion of calcareous matter as to justify the bed being referred to as chalk.

The relations of the Chloritic Marl have been discussed by Messrs. Barrois, Parkinson, Meyer, Jukes-Browne, and others. By M. Barrois it was grouped with the Chert Beds and the

freestone as the zone of *Pecten asper*, which fossil is recorded
from the freestone. Mr. Parkinson, however, denies that *Pecten
asper* occurs in the Isle of Wight at any other horizon than in
the Chloritic Marl, and that there it only appears as a derived
form,* in a layer of broken specimens at the base of the uppermost bed of this subdivision near Ventnor. This *Pecten* being
a characteristic Upper Greensand form, its occurrence as a
derived fossil only in the Chloritic Marl seems to indicate that
this bed is in part made up of the reassorted materials of the
Upper Greensand. All the phosphatised fossils which occur in
the Chloritic Marl are also of Upper Cretaceous type, and,
though they appear to have been phosphatised in a matrix similar
to that in which they are now imbedded, namely a glauconitic
sand, they have all been broken and many have been rolled.

Near Ventnor and St. Lawrence the Chloritic Marl is divisible
into two or more bands, the uppermost of which contains the
numerous phosphatic casts before alluded to.† According to
Mr. Meyer,‡ there are included under this title of Chloritic Marl,
as first applied, "two sets of strata with, in time at least, a gap
between them," the (local) top of the Upper Greensand, and the
(local) bottom of the Chalk Marl, the lower including in its fauna
*Pecten asper, Terebratella pectita, Catopygus carinatus (colum-
barius), Echinoconus (Galerites) castanea, &c.*, the upper, *Ammonites, Scaphites, Turrilites, &c.*, mostly phosphatic. These two
sets he correlated with the beds overlying the Chert Beds at
Warminster, for which he had previously used the name of the
Warminster Beds.

But Mr. Jukes-Browne§ remarks that the Warminster fossils
occur only in a *remanié* form in the Chloritic Marl of the Isle
of Wight, and that the small indigenous fauna differs very little
from that of the Chalk Marl, but is quite distinct from that of the
zone of *Pecten asper*. The Chloritic Marl is therefore regarded
by him as the natural base of the Chalk.

Not only, however, is it impossible to recognise sub-divisions in
the Chloritic Marl throughout the Island, but it is almost as
difficult to fix on a definite line between it and the Chert Beds.
While palæontologically it forms the base of the Chalk,∥ lithologically it is Upper Greensand, and the only line which can be
traced across country is that which runs at the foot of the Chalk
Downs, and marks the position of the lowest bed of chalk. Over a

* *Quart. Journ. Geol. Soc.*, vol. xxxvii. p. 372. 1881.
† This upper band seems to have constituted the Chloritic Marl of Captain
Ibbetson, who gives a thickness of 1 to 3 feet only to the bed. "Notes on the
Geology, &c. in the Isle of Wight," p. 24 (but see also p. 21 where he speaks of it
as consisting of two portions, the upper exhibiting a conglomerate of pebbles and
small boulders).
‡ *Geol. Mag.* for 1878, pp. 547–551. See also *Quart. Journ. Geol. Soc.*, vol. xxx.
p. 369. 1874.
§ *Geol. Mag.* for 1877, p. 357.
∥ When the geological mapping of Dorset was undertaken by H. W. Bristow, E.
Forbes, the palæontologist, pronounced that the Chloritic Marl, containing *Scaphites
æqualis*, constituted the lowest bed of the Chalk, of the fossils of which this formed
the earliest appearance.—H. W. B.

great part of the Island the outcrop of the Chloritic Marl is so
narrow that a single line suffices on the one-inch map to cover it ;
but around the Southern Downs and near Gatcombe, where the
dip is gentle, the Chloritic Marl runs up the dip-slope of the
Chert Beds considerably beyond the foot of the Chalk Downs,
ending off along an irregular line marked neither by feature nor
change of soil. In such cases, the line at the base of the Chalk
has been engraved, as the only boundary capable of being traced
with any accuracy.

In Compton Bay, the Chloritic Marl, 13 feet thick, consists
of marly sand with much glauconite and numerous pale-brown
phosphatic nodules, most of which are the rolled casts of
Ammonites (chiefly *A. varians*), *Turrilites Bergeri* and bivalves.
Some lines of irregular-shaped concretionary masses in it may
possibly be imperfectly formed chert. The same subdivision is
again well exposed in the road-cutting above Brook (*see* p. 70),
where it is 11½ feet thick, and contains abundant rolled casts of
Ammonites varians. The same description will apply also to the
section in the chalk-pits on the Brixton and Calbourne road.

In the Undercliff the Chloritic Marl is well exposed, some of
the best sections being on the top of Gore Cliff, and on the cliff
above St. Lawrence, in the zig-zag road at Ventnor, at the
railway station and 100 yards east of it, and in a pit by the road-
side near the Pulpit Rock above Bonchurch. It is about 7 feet
thick, and consists in the upper part of marly sand with
glauconite and many phosphatised casts of fossils, and in the lower
part of laminated sand of a darker tint, with broken shells of
Pecten asper, while between the two bands there runs a line of hard
white stony lumps. The old coprolite diggings, before alluded
to, were in the upper part of the Chloritic Marl and may still be
distinguished on the edge of Gore Cliff, on either side of the
township boundary.

In Culver Cliff, the upper limit of the Chloritic Marl is difficult
to fix. If it is taken at the base of the lowest bed that can be
fairly called chalk, the thickness obtained for the Chloritic Marl
is 15 feet, the section being as follows : —

		FT.	IN.
Chalk Marl (see p. 89).			
Chloritic Marl.	Green marly sand with lines of grey concretions, with *Plocoscyphia*, and a few scattered phosphates - - - -	5	0
	Do. with phosphatised *Ammonites* - -	5	6
	Line of large lumps of very hard grey stone	1	6
	Very green sand, with pipe-like markings, and a few phosphates - - -	3	0
Chert Beds (see p. 70).			
		15	6

Inland the Chloritic Marl being very soft is usually hidden,
but sections of it may be seen 100 yards south-east of Garstons, in
the roadside by the Convent at Carisbrook, and at Frogland,

where a fine spring issues from its junction with the Chalk Marl.

At Punfield, the Chloritic Marl is 3 feet 6 inches thick and contains phosphatic casts of *Ammonites, Nautilus,* and an *Exogyra* which seems indigenous. The Chert Beds below it, in which the chert occurs only as cherty lumps, contain *Exogyra conica* in great abundance, with *Siphonia tulipa,* Zittel, *Pecten asper,* Lamk., *P. orbicularis,* Sow., *P. quinquecostatus,* Sow., *Pleurotomaria,* and *Ammonites varians,* Sow.

UPPER, MIDDLE, AND LOWER CHALK.

1. *Compton Bay, along the Central Downs, to Culver Cliff.*

In proceeding westwards from the Needles along the coast, we find the Middle Chalk first coming in at the foot of the cliff at Oldpepper Rock, 700 yards east of Sun Corner. Up to this point the cliff, which is over 400 feet in height, is vertical and descends sheer into the sea, but, where the Middle Chalk rises from beneath the water, is fringed with a rough beach of fallen blocks. Oldpepper Rock is an outstanding mass of Middle Chalk, still *in situ.* After 500 yards the top of the Middle Chalk descends again beneath the sea, and the cliff becomes once more vertical. At a point 800 yards west of the Beacon (or about $1\frac{1}{4}$ mile east of Sun Corner), known as New Ditch Point, the Middle Chalk rises again into the cliff, and so continues for a little over a mile, when it once more sinks below the sea. The same change in the character of the cliff is observable here also, and the vertical walls of chalk and remarkably picturesque range of caves are conveniently situated for examination. The Middle Chalk rises finally about 600 yards east of the easternmost point of Freshwater Bay. Thence it slants gradually up the cliff to a cutting in the Military Road on Afton Down, where the following section occurs :—

Military Road Cutting, Afton Down.

		FT.	IN.
Upper Chalk	Chalk with flints - - - -		
	Nodular chalk, without flints - -	6	0
	Marl - - - - -	0	1–2
	Nodular chalk, without flints - -	8	0
	White shaly marl - - -	0	1–2
	Nodular chalk - - - -	6	0
	Nodule Bed (Chalk-rock), green-coated nodules in the top 3–6 inches -	1	6
Middle Chalk -	Massive chalk, weathering into small fragments, but not nodular, with bands of marl at 4–10 feet intervals. Pyrites and *Terebratula semiglobosa* - - -	60	0+

The Lower Chalk first rises from beneath the sea, at a point on the beach nearly midway between the easternmost point of Freshwater Bay and the path down the cliff on the outcrop of the

Gault in Compton Bay. A poor representative of the Melbourn Rock was detected here by Mr. Whitaker, the sequence being as below :—

Middle Chalk { Massive thick-bedded chalk traversed by straight joints.
Melbourn Rock, hard thinly bedded chalk with layers of marl, about 8–10 feet.

Lower Chalk - Softer chalk, traversed by curving joints, producing 'conchoidal fracture' on a large scale.

A small fault throws the beds about 6 feet down to the north-west, at the point where the Melbourn Rock comes down to the beach. Downwards the Lower Chalk passes so gradually into the Chloritic Marl that it is difficult to fix its base. The following section, which forms the continuation of that given on p. 68, was obtained by climbing a short distance up the cliff.

FT. IN.

Alternations of chalk and marl in beds of 1–2 feet thick.
Chalky sand, with glauconite, and containing rolled *Ammonites, Turrilites,* &c. at base. The bed looks like chalk at first sight, but contains perhaps more sand than chalk - 8 0
Pale-blue marl and chalk in alternations - - - 7 0
Chloritic marl (see p. 81).

Along all this part of the coast, from Compton Bay to Sun Corner, a line of rocks may be seen under the water when the sea is smooth and clear, running nearly parallel to the foot of the cliff, and still more nearly parallel to the line marking the top of the Middle Chalk, as traced above. This line of rocks marks the submerged outcrop of the Chert Beds, for further east it joins a reef formed by these beds, which is bare at low water in Compton Bay. It shews no deviation from its course opposite Freshwater Bay, whence we may infer that no fault runs along this valley, where a fault might have been suspected from the course taken by the topmost beds of the Chalk.

Following the Downs eastwards, we find the next sections at the south-eastern corner of Shalcombe Down. Here there are two pits, the upper of which was described by Mr. Whitaker.* The section seen in 1887 was as follows :—

Pit at the south-eastern corner of Shalcombe Down.

FT. IN.

Chalk with flints - - - - - -
Rough nodular chalk without flints - - - - 10 6
Black clay or shale - - - - - - 0 1–3
Rough nodular chalk - - - - - 6 0
Nodular bed (Chalk Rock), the nodules in the upper 3 inches green-coated - - - - - 1 3
Massive thickly bedded chalk with two or three seams of marl about 10 feet apart - - - - - 20 0+

The lower pit is in the lower beds of the Middle Chalk and seems to touch the Lower Chalk, but the Melbourn Rock could

* *Quart. Journ. Geol. Soc.,* vol. xxi. p. 402. 1865.

F 2

not be clearly distinguished. *Inoceramus mytiloides* is abundant in the Middle Chalk. In the same neighbourhood a deep cutting for the coach-road shows in the upper part:—

	Ft.	In.
Alternations of chalk and marl, top not seen - - -	120	0
Rocky chalk, very impure, with glauconite; passing down into the - - - - - - -	5	8
Chloritic marl (see p. 70) - - - - -	11	6

Large specimens of *Ammonites rhotomagensis*, Defr., occur here, and *A. varians*, Sow., is common but badly preserved.

The Middle and Lower Chalk are both seen in a pit north of Mottistone, where the latter has been worked near its base; but the junction between the two subdivisions is obscured. This seems to have been the pit alluded to by Mr. Whitaker,* and the layer of hard yellowish nodules seen by him may have been the Melbourn Rock. The pits do not reach up to the horizon of the Chalk Rock.

At the west end of Brixton Down a fine series of pits extends from the Upper Chalk to the Malm Rock of the Upper Greensand. The Upper and Middle Chalk are seen in the uppermost pit on the north side of the Calbourne and Brixton road, the section, which was measured in company with Mr. Whitaker, being as below:—

West end of Brixton Down.

		Ft.	In.
Upper Chalk	Chalk with flints, seen up to about - -	20	0
	Rough nodular chalk, without flints -	20	0
	Nodular chalk (Chalk Rock), the nodules green-coated - - - -	1	3-6
	Rough nodular chalk - - -	1	6
Middle chalk -	Smooth massive chalk - - -	6	0+

The lower beds of the Middle Chalk are seen in a pit a few yards further south, but the Melbourn Rock is not now exposed. Another and larger pit in the Middle Chalk has been opened about one third of a mile further west in Mottistone Down, and seems from the character and curvilinear jointing in the lower part to have reached the Lower Chalk, but the Melbourn Rock is not distinguishable. *Holaster subglobosus* occurs in these lower beds.

The Brixton Down pit was visited in 1865 by Mr. Whitaker, and figured on p. 403 of the paper already quoted. At that time a line of clay was visible, which seemed to shew an unconformity (or perhaps false bedding) in the Chalk; for southwards it was further from the nodular bed, whilst northwards the latter was not seen, but seemed to be cut off. This line of clay, however, runs persistently through the Island at a scarcely varying distance above the Chalk Rock; it was in fact selected by M. Barrois as the base of his Chalk-with-flints. Moreover, it was figured by him

---

* *Quart. Journ. Geol. Soc.*, vol. xxi. p. 402. 1865.

again in 1875,* and described as occurring at about its proper distance above the Chalk Rock. Presumably therefore its irregular appearance in the Brixton pit was the result of squeezing. The dip ranges from 27° to 30°. Southwards from this pit the high road passes steep banks of thin-bedded chalk and marl, which become very impure and sandy in the lower part, and so merge into the Chloritic Marl (p. 81).

Near Coombe Tower, north of Brixton, several large pits in the Lower Chalk reach upwards into more massive beds which seem to be the Middle Chalk, but the Melbourn Rock is not distinguishable.

The large pit at Shorwell exposes this rock, but unfortunately in a position wherein it is quite inaccessible at present. The strata consist of thin-bedded chalk and marl (Lower Chalk) but the top of the vertical wall of the pit is formed of a hard flaky chalk, underlain by a thin seam of marl, the appearance of the beds, as studied at a distance of a few feet, being the same as that of the Melbourn Rock near Arreton and Yarbridge (*postea*, pp. 88, 89).

The next sections are found in the projecting promontory of Chalk of Chillerton Down. The uppermost pit touches the Chalk Rock, but starts a few feet below the Chalk-with-flints.

Chillerton Down.

							Fr.	In.
Upper Chalk	Massive chalk	-	-	-	-		4	0
	Grey marl	-	-	-	-		0	1
	Massive chalk	-	-	-	-		4	0
	Green-coated nodule bed (Chalk Rock)				-		0	8
Middle Chalk	Rough knotty chalk	-	-	-			8	10
	Smoother flaky chalk	-	-	-			6	0+

The Middle Chalk, which has been worked, is now overgrown; but the Lower Chalk, presenting its usual character of thin-bedded chalk and marl is worked in the lowest pit. At least three faults are visible in the pit, in each case with slickensided faces, coated with a film of blue marl. Two of them range nearly north-east, throwing down a wedge between them, while the third runs north-west with a downthrow to the south-west. The dip is at 7°—9° to the north, but decreases to 2° or 3° further north towards Gatcombe, and changes in direction to north-west. The faults may be connected with this change.

The boundary of the Upper Chalk is shown upon the map, as running across the Downs from near Shorwell to Carisbrook, but the hills lying outside this boundary are believed to be capped with outliers of Upper Chalk. The existence of outliers there is not quite certain, because of the uniform sheet of flint gravel covering the tops of all the hills, but it is inferred from the position of the Chalk Rock in the pit last described. They must, however, be exceedingly thin, the Chalk-with-flints having nearly all mouldered down into flint gravel. The few pits, which are open round the brows of these hills, expose the massive beds of the Middle Chalk. The best sections in the Lower Chalk are to be found 500 yards south of Newbarn and at Garstons.

---

* Description Géologique de la Craie de l'Ile de Wight, p. 18.

The evidence on which the base of the Upper Chalk has been traced across the Downs to Carisbrook is somewhat scanty. Middle Chalk is seen in the road at the top of Shorwell Shute, in a pit at Cheverton, at Rowborough, and in a pit at Bowcombe, while the Upper Chalk is exposed in pits on the southern and eastern brows of Idlecombe Down. In one spot only, namely a cart-road running northwards from Rowborough, a poor exposure of the Chalk Rock may be seen. There are sections of Upper and Middle Chalk close together in a lane leading up the hill to the north-west from Bowcombe, but no section of the Chalk Rock. In Carisbrook, however, this latter subdivision is well exposed. We first see it in a cutting where three lanes meet near Clatterford. Thence it runs along the south front of the hill on which the castle stands (the hill being a portion of the escarpment of the flinty Chalk), to a quarry near the Convent, where the following section occurs :—

Quarry east of Carisbrook Castle.

		FT.	IN.
Upper Chalk {	Nodular chalk with grey flints.		
	Do. without flints - -	4	0
	Smoother chalk - - -	1	0
	Marl - - - - -	0	1
	Rough hard chalk - - - -	7	6
	Dark marl - - - - -	0	1
	Hard chalk - - - - -	6	0
	Chalk { Line of green-coated nodules	0	3
	Rock { Nodular chalk - - -	2	4
Middle Chalk {	Smooth chalk - - - -	4	0
	Fault		
	Thick-bedded smooth chalk with partings of		
	marl at 2 to 4 ft. intervals - -	60	0+

The occurrence of the nodular bed here was first noted by Mr. Whitaker in 1865,* but, the Chalk-with-flints not being exposed at that time, he was unable to correlate it positively with the Chalk Rock. The fault mentioned above runs W. 15° S., very nearly along the strike of the strata which it throws down to the north, its effect being to depress out of sight an unknown thickness of the upper beds of the Middle Chalk. The dip points a little west of north at 42°.

The Middle Chalk and Melbourn Rock are exposed in an old pit half a mile further east, on the west side of the Shide and Gatcombe road, the section being as follows :—

Pit on the east side of Mount Joy.

	FT.	IN.
Chalk in beds of 2 to 3 feet thickness, with bands of marl,		
top not seen - - - - - -	30	0
Thin-bedded chalk with bands of greenish marl, about -	4	0
Chalk with yellowish nodules (Melbourn Rock), about -	2	0
Marl (? Belemnitella Marl) - - - -	2	0+

The pit is now occupied by farm-buildings, and the section somewhat obscured. The nodular bed was first noticed and

---

* *Quart. Journ. Geol. Soc.*, vol. xxi. p. 403.

described by M. Barrois in 1875.* He obtained from it *Inoceramus labiatus*, *Rhynchonella Cuvieri*, and *Cidaris hirudo*.

The Chalk Marl appears on the east side of the Gatcombe road, and in an old pit midway between the two described above. On the east side of the valley a large pit exposes the Lower Chalk; the Middle and Upper Chalk are seen, but not well, in the side of the road.

Some fine sections occur at the east end of Arreton Down. On the west side of the high road, in the bottom of a disused pit, Mr. Whitaker found the Chalk Rock. It is now overgrown, but the beds above it are seen as follows :—

	Ft.	In.
Nodular chalk with a few grey flints.		
Smooth chalk with *Terebratula semiglobosa*, Inoceramus, &c.	2	6
Rough nodular chalk · - - - -	6	0+

Fifty yards east of this pit, and on the opposite side of the road, a marl-pit exposes a good view of the Chalk Rock, the section being as below :—

		Ft.	In.
Upper Chalk	Smooth chalk - - - -	4	0
	Black clay - - - - -	0	1
	Rough chalk - - . -	8	6
	Chalk ⎰ Line of green-coated nodules -	0	4
	Rock ⎱ Rough nodular chalk -	2	2
Middle Chalk	Smooth chalk - - - -	2	8
	Marl - - - -	0	1
	Smooth chalk - - - -	2	6
	Marl - - - -	0	0½
	Smooth chalk - - - -	10	6
	Marl - - - -	0	3
	Smooth chalk - - - -	2	0+

Following the foot of the Down eastwards we find a large pit 300 yards north-west of Heasley Lodge, in the upper part of which a band of rough chalk, nodular in parts, is no doubt the Melbourn Rock. The section is as follows :—

Pit on Mersley Down.

		Ft.	In.
	Massive chalk with marly partings - -	60	0
Melbourn Rock.	Nodular chalk, the top concealed, seen up to	2	0
	Thin-bedded chalk with partings of greenish marl - - - - -	4	0
	Hard chalk, nodular at the base - -	3	6
	Alternations of chalk and marl - -	3	0
? Belemnitella Marl.	Laminated marl - - -	2	0
	Marly chalk with curving joints - -	2	6

The pit is worked deep into the Chalk Marl, but the rest of the section is obscured by talus. There is a large pit in the same beds by the side of the Ryde and Newchurch road, but the Melbourn Rock was not to be found. The ·Chalk Marl is well seen in a large pit north of Kern.

---

* Craie de l'Ile de Wight, p. 16.

At Yarbridge all the subdivisions of the Chalk are exposed. Two large pits are situated on the side of the road to Alverstone, the upper one wholly in the Chalk-with-flints, the lower one partly in this and partly in the Middle Chalk.

Pit half a mile west-north-west of Yarbridge.

		FT.	IN.
Upper Chalk {	Chalk with a few grey flints - - -	8	0
	Rough nodular chalk, with lumps slickensided and weathering yellow; fossiliferous in the lower part - - - -	6	0
	White marl parting - - - -	0	1
	Rough chalk with *Terebratula semiglobosa* -	8	0
	Black clay - - - - -	0	1
	Rough chalk - - - - -	4	0
	Smoother chalk - - - -	4	0
	Chalk Rock, a single line of green-coated nodules lying on - - -	0	1
	Rough nodular chalk - - -	2	6
Middle Chalk {	Smooth chalk - - - -	5	6
	Marl - - - - -	0	1

The same beds were formerly exposed in Yarbridge in some pits which are now partly hidden by building. Mr. Whitaker noted the following section* :—

> Chalk with a few nodular flints (shown only at the northern end of the quarry, where it is 20 to 30 feet thick).
> Thin seam of dark-grey clay.
> Chalk, about 8 feet.
> Inconstant layer of irregularly-shaped green-coated nodules (Chalk-rock ?)
> Evenly and massively bedded chalk, without flints, but with seams of marl.

The Middle and Lower Chalk are well exposed in a pit about 200 yards west of the upper road in Yarbridge, which shows the following section :—

Pit west of Yarbridge.

		FT.	IN.
	Massive chalk in beds of 2–3 feet, iron pyrites - - - -	30	0
	Thin-bedded chalk in beds of 6–8 inches, with partings of greenish marl - -	3	0
Melbourn Rock. {	Hard nodular bed - - -	2	0
	Laminated greenish marl - - -	0	3
	Hard nodular bed - - - -	4	6
? Belemnitella Marl. {	Smooth earthy chalk with curvilinear jointing passing down into - -	2	0
	Grey or greenish marl with curvilinear jointing passing down into - -	2	6
	Hard chalk - - - - -	7	0
	Marl - - - - -	0	1
	Alternations of marl and blocky chalk -	21	0

We now reach the great section afforded by Culver Cliff. There the sub-divisions are not only well exposed, and the different

---

horizons identifiable, but by choosing the least steep parts of the cliff we have found it possible to take an unbroken series of measurements from the base of the Upper Chalk downwards, thus continuing the measurements which have already been given for all the beds down to the base of Lower Greensand. The total thickness of beds measured in this section amounted to 1,218 feet, as shown drawn to scale in Plate III. The section in the Chalk is as follows :—

Culver Cliff.

			Ft.	In.
Upper Chalk		Chalk with grey flints - -		
		Smooth chalk with *Holaster* -	4	0
		Chalk, splitting up into nodular masses along wavy dark lines; fossils - - -	3	0
		Marl - - - -	0	0⅓
		Chalk as above - - -	2	0
		Beds, obscured by talus -	16	0
	Chalk Rock.	Hard grey chalk, with a line of green-coated nodules at top -	1	2
Middle Chalk, 180 ft. 3 in.		Thick-bedded white chalk with partings of marl - -	164	0
		Shaly chalk - - -	2	0
	Melbourn Rock.	Chalk with yellow-coated nodules - - -	0	3
		Chalk split up by partings of greenish marl - -	6	0
		Chalk with yellow-coated nodules - - -	2	0
	? Belemnitella Marl.	Bluish marl, about - -	6	0
Lower Chalk, 206 ft.		Massive smooth chalk - -	86	0
	Chalk Marl.	Thin-bedded grey chalk and marl in numerous alternations, passing down - -	50	0
		Similar beds, but rather bluer and more marly; the chalk bands very hard and lumpy, and containing *Ammonites varians* and sponges abundantly - -	70	0

Chloritic Marl (see p. 81).

An abstract of this section may be arranged as follows :—

Abstract of the Section of Middle and Lower Chalk in Culver Cliff.

		Ft.	In.
Middle Chalk, 180 ft. 3 in.	Thick-bedded chalk - - -	166	0
	Melbourn Rock - - -	8	3
	Belemnitella Marl - - -	6	0
Lower Chalk, 206 ft.	Massive chalk - - -	86	0
	Chalk Marl - - -	120	0
		386	3

The thicknesses of these sub-divisions at Punfield compared with those given above, show a westerly attenuation of the Chalk as of the other Secondary Rocks. The Upper Chalk becomes devoid of flints but very nodular in the lower 20 feet, and has as its base a conspicuous band of green-coated nodules, about 4 inches thick (Chalk Rock), below which the section runs as follows :—

Near Punfield Cove.

		Ft.	In.
Middle Chalk, 111 feet.	Hard, rough, and lumpy chalk	6	0
	Smoother chalk, thick-bedded, with partings of marl	90	0
	Melbourn Rock	6	0
	Smooth chalk with conchoidal fracture, with several partings of marl	3	0
	Fine marl (? Belemnitella Marl)	9	0
Lower Chalk, 132 feet.	Alternations of chalk and marl in beds of 1 to 2 feet, with an occasional line of nodules, some of which are green like those of the Chalk Rock	132	0
		243	0

2. The Southern Downs.

The outliers of chalk, which cap these hills, consist of the Lower, Middle, and a mere film, if any, of the Upper Chalk, the Chalk-with-flints (and according to M. Barrois the whole of the Middle Chalk) having been denuded away. The tops of the hills, however, are so thickly overspread with flint gravel, a residue of the mass of beds that have been removed by subaërial agencies, that it is not possible to say what is the highest bed present beneath this covering.

In the outlier of St. Catherine's Down the dip is at a gentle angle to the east-south-east—that is, about the same as that of the Lower Cretaceous Rocks seen in the coast.* The thickness of chalk forming the outlier amounts to about 180 feet, and must therefore belong wholly to the Lower Chalk. But it is noticeable that the hill is capped with flint gravel, a relic of the Upper Chalk, that must have been slowly let down from above by the dissolving away of the chalk. The best exposures of the beds are to be met with in a large marl-pit at the north end of the outlier. They consist of alternations of chalk and marl generally in thick beds, and are traversed by a small fault running about E. 10° N. with a downthrow to the south.

A second outlier, scarcely separated from the first, occurs on the brow of Gore Cliff. The beds, well exposed along the cliff, with the underlying Chloritic Marl, are very fossiliferous. This outlier evidently forms the northern flank of a chalk-hill, of

---

* It was stated by Captain Ibbetson that an unconformity between the Upper and Lower Cretaceous Rocks was visible in the Isle of Wight (*Quart. Journ. Geol. Soc.* vol. iii. p. 315. See also Judd on the Punfield Formation, *ib.* vol. xvii. p. 221, 1871). This statement was founded on a mistaken idea that the Chalk of the Southern Downs is horizontal, while the easterly dip of 2° of the lower rocks, as seen in the cliff section at Atherfield, was supposed to be maintained beneath them. Neither supposition is correct.

which the only other traces left are masses of fallen chalk in the Undercliff. Some of the rain-wash, however, from the slopes of this vanished chalk-down forms a conspicuous bed on the brow of the cliff (see *postea*, p. 237). There is a small pocket of flint-gravel in this chalk.

The same description will apply also to the chalk which caps the cliff east of St. Lawrence. The Chalk Marl only is seen, but it is possible that the tops of the hills touch the more massive upper beds of the Lower Chalk. The base of the Chalk Marl occurs in St. Lawrence Shute and in the footpath leading up the cliff to Whitwell. The dip is southerly and south-easterly at a gentle angle.

In the high down which extends northwards to Appuldurcombe Park, there is a thickness of about 270 or 280 feet of chalk at a point between Week Farm and Rew Farm, and there must therefore be from 60 to 70 feet of Middle Chalk on this hill, underneath the gravel. Numerous old pits have been opened in the Chalk Marl around Stenbury and Appuldurcombe Downs, and a pit is now worked near Ventnor Cemetery, in a more massive chalk, apparently the upper part of the Lower Chalk (the Grey Chalk). Mr. Norman remarks that a portion of the head and jaws of a large fish was dug up in the Cemetery, but unfortunately not preserved.[*]

The junction of the Chalk Marl and Chloritic Marl is seen on the brow of the cliff 900 yards east of St. Lawrence Shute, and in the side of the zig-zag road leading up the cliff above the Royal Hotel, Ventnor. It is exposed also in the cutting at Ventnor Station, but is more accessible by the road-side, 150 yards east of the Station, and in a road-side 300 yards east of St. Boniface Well.

St. Boniface Down forms the highest ground in the Island, reaching a height 787 feet above Ordnance Datum. The base of the Chalk on the north side of the Down is about 450 feet above the sea, and on the south side about 300 feet, the distance across being 1,320 yards. From these data it may be calculated that the southerly dip amounts to 1 in $26\frac{1}{2}$, or a little less than $2°$, —a result which agrees with that obtained in the tunnel (p. 72). From the same data it may be calculated that the thickness of chalk and gravel under the highest point of the Down must be about 430 to 440 feet. But it will be remembered that the united thicknesses of Middle and Lower Chalk at Culver Cliff amounted to only 386 feet. Above these there were 26 feet of Chalk Rock and flintless chalk, making a total of 412 feet of chalk below the lowest band of flints. If to this we add 20 feet for the estimated thickness of flint-gravel on St. Boniface Down, we obtain a total of 432 feet. It would seem then that, though the lowest bed of the Upper Chalk may be present, there is not room for any of the Chalk-with-flints, or at most for more than a mere film of it beneath the gravel.

---

[*] Geological Guide to the Isle of Wight, p. 99.

No section, however, occurs of the higher beds forming the Down, with the exception of a small hole on the east side of Shanklin Down, which seems to be in the massive beds of the Middle Chalk. On the very steep slope of chalk over Ventnor a small spring rises, known as St. Boniface's Well. It was remarked by Sir H. Englefield (*op. cit.* p. 37) that "a spring at this height, is a most remarkable circumstance, and the only instance of the kind in the whole island. It indicates some stratum within the hill differing from the chalk, which certainly would let the water sink through its substance here, as it does everywhere else." This spring occurs at about the height at which it may be calculated that the Melbourn Rock and Belemnitella Marl should occur.

DIVISION OF THE UPPER CHALK INTO ZONES.

The inland section of the Chalk-with-flints presents a remarkable uniformity in lithological character. The sub-division of this great mass by M. Barrois depended therefore principally on the evidence of the fossils, which he collected himself. The following account of the four zones is an abstract of the description published by him in 1875.* The thickness of the various zones are given by M. Barrois in round numbers of *métres*. The conversion of *métres* into feet gives a misleading impression of minuteness of measurement. The zones are taken in ascending order.

Zone of Holaster planus.

For the base of this zone the seam of black clay, described on pp. 87, 88, was chosen by M. Barrois. The zone is seen in the Military Road cutting near Freshwater, as a very hard nodular chalk about 65 feet thick. The nodules are of a yellowish-white and very hard, so that it is difficult to detach some urchins, which occur in them. The rock enclosing the nodules is softer, and of a greenish-grey colour; and numerous layers of homogeneous white chalk with nodules are intercalated. Tabular layers of flint are abundant, and the zone is rich in fossils. At Watcombe Bay, near Freshwater, where the rocks are continually being scoured by the waves, there may be seen in every square yard of the cliff all the fossils characteristic of the lower part of the white Chalk.

Zone of Micraster cor-testudinarium.

This zone is exposed in parts of the cliffs scarcely accessible, and is rarely quarried inland. It forms the central part of the range of Chalk Downs. The thickness is 160 to 170 feet, but is difficult to estimate. The zone is exposed in pits at the west of Bembridge Down, south-east of Brading Down, in the road to the south of the great quarry on Arreton Down, in the road

---

* Craie de l'Ile de Wight, pp. 22–29.

from Compton Bay to Freshwater and in the cliffs known as the Nodes and the Main Bench.

Zone of Micraster coranguinum.

This zone has furnished but few fossils; and differences in fauna were not therefore relied upon by M. Barrois in making this sub-division of 500 to 550 feet of chalk. He correlates it with the two divisions established by Mr. Whitaker in the Chalk of the Isle of Thanet, namely the Margate Chalk above, and the Broadstairs and St. Margaret's Chalk below. In this lower division in the Isle of Thanet he has obtained many specimens of *Micraster coranguinum*, and in the upper, a great abundance of *Belemnites verus*, Miller, *Marsupites Milleri*, Mant., *M. ornatus*, Miller, which, according to M. Hébert, are characteristic of the upper part of the zone of *Micraster coranguinum*. The upper or Margate zone also contains but few flints, while the lower or Broadstairs zone contains a great number. These two zones he considers to be recognisable in the Isle of Wight. To the Margate zone he attributes the chalk of the great quarry on Arreton Down, and of that to the east of Mersley Down; while the Broadstairs and St. Margaret's type is seen in the small quarry of Bowcombe Down.

Zone of Belemnitella.

The great quarry to the north of Shalcombe Down shows, in the lower part, white chalk with many large black flints, and, in the upper part, softer chalk with smoke-grey flints. These correspond respectively to the zones known in France as those of *Belemnitella quadrata* and of *B. mucronata*. There are many quarries along the north side of the Downs, all in the zone of *Belemnitella*, but the deepest only reach the horizon of *B. quadrata*. The flints of the zone of *B. mucronata* are often grey as at Shalcombe and the Needles, but sometimes black, as at Alvington and Mottistone. In the upper part of the lower zone (that of *B. quadrata*), *Magas pumilus* is abundant. The united thickness of these zones of *Belemnitella* is 260 feet.

The junction of the *Belemnitella* zone and the zone of *Micraster coranguinum* may be observed on Arreton Down, but, except in their palæontological characters, there is little difference between them. They are distinguishable only by the relative abundance of flints in the *Belemnitella* zone, and their almost entire absence in the upper part of the *Micraster* zone.

M. Barrois alludes also to the road-cutting near Apes Down, which extends for some three hundred yards along the junction of the Chalk and Plastic Clay. The section has now become somewhat obscured by talus and vegetation, but the contrast between the red clay of the north, and the white chalk of the south side of the road, is still sufficiently striking.

CHAPTER VIII.

EOCENE.

INTRODUCTION.

THE Eocene strata of the Isle of Wight may, as a whole, be more conveniently studied in the cliffs in Alum Bay* than in any other part of the Island.

In this remarkable section the whole of the strata from the Chalk to the Fluvio-marine formation are displayed in unbroken succession, and that too in a manner the most favourable for close examination, in consequence of their being thrown into a vertical position by the action of the same elevatory force which has caused the Chalk to assume its present high inclination.

When the face of the cliffs has been laid more than usually bare, and the colours of the various beds have been heightened by heavy rains, the aspect of the bay, always beautiful, is rendered still more striking. Every bed is then revealed to the eye from the base of the cliff to where it crops out at its summit, and while some of the beds attract the attention by their contrast in colour, others, like the coals in the Bracklesham series, the conglomerate bed dividing that series from the overlying Barton Clay, and the bed of white pipeclay in the Lower Bagshot series which is so crowded with vegetable remains, are not only rendered conspicuous by their different colours, but, standing out from the rest of the strata, they become useful by enabling the observer more readily to perceive from a distance the positions and limits of the various formations.

No drawing without the appliance of colour can do justice to the section, and even then no artist is capable of rendering a faithful and characteristic representation of it, who does not (like the late lamented Edward Forbes) combine with a dexterous use of the pencil a thorough knowledge of the geological structure of the scene he wishes to delineate.

READING BEDS.

THE lowest member of the Tertiary Group in the Isle of Wight is the Reading Series of Prof. Prestwich, formerly called the "Plastic Clay" from the occurrence in it of beds used in the manufacture of tiles and coarse earthenware. Owing to the strata being nearly vertical throughout the Island, this division can only be examined at Alum and Whitecliff Bays. Formerly there were pottery works at Newport in the red clays, but the pits are now filled up and overgrown. The only other inland sections now visible are near Brading; in a railway cutting at Ashey; and at Downend Brickyard, near Arreton. The last has been opened since the

---

* So called from the quantities of alum formerly manufactured there.

new Survey was complete, and there has been opportunity of examining it.

In the Isle of Wight the Reading Beds consist almost entirely of mottled clays, in which shades of red and purple predominate. These rest on a slightly eroded surface of the Chalk, and contain at their base small rolled flint pebbles. (See Fig. 16, from a sketch by Sir Andrew Ramsay.)

FIG. 16.

Junction of the Chalk and Lower Tertiary Beds, in Alum Bay.

The following section was measured, with the assistance of Mr. Richard Gibbs, in 1852.

Section of the Reading Beds in Alum Bay.

	FT.	IN.
Red and white mottled clay, with a ferruginous parting at 4 feet - - - - - - - -	25	0
Ferruginous-brown clayey sand - - - -	14	0
Bright-red and white mottled clay (pipeclay) - -	20	0
Brown and grey sandy clay (with a bed towards the middle of dark-red clay 3 feet thick) ; most sandy in the upper 5 feet - - - - - - -	10	0
Tenacious, wet, red and white mottled clay - -	3	0
Tenacious blue and brown ferruginous clay - -	8	0
Brown sand covering an uneven eroded surface of Chalk	3 to 4	0
	84	0

As the strata are traced eastward their thickness increases to 110 feet at Downend, 92 feet at Ashey, 140 feet at Brading, and 163 feet at Whitecliff Bay. At the last-named locality they consist principally of mottled clay, but are so hidden by landslips and mud-streams that their details cannot at present be noted and the total thickness here given is taken from the original measurement made in 1852.

The section in the railway cutting at Brading is now entirely overgrown, but a sketch and description, made by Mr. Whitaker during the construction of the line in 1878, is here given. (Fig. 17.)

Fig. 17

Railway Cutting just south of Brading Station (higher, or western, side).

Scale, horizontal and vertical, 30 feet to an inch.

a. Wash.

Reading Beds.
- *b.* Buff and brown fine clayey sand, with layers of clay.
- *c.* Purple-grey plastic clay, mottled crimson.
- *d.* Buff and light-grey fine sand.
- *e.* Grey and purple plastic clay.
- *f.* Crimson and grey plastic clay.
- *g.* Greenish and crimson plastic clay.
- *h.* Red plastic clay, mottled grey.
- *i.* Grey and crimson plastic clay.
- *k.* Brown grey and purple plastic clay.

l. Chalk with flints; showing a dip of 60°, and on the other (eastern) side of the cutting as much as 80°.

The junction of the Reading Beds and the Chalk was not clearly shown, but must be somewhat irregular. Some huge roundish flints occur at the junction, as in Alum Bay, partly projecting into the clay. The cutting is deeper further south, and in Chalk.

Some caution is needed in estimating the true thickness of the Reading Beds in the Isle of Wight; for it must not be forgotten that the strata are nearly vertical and have been subjected to violent pressure, varying in direction and amount according to their proximity to the sharp monoclinal curve which forms such a conspicuous feature in the geology of the Island. Where the Chalk is thrust northward, beyond the ordinary line of the Downs, the compression of these lower Tertiary strata is also greatly exaggerated, but where the Downs recede slightly to the southward the thickness of the Reading Beds increases considerably. Allowing for this compression, and taking into account the measurements obtained on the mainland, it seems probable that the thickness we might expect to find in wells sunk beyond the limits of the most violent disturbance would be from 100 to 120 feet.

The only fossils this series has yet yielded in the Isle of Wight are fragments of plants; and though the beds are probably in the main of freshwater origin, there is little direct evidence in the district. On the mainland the principal fossils found in Reading Beds of this type consist of leaves of plants and other vegetable remains, showing, according to Sir J. Hooker and Mr. J. Starkie Gardner, a temperate climate. In similar beds at Lancing, however, the mottled clays are not entirely freshwater, for they contain a line of ironstone nodules with casts of marine shells.

LONDON CLAY

Like the Reading Beds, the London Clay forms a narrow belt extending across the Island, between the west and the east coast, from Alum Bay to Whitecliff. In consequence of the highly inclined position of the strata between these points, the width of the out-crop of the London Clay, or the space occupied by it at the surface, is frequently very little more than the actual thickness of the formation. The only places where it can be thoroughly examined are on the coast.

The junction of the Reading Beds and the London Clay is sharp and well defined. In Alum and Whitecliff Bays the highest part of the older deposit consists of red mottled clays, while the base of the newer one is ferruginous or blue sandy clay. At both localities the division between the two formations is indicated by a band of flint pebbles, sometimes mixed with pebbles of the underlying red clay, representing the Basement Bed of Professor Prestwich. In Alum Bay, however, this seam of pebbles is not perfectly continuous. Inland, the Basement Bed is better represented by an impersistent bed of fine sand, seen in the road cuttings between Calbourne and Swainstone, and dug near Ashey Chalk Pit and close to Ryde Waterworks. This sand appears nowhere to exceed 10 or 12 feet in thickness. There is nothing especially characteristic in the fauna of these basement beds in the Isle of Wight, all the species being also found in higher zones.

Junction of London Clay and Reading Beds at Alum Bay.

(Observed by Mr. Whitaker in 1865.)

London Clay
- Grey and brown sandy clay, with here and there a small flint-pebble;—passing down into the next bed.
- Basement-bed.
  - Grey and brown loam or clayey sand partly with clay-lines and green grains, shells, and hard masses (sometimes concretionary clayey limestone and ironstone 6 inches thick) 4 or 4½ feet.
  - Coarse pea-iron-ore 3 inches to a foot or more.

Reading Beds.   Grey and crimson plastic clay.

Fossils from the Basement-bed of the London Clay in the Isle of Wight.

P = first noted by Prestwich.
W = „   „   „   Whitaker.

—	Alum Bay.	Whitecliff Bay.
Lamna, teeth - - - - -	P	P
Aporrhais Sowerbyi, *Mant.* - - - -	W	P
*Calyptræa ? - - - - -	W	—
*Fusus - - - - - -	W	—
Natica labellata, *Lam.* - - - -	W ?	P
*Pleurotoma - - - - - -	W	—
Pyrula tricostata, *Desh.* - - - -	- - -	P
Rostellaria (? = Aporrhais Sowerbyi) - -	W	—
*Solarium - - - - - -	W	—
*Arca - - - - - - -	W	—
Cardium plumsteadiense, *Sow.* - -	W ?	P
Corbula - - - - - -	- - -	P
*Cyprina Morrisii, *Sow.* - - - -	W	W ?
Cytherea obliqua, *Desh.* - - - -	W	P
* „ orbicularis, *Edw.* - - - -	W	—
*Glycimeris ? - - - - - -	W	—
*Nucula - - - - - -	W	—
Ostrea - - - - - -	- - -	P
*Panopæa - - - - - -	W	—
Pectunculus brevirostris, *Sow.* - -	W ?	P
Ditrupa plana, *Sow.* - - - - -	P	P
Wood, &c. - - - - - -	- - -	P

* Here recorded for the first time (from the Isle of Wight).

The following section was measured in July 1888 with the assistance of Mr. Henry Keeping. It continues the upward succession given at p. 95.

Section of the London Clay in Alum Bay.

	FEET.
Dark blue loamy clay, with ironstone nodules. Becomes sandy in the upper part - - - - -	46
Laminated dark grey loam - - - -	13
Loam, passing upward into fine sand - - - -	23
Blue clay, becoming more loamy above - - -	17
Line of large septaria full of *Cardita Brongniatii* (a conspicuous bed)	
Dark blue clay - - - - - -	62
Loam with scattered small flint pebbles. *Panopæa intermedia, Tellina, Cassidaria, Fusus, Turritella imbricataria, Natica labellata* - - - - - - - -	2
Brown and bluish clay, with lines of septaria - - -	35
Septaria full of *Pinna affinis* (a conspicuous bed) - -	
Brown and bluish clay, sandy in places, with lines of septaria	20
Basement Bed—Sandy glauconitic loam with a little pyrites. *Ditrupa* at the base - - - - -	15
Total - - -	233

Other measurements made the total 200 feet and 220 feet.

Here again it must be observed that no reliance can be placed on the minute accuracy of the measurements, for the top of the cliff will give a different result from its base. If the monoclinal curve of the Isle of Wight be carefully plotted and measured, it will be seen that the upper and under surface of any bed affected by the disturbance cannot always be parallel, but that the thickness will vary according to the part of the curve at which it is taken, and also according to the hardness or softness of the beds affected.

At Whitecliff Bay the basement pebble-bed, two inches in thickness, is overlain by eighteen inches of buff-coloured sands, above which there lies a bed of hard sandstone, abounding in *Ditrupa plana*, that appears on the shore and may be seen stretching out to sea, for a considerable distance, at low water. About thirty-five feet above the basement bed there occurs a zone of *Panopæa intermedia* (Fig. 19), and *Pholadomya margaritacea* (Fig. 18),

FIG. 18.

Pholadomya margaritacea,
Sow.

with their valves closed; at fifty feet another band of *Ditrupa plana* (Fig. 20) comes in, and at about eighty feet there is a well-marked band of *Cardita.*

The remainder of the section in Whitecliff Bay consists, in ascending order, of lignite in dark-grey clayey sand, aluminous and weathering to a brown colour; ferruginous-brown sands; clayey sand or sandy clay as before, but darker, harder, and more clayey than the beds below, and containing *Panopæa intermedia,*

FIG. 19.

Panopæa intermedia, Sow.

FIG. 20.

Ditrupa plana, Sow.

with their valves joined, lying in the positions they occupied when
alive.   Succeeding these, are similar beds with sandy alternations
and laminæ, and a layer of large septaria.   *Pinna affinis* (Fig. 21)
is found in the septaria.*   The total thickness of the London
Clay amounts to about 320 feet.   A bed of flint-pebbles is found
at 255 feet above the base.

FIG. 21.

Pinna affinis, Sow.

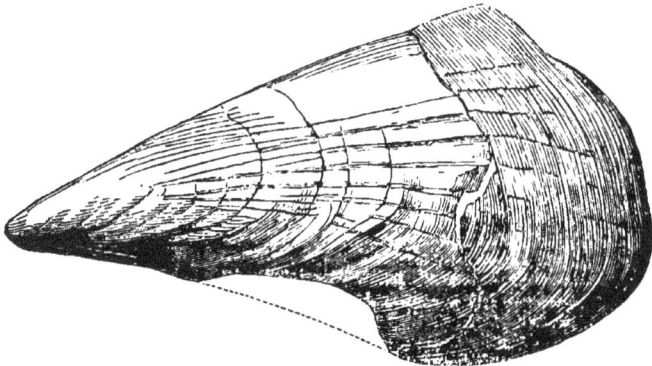

No inland sections of the London Clay are now visible in the
Island, unless the cutting at Ashey is partly in this division.
Probably, however, the clays there exposed belong almost entirely
to the Bracklesham Beds, nearly the whole of the London Clay
being cut out by a strike fault.

The fossils of the London Clay (*see* Appendix) have not yet
been fully collected in this district; but as far as they go they
indicate a subtropical climate, as in the London Basin.   The
occurrence of occasional scattered lines of flint-pebbles in the
clay is noteworthy.   This and the more sandy nature of the strata
seem to point to a gradual shoaling of the sea towards the south,
at the time when the London Clay was in course of being
deposited.

---

* See also Caleb Evans, On the Geology of the neighbourhood of Portsmouth and
Ryde.   *Proc. Geol. Assoc.,* vol. II. p. 70.   (1871.)

LOWER BAGSHOT BEDS.

101

LOWER BAGSHOT BEDS.

In 1847 Professor Prestwich[*] pointed out that the series of sands and clays between the London Clay and the Oligocene Beds in the Isle of Wight is the equivalent of the Bagshot Beds on the mainland. He also showed that in the Isle of Wight there is a similar three-fold division—into Lower Bagshot, without fossils; Middle Bagshot, with marine fossils like those found at Bracklesham; and Barton Clay and Sands, the last two perhaps being equivalent to the Upper Bagshot of the London Basin, perhaps in part (the Barton Clay) dying out northward, or passing into the middle division.

Subsequent research—especially the observations of the Rev. Osmond Fisher—has added largely to our knowledge of these strata and their fauna; but there is still considerable doubt as to the exact limits of the divisions, which in fact pass almost imperceptibly into each other. Recent observations have also indicated that the Upper Bagshot Beds in the London Basin are probably the equivalent of the lower part of the Barton Clay in the Hampshire area; and that the glass-sands (the so-called Upper Bagshot Series of the Isle of Wight) belong to a higher zone, apparently unrepresented north of Hampshire.

Owing to the Bagshot Beds being nearly everywhere vertical, it has been found impracticable to trace their subdivisions on the map, especially in the absence of fossils. The whole series has therefore been grouped together, represented by one colour, and indicated on the map by the letters i 4 to i 7. In this Memoir the term 'Bagshot' is only applied to the plant-bearing pipe-clays and sands formerly called 'Lower Bagshot.'

These Lower Bagshot Beds are highly developed in the Isle of Wight, attaining a thickness of 660 feet in Alum Bay. But it may be well at once to point out that part of this great thickness of sparingly fossiliferous beds may be the equivalent of the lower part of the marine Bracklesham Beds, which appear to thicken so greatly towards Whitecliff Bay.

Lower Bagshot Beds in Alum Bay.

	FT.	IN.
Very thinly laminated pale yellow sand	10	0
White crimson, and rose-coloured variegated sand passing into pale brownish-yellow sand	50	0
Thinly laminated light grey pipeclay	1	6
Pale yellow sand and white laminated clay, with crimson streaks.		

Details of the upper part of this subdivision:—	FT.	IN.		
Yellow sand	14	6	}	104 0
Pipeclay parting	—			
White sand	11	6		
Yellow sand	12	0		
White and crimson sand				

[*] *Quart. Journ. Geol. Soc.*, vol. iii. p. 386.

	Ft.	In.

Thinly laminated clay, chocolate-coloured in the upper part.
  Details :—   Ft. In.
    Clay - - - - - - 27 0
    Lignite (very bituminous) - - - 0 6
    Clay, with a band of lignite 5 or 6 feet from the
      base - - - - - - 44 0   99 0
    Thinly laminated yellow sandstone, with much
      carbonaceous matter - - 4 inches to 0 6
    Clay ; white, hard and marly - - - 27 0
Tawny, variegated, pink and white sands, with brown
  laminæ ; white sand predominates. - -
(Iron bands 1 inch thick occur at 52 feet and 79 feet from   90 0
  the bottom) - - - - -
Pale grey and yellowish-brown sands, with thin laminæ of a
  darker grey clay, containing pyrites and carbonaceous
  matter - - - - - -
(Some of the laminæ, when newly broken, are of a greenish   60 0
  colour. These beds are darker and most laminated in the
  lower part, and are most sandy towards the upper part) -
Light grey sandy clay, with vegetable matter lying across
  the bedding - - - - - -   2 0
Fawn coloured and whitish sands, slightly variegated with
  red : the upper 10 feet slightly laminated.
  Details :—   Ft. In.
    Slightly laminated white sand - - 9 5
    Irony band - - - - - 0 1   40 0
    White, pink and yellow laminated sand, with
      veins of white pipeclay and bright red
      laminæ of iron - - - - 7 6
    Fine light yellow sand - - - 23 0
Pipeclay (full of leaves) between yellowish-white and varie-
  gated laminated clays. The lower 2 inches are composed
  of sandy white pipeclay, with laminæ of yellow and
  crimson sand, becoming thicker towards the upper part of
  the cliff - - - - - - - -   6 0
Bright yellow sand, with thin laminæ of blue clay - 13 0
Iron band - - - - - -   2 0
Grey and yellow sands.
  Details :—   Ft. In.
    Yellow and grey sands - - - - 15 0
    Grey laminated sands and clays; mostly sands 18 0
      Do.   nearly all   45 0
    clay : very carbonaceous - - - 11 0
    Grey laminated sands and clay ; clay predomi-
      nating - - - - - - 3 6
Iron sandstone band and tawny ironsand with ferru-
  ginous veins and strings, and pebbles of quartz - 0 6 to 3 0
Grey sands, &c.
  Details :—
    Pale yellow and bluish white sand, darker in
      the upper part and with a few laminæ of clay 16 0
    Blue clay with thin (¼ inch) sandy laminæ ;   104 0
      carbonaceous matter - - - 27 0
    Grey and yellow sands, with thin laminæ of
      blue clay ; much pyrites and carbonaceous
      matter - - - - - 61 0
(N.B.—These beds have a slightly reversed dip towards the
  top of the cliff.)
Bright yellow and white sands, more laminated and clayey
  than the bed above, and containing much carbonaceous
  matter. The lower 5 feet sand - - - - 23 0
Iron sandstone - - - - -   3 0
Parting of pale clay of variable thickness - - - 0 2

	FT.	IN.
Very thinly laminated white and yellow sand - -	1	10
White sand and blue clay, becoming more clayey towards the lower part.	5	0
[On London Clay.]		

	662	6

At the eastern end of the Island the Bagshot Beds present a different aspect. The mass of white pipeclay has there disappeared, and the beds have either thinned from 600 feet to about 100 feet, or the upper portion has become somewhat marine and is inseparable from the Bracklesham Beds.

The junction between the London Clay and the Bagshot Beds is clearly shown in Whitecliff Bay, the former being represented there by ferruginous brown clay, and the latter by pale grey sands weathering nearly white and containing occasional thin laminæ of pipeclay. Thirty-seven feet of these sands, clays, and pipeclays intervene between the upper part of the London Clay, and a band of sandstone that runs out to sea at the base of the yellow micaceous sands which constitute the greater proportion of the Lower Bagshot series there. Above them there is an 18-inch band of flint pebbles, taken by Mr. Fisher as the base of the Bracklesham Series, for in the clay immediately above marine shells occur.

The inland sections are of little interest, none of them being fossiliferous or showing satisfactorily their relation to the over or underlying deposits. Commencing at the west end of the Island, we find the sands well exposed in pits around Freshwater, especially in one close to Easton, and another on the opposite side of the marsh near some new houses. At the latter there are seams of pipeclay. The road cutting south of Farringford House also shows a good section of ferruginous sand.

Continuing eastward, we learn that pipeclay was formerly dug in a piece of rough ground half a mile east-south-east of East Afton. Due north of this old pit sandy white clay is again seen in the deep channel cut by a small stream north of the high road. This is probably a higher seam—perhaps in the Bracklesham or Barton Series.

About a quarter of a mile east of Chessel a pit has been dug in sand with the bedding vertical. Between this pit and the London Clay a number of flint pebbles are ploughed up in the field, but it is not at all clear from what bed they are derived, though they seem to occur low down in the Bagshot Series, possibly at its base.

Continuing along the high road, we come to a deep cutting in sand with seams of pipe clay between the two entrance lodges belonging to Westover. Similar beds occur in the road to Shalfleet, about a quarter of a mile north of Calbourne. Higher beds are exposed in a small pit half a mile north-east of Calbourne, where sand with a dip of 40° is overlain by a bed of pebbles, and that again by clay. Probably this pebbly bed marks the base of the Bracklesham Beds. A few chains further north there are a number of old sand pits close to Five Houses. These were

probably opened in the glass sands of the Barton Series, but no
section can now be seen.

From this point eastward no sections occur till Newport is
reached. Here the brick-yard near St. John's Church shows at its
southern end sand, with the bedding vertical. Wells in Elm Grove
reach the same bed and a house at the corner of Elm Grove and
the main road, is built on the site of an old sand pit.

From Newport to Downend nothing is seen of the strata, the
slope being much masked by a wash of clay and flints from the
higher ground to the south. At Downend, however, the beds
were well seen in a small pit in Saltmoor Copse, where clay
rests on a bed of pebbles overlying fine buff and red sand, the
whole dipping north-north-east at 80°. The pebble bed, which
perhaps forms the base of the Bracklesham Beds, is apparently
only 150 feet above the London Clay. The Bagshot Beds must
therefore have rapidly thinned out eastward, or else the beds of
pebbles come in on different horizons in different parts of the
Island. As the position of this pit necessitated the cartage
uphill over a bad road of the sand needed in the brick-yard, it was
pointed out by one of the writers that the same bed would be
found close to the kilns, underlying the brick-earth. The pro-
prietor has consequently opened a new sand pit since the survey
was made, and probably the section above described will now be
overgrown.

At Brading Station the sands are again seen, and they re-appear
in the bluffs on the eastern side of the Yar, but without any clear
section. A few chains further east, close to Longlands, a pit
shows a dip of 95°—*i.e.* reversed 5°—to the north-east.

Very little is yet known of the fossils of the Lower Bagshot
Beds in the Isle of Wight, except the plants, for it is doubtful
whether any other organic remains besides elytra of beetles have
been found in this series.

On the Flora of Alum Bay. By Mr. J. Starkie Gardner, F.L.S., F.G.S.

The plant remains were found in a pocket or lenticular
thickening of a seam of fine white pipe-clay in the midst of the
Lower Bagshot Sands. They consist principally of most delicate
impressions of leaves, rarely presenting traces of colour, and giving
little indication of their texture when living. They lie with the
planes of bedding and are rarely twisted or rolled. The leaflets
of compound leaves, of which there are many, are almost always
detached, though a few specimens exist in which they still adhere
to the axis. With the leaves are twigs of a conifer, shreds of
fan-palm and reed, small leguminous pods, drupes and other bodies
too decomposed for identification, and very rarely, a flower like

Porana or *Kydia*, and the detached elytron of a beetle. All bear the appearance of long immersion and tranquil deposition, and the sediment is so fine that the disturbance in it caused by the formation and passage of gas bubbles is distinctly visible. Every trace of carbon has been chemically removed.

This pocket must have been of considerable size, for it was known to Mantell as far back as 1844, and it continued to yield specimens of leaves abundantly down to about 1883, when they became rare, while at present scarcely any vestige of leaf-bearing pipe-clay can be found.

The number of species obtained from this pocket has been variously estimated. The first critical examination of the flora was by De la Harpe in 1856, when out of 48 species seen, 43 were pronounced determinable and named specifically. Of these 21 of the most important were figured in the former edition of this work. Heer added a species in 1859.* Ettingshausen in 1879 spent a winter in studying collections from Alum Bay, and announced† that the flora comprised 274 species divided among 116 genera and 63 families. Like Heer, he found considerable affinity between these and the flora of Sheppey, and further called attention to the community of more than 50 species with the floras of Sotzka and Häring. We are not able to reconcile this estimated richness with our knowledge of the flora, and surmise that fossil plants from other localities must have been inadvertently included.

The flora appears indeed, very restricted as to species, as we might reasonably anticipate, since we are limited to the leaves which drifted waterlogged into a single pool. The most conspicuous and typical of these are unquestionably the *Ficus Bowerbankii*, De la H., *Aralia primigenia*, Heer, *Dryandra acutiloba*, Sternb., *D. Bunburyi*, De la H., *Cassia Ungeri*, Heer, and the fruits of *Cæsalpinia*. It is not certain that these determinations are generically accurate, and indeed one of the latest specimens discovered proved conclusively that the *Dryandra acutiloba* is actually a *Comptonia*; but they are all well-defined species, and as such form exact bases for comparison. These, with a number of less common but scarcely less conspicuous forms, unite to give the flora of which they are the chief elements, a very special and singularly early impress, so much so that Prof. Newberry would regard them as Cretaceous, if their horizon were not stratigraphically defined. The floras which it chiefly resembles are, firstly, that of Monte Bolca, as already noticed by Heer, and secondly, in a far higher degree, the flora of the Grès du Soissonnais, which though resting on the lignites of Woolwich age in the Paris Basin, are really unconformable and doubtless contemporary with our Lower Bagshot.

The chief cause of the highly distinctive and interesting character of the Alum Bay flora, lies in the fact that it is the

---

* Flora Tertiaria Helvetiæ, vol. iii., fol. *Winterthur*. (p. 315, *Drepanocarpus Dacampii*, Mass.)

† Proc. Royal Soc., vol. xxx. p. 228. 1880.

most tropical of any that has so far been studied in the northern hemisphere. Following so immediately the flora of Sheppey, with its wealth of Palm fruits, some denoting the largest species, it presents us probably with an insight into the dicotyledonous vegetation which accompanied them. Sifted as they have been by the agency of water, only those leaves and bodies endowed with certain powers of flotation were able to drift to that point; the heavy palm leaves and fern fronds, and the large leguminous pods which give the Lower Bagshot flora its tropical aspect, have been eliminated. These were left in higher reaches of the stream, and we meet with them at Studland, where large quantities of Fern and Palm are massed together, and at Creech Barrow near Corfe, where the most magnificent opportunities for collecting fossil plants have passed away, never perhaps to recur.*

The Reading flora has an exceedingly temperate facies, and thus presents to us a relatively recent aspect. The Woolwich flora is less temperate, for Palmettos appear in it. The Lower Bagshot flora is like that of the London Clay, decidedly the most tropical. The Middle Bagshot flora begins to lose its tropical elements, and these appear to drop out very gradually and without any sudden changes, down to the close of the Hamstead period, when all traces of Eocene plants disappear from this country. Allowance must be made for the fact that local accumulations will of course present very different appearances and plant remains derived from a sheltered and swampy station will appear luxuriantly sub-tropical, which are not so, and conversely, leaves blown from an arid spot may seem to indicate a harsher climate than actually prevailed.

The break between the London Clay flora and those which preceded it, is very great, and obviously due to a considerable increase of temperature. The connection between that of Sheppey and of Alum Bay, though probably a good deal over-estimated, is likewise due, it appears, to the high temperature having been maintained, bringing in a vegetation that had not been able to exist so far north since the close of the Cretaceous period; whence the Cretaceous aspect that has struck so many observers. The break, which is very great indeed, between the floras of Alum Bay and Bournemouth, deposited as they must have been under very similar conditions, is far less easy to explain. It is not one altogether of temperature, because there are still many large palms in the latter, as *Iriartea, Phœnix, Calamus, Nipa*, with decidedly sub-tropical ferns. Some break or change must have driven the then indigenous flora almost completely away and brought in the new set of plants which

---

* There are still fragments, some of them two feet in diameter, of enormous leaves of fan palms, which might easily have been extracted entire, and parts of huge pods of Cassia and Acacia, preserved in the Dorchester and Jermyn Street Museums and in private collections; but for upwards of 20 years no leaf deposits of Lower Bagshot age have been found. The beds at Creech are much folded and leaf beds of Middle Bagshot age are preserved in the folds, from one of which the large series in the Oxford Museum must have been obtained, and from others I have more than once myself been able to collect.—J. S. G.

maintained themselves and spread over central Europe, only dying out or giving way in late Miocene times. This is why the Flora of Alum Bay is of such immense interest and importance, why its composition is so different from other Eocene floras, and why it is confined to a single horizon. Misled by its striking facies, together with that of the flora of Monte Bolca, which resembles it, and being unacquainted with any other type of Eocene flora, Heer set it up as a sort of test flora, determining according to the degree in which other floras resembled it, whether they should be classed as Eocene or not. Thus the floras of Mull and Bovey were discarded from the Eocene, as those of Reading and Bournemouth would have been had they been adequately known at the time. For the same reason the representatives of the Bournemouth flora on the Continent, became his type of a Lower Miocene (now Oligocene) flora.

In the present state of our knowledge no real analysis of the Alum Bay flora is possible. It is remarkable for the absence of any well authenticated ferns, except the pinnæ of a still somewhat doubtful *Marattia*. *Anæmia subcretacea*, Sap., has been recorded only as *Asplenium Martinsi* by Heer. As it is common in the Reading Beds and again in the Bournemouth Beds and could evidently support a high temperature, its occurrence would not be extraordinary in the Lower Bagshot Beds, but requires confirmation. *Chrysodium lanzæanum*, Visiani, which abounds in the corresponding pipe-clays of Studland, has also been recorded, probably erroneously, from Alum Bay. Of Gymnosperms the *Cupressites elegans* of our former edition has been transferred to the genus *Podocarpus*. Two specimens have revealed traces of fruit, but of too indistinct a character to be very reliable. The foliage greatly resembles that of *Glyptostrobus* which occurs plentifully in the Reading Beds beneath and the Bournemouth Beds above. There appear to be no other Coniferæ in the flora. Of Monocotyledons none whatever are determinable unless it be a very doubtful and unique orbicular leaf something like a *Smilax*. Palms are represented by a few macerated fragments that may have come from the fringe of a leaf such as *Sabal*, and Reeds by almost equally unsatisfactory fragments of sword-shaped leaves. The Dicotyledons are probably between 40 and 50 in number, of which almost all the most characteristic are absolutely confined to the Lower Bagshot horizon in this country. A dwarf leaf of a similar *Aralia* was once found in the highest Woolwich beds at Lewisham, and twice the *Dryandra* (*Comptonia*) *acutiloba* has been found in a small patch of pipe-clay low down in the Bournemouth beds, on the last occasion in the presence of that distinguished palæobotanist M. de Saporta. Some of the most ordinary types of leaves look as if they may be common to other formations, but no importance attaches to them, and with the exceptions just alluded to no strikingly well-marked leaf of either the Woolwich, Reading, or Bournemouth series is known to be common to the Alum Bay flora. The wealth, greater than is supposed, of leguminous plants is one of its chief characteristics, and next in order, are the large leaves ascribed to *Ficus*. The abundance of the single species of *Aralia* and of a larger *Acer* furnish a higher proportion of palmate leaves than we are accustomed to in later Eocene strata.   There are the usual simple laurel and willow-looking leaves, most of which afford no characters on which we can ever base any valid determinations. The question as to whether there are any true Proteaceæ in the flora is still in suspense.   There are several forms of leaves in this remarkable family which are quite unmistakable, but none of these have been found fossil in Europe.   Nor have any unmistakably proteaceous fruits yet been discovered, even among the tens of thousands that have been collected at Sheppey, where they most certainly must have been met with, for the supposed *Petrophiloides* is proved to be an Alder.*   The Australian elements in the Tertiary at one time thought to be so preponderant, grow more and more doubtful when critically examined, and it appears that it is rather to Central America on the one hand, and the Malayan Archipelago on the other, that we must look for species nearly related to those of our Alum Bay and Bournemouth floras.   That there are some Australasian species cannot be questioned in presence of the Bournemouth *Araucaria*, and the Hordwell *Athrotaxis*, but these Gymnosperms may well be of immense antiquity and once perhaps universal, so that their occurrence here or in Australia is of little importance.   The study of Dicotyledons would alone show whether any part of the existing Australian flora had ever migrated across Europe or America, as the existing Japanese flora has most certainly done, and that study, too long postponed, will, it is to be hoped, shortly be continued in the pages of the Palæontographical Society.

PROVISIONAL LIST of the FLORA of the PIPE-CLAY of ALUM BAY (revised by J. STARKIE GARDNER).

Apeiobopsis Symondsii, *De la Harpe.*
Aralia primigenia, *De la Harpe.*
Cæsalpinia æmula, *Heer.*
———— Bowerbankii, *De la Harpe.*
———— brevis, *De la Harpe.*
———— mollis, *De la Harpe.*
———— Salteri, *De la Harpe.*
———— phaseolites, *Unger.*
———— Ungeri, *Heer.*
Ceropetalum myricinum, *De la Harpe.*
Chrysodium lanzæanum, *Visiani.*
Cluytia aglaiæfolia, *Wess. & Web.*
Comptonia acutiloba, *Brong.*
Cornus, sp.
Cupania, sp.
Dalbergia Salteri, *De la Harpe.*
Daphnogene anglica, *Heer.*
———— veronensis, *Massal.*
Diospyrus, sp.
Drepanocarpus Dacampii, *Massal.*
Dryandra Bunburyii, *De la Harpe.*
Elæodendron Heerii, *De la Harpe.*

Ficus Bowerbankii, *De la Harpe.*
———— Forbesii, *De la Harpe.*
———— Granadilla, *Massal.*
———— Morrissii, *De la Harpe.*
Grevillea La Harpii, *Heer, MS.*
Juglans Sharpei, *De la Harpe.*
Laurus Forbesii, *Unger.*
———— Jovis, *Unger.*
———— primigenia, *Unger.*
———— Salteri, *De la Harpe.*
Marattia Hookeri, *Ett. & Gardner.*
Podocarpus elegans, *De la Harpe.*
———— eocenica, *Unger.*
Quercus eocenica, *De la Harpe.*
———— lonchitis, *Unger.*
Rhamnus densinervis, *Heer.*
———— 3 sp.
Sapindus, 2 sp.
Smilax, 2 sp. n.
Zizyphus integrifolius, *Heer.*
———— vestustus, *Heer.*

* J. S. Gardner, On *Alnus Richardsoni, Journ. Linn. Soc.*, vol. xx. p. 417.

CHAPTER IX.

EOCENE—*continued.*

BRACKLESHAM AND BARTON BEDS

ABOVE the Lower Bagshot Beds a variable series of sands and clays with lignite attains a thickness of about 700 feet. There is no clear line of division between this series and the underlying leaf-bearing beds, but the separation is often made at the point where a pebble bed occurs, or at the lowest point where marine fossils have been found. It should be remembered, however, that there is no evidence of any real break, and that the change is so gradual that it is very doubtful whether we have really taken the boundary even approximately at the same horizon at opposite ends of the Island. The difficulty of following the beds inland makes it impossible to connect the sections by tracing the boundaries on the Map.

The beds now to be described are often known as the Middle and Upper Bagshot series, but recent observations have shown that the Upper Bagshot Beds of the London Basin are probably the equivalent of the Barton Clay (*i.e.* of the so-called Middle Bagshot of the Hampshire Basin). It has therefore been thought safer to drop these names and simply to call the groups—for the present at any rate, and having regard only to the Isle of Wight— Headon Hill Sands, Barton Clay, and Bracklesham Beds.

BRACKLESHAM BEDS.

In 1847, Prof. Prestwich showed that the marine bands over-lying the unfossiliferous Lower Bagshot Beds of Whitecliff Bay were probably equivalent to the fossiliferous Bracklesham Beds so well seen near Selsey.* Subsequently the Rev. Osmond Fisher worked out the palæontology of the beds in greater detail, and the following account of the sections at the two extremities of the Isle of Wight is mainly taken from his paper.†

The Bracklesham Beds are represented in Alum Bay by clays and marls in the lower part, by white, yellow, and crimson sands in the middle portion, and by dark sandy clays with numerous impressions of fossils in the upper part. The latter alone have been attributed to the Bracklesham Beds in Mr. Fisher's Memoir. The lower beds are remarkable for the quantity of vegetable matter contained in them, not, however, in the shape of leaves, as is the case in some of the Lower Bagshot Beds, but in the form of coal (lignite), constituting solid beds from fifteen inches to two feet three inches thick. Four of these beds, when

---

* *Quart. Journ. Geol. Soc.*, vol. iii. p. 385. (1847.)
† *Ibid.*, vol. xviii. p. 65. (1862.)

fully displayed, are conspicuous objects in the cliff, where they project out of the softer strata, and on the shore, owing to their black and coal-like appearance.

At the time of our survey these beds of coal were more than usually well displayed in consequence of the prevalence of long continued wet weather having worn away the soft intervening strata in which they are imbedded. On examining them during a brief visit made to the Island, in company with Sir A. Ramsay, during the autumn of 1860, it appeared evident that the beds in question occur in the manner of ordinary coal. Like true coal, each bed was based upon a stratum of clay, containing, apparently, the rootlets of plants, as in the underclay of the Coal Measures. The underclays, which occur beneath beds of coal of Carboniferous date, are thought to have been soil that supported the vegetation which, by certain chemical changes, became subsequently converted into coal : it is reasonable, therefore to infer from the presence of similar underclays beneath the coal in the Bracklesham Beds at Alum Bay, that the plants out of which that coal was formed grew on the spot, and were not drifted from elsewhere, as was the case with the vegetable remains in the pipe-clay beds of the Lower Bagshot Series.

A similar underclay was visible in Whitecliff Bay in December 1886, but, owing to the coal having been worked a few years before as far as it could be conveniently reached, the seam itself could not be examined or measured, though a sketch of the roots was made.

On comparing the section of the Bracklesham Beds in Whitecliff Bay with the corresponding section in Alum Bay, it will be seen that the beds are much better developed in the former locality than in the latter. It is, therefore, at the eastern extremity of the Island that these deposits may be studied to the most advantage. Indeed, this is the only locality in the country where the entire series can be seen exposed to view. The following section is taken from Mr. Fisher's paper.*

Section of the Bracklesham Beds at Whitecliff Bay.

No. I. is the lowest of the series occurring towards the south end of the Bay, and No. XIX. the highest of the series further to the north. The letters *a b c*, &c., denote the more important fossil-beds.

Nos. in Fisher's Section.	Nos. in Prest-wich's Section.†	—	Feet.
XIX.	(17)	*a* Greenish and blue clays - - - - At 24 feet from the top is a band of small shells imperfectly exhibited. Ostrea flabellula. Cardita, a small species like Mytilus, a small species. *C. oblonga*.	162

* *Quart. Journ. Geol. Soc.*, vol. xviii. p. 67. (1862.)
† *Ibid.*, vol. ii. p. 223. (1846.)

Nos. in Fisher's Section.	Nos. in Prest- wich's Section.		Feet.
XVIII.	(16)	Dark-blue clay, weathering brown - - -	22
XVII.	—	*b Nummulites variolarius* in blue clay. The clay is crowded with Nummulites, which are often black	10
		Turbinolia sulcata.	
		Nummulites variolarius.	
		Quinqueloculina Hauerina.	
		Alveolina sabulosa.	
		Rotalia obscura.	
		Fusus longævus.	
		—— pyrus.	
		Mitra parva.	
		—— —— var.	
		——. labratula.	
		Turritella sulcifera.	
		Dentalium politum.	
		—— striatum ?.	
		Rissoa cochlearella.	
		Pecten corneus.	
		Cassidaria nodosa.	
		Pleurotoma inflexa.	
		—— plebeia.	
		—— scalarata.	
		—— Fisheri.	
		Voluta nodosa.	
		Cardium parile ?.	
		Lucina ?.	
		Cardita planicosta.	
		Crassatella (the species found also at Brook).	
		Corbula pisum.	
		—— cuspidata.	
XVI.	(15)	*c* Light-coloured sand, with two beds of sand-rock. *Tellina* and small Univalves in the bottom of the lower rock - - - - - -	6
		Natica. Tellina donacialis. T. plagia.	
		('This stratum forms a good horizon of reference being distinct in character and noticeable.)	
XV.	(14)	Sandy clay, passing into lead-coloured compact clay	10
		Echinoderm in sand. Ancillaria canalifera in clay.	
XIV.	—	*d* Dark sandy clay, with grains of black sand, full of *Corbula pisum* in the upper part, and with numerous shells below; passes into dark clayey sand with *Pecten corneus* - - - - -	3
		Nummulites variolarius (common).	
		Rostellaria sublucida.	
		Murex asper.	
		Fusus pyrus.	
		Strepsidura turgida.	
		Cassidaria nodosa.	
		Pleurotoma plebeia.	
		Voluta nodosa.	
		—— Selseiensis.	
		Cerithium tritropis, *Edw. MS.*	
		Calyptræa trochiformis	
		Turritella imbricataria.	
		—— sulcata.	
		Ditrupa plana.	
		Pecten corneus.	
		Pinna margaritacea.	
		Nucula Dixoni, *Edw. MS.*	
		Leda.	
		Crassatella (the Brook species).	
		Corbula pisum (abundant).	
		—— costata.	
		Cytherea lucida.	
		Cultellus.	
XIII.	—	Beds not exposed; apparently clays - -	39
XII.	—	Streaked, whitish-yellow, and foxy sands - -	10
XI.	—	*e* Sandy clays, weathering grey and yellow. There is a layer of casts of shells where it passes into the next bed, *Sanguinolaria Hollowaysii* being extremely abundant - - - - - - -	4
		Turritella sulcifera.	
		Pecten corneus.	
		Pectunculus pulvinatus.	
		Cytherea lucida.	
		Sanguinolaria Hollowaysii.	
		Solen obliquus.	
X.	—	Sand, weathering yellow and grey - - -	7

Nos. in Fisher's Section.	Nos. in Prestwich's Section.†	—	Feet.
IX.	(13)	*f* Brownish sandy clay, with shells and pebbles at the bottom. The shelly layer appears to be a lenticular mass, and not to be persistent - - -	6
		Nummulites variolarius. Ostrea zonulata ?.	
		Murex minax. Arca.	
		Voluta nodosa. Pectunculus pulvinatus.	
		Turritella imbricataria. Chama gigantea.	
		—— sulcifera. Crassatella compressa.	
		Natica labellata ?. Cardita planicosta.	
		Nucula subtransversa ?. Corbula pisum.	
		Tellina plagia ?. Sanguinolaria Hollowaysii.	
		Pecten 30-radiatus.	
VIII.	(12)	Foliated, dark, sandy clays, weathering brown ; with vegetable matter interspersed. There is a layer of casts of shells at the junction with the next bed -	46
VII.	—	*g* Green sand, in which *Sanguinolaria Hollowaysii* is very abundant - - - - - -	15
		(*Nummulites lævigatus* occurs in a mass four feet from the bottom.)	
		Nummulites lævigatus. Sanguinolaria Hollowaysii.	
VI.	(11)	*h* Light- and dark-coloured green sands, with many shells in the upper part. (A spring at the base of the cliff) - - - - - -	62
		Nummulites lævigatus. Pecten corneus.	
		Fusus longævus. Mytilus.	
		—— pyrus. Nucula.	
		Voluta nodosa. Leda.	
		—— spinosa. Lucina.	
		Pleurotoma dentata. Cardita planicosta.	
		Natica (small). Tellina plagia.	
		Turritella sulcata. Sanguinolaria Hollowaysii.	
		—— sulcifera. Solen obliquus.	
		—— terebellata. Corbula (? Gallica).	
		Calyptræa trochiformis. —— pisum.	
V.	(10)	Laminated grey clay, with some beds of calcareous green-sand, and a few beds of lignite - -	76
IV.	(9)	*k* Calcareous, clayey, green, and iron sand, with numerous shells in seams. The base seems washed into the next bed - - - - -	52
		Nummulites lævigatus Calyptræa trochiformis.	
		(rare). Ostrea flabellula.	
		Fusus pyrus. Cardita planicosta.	
		Metula (Buccinum) Cytherea lucida.	
		juncea. C. suberycinoides.	
		Pleurotoma (small). Tellina.	
		Voluta nodosa. Panopæa.	
		Natica. Corbula pisum.	
		Turritella imbricataria ?.	
III.	(8)	Alternating beds of green sand and finely laminated clay, weathering grey and brown ; with thin seams of lignite - - - - - -	18
II.	(7)	Yellow sand - - - - - - -	10

Nos. in Fisher's Section.	Nos. in Prestwich's Section.	———	Feet.
I.	(6)	Sandy clay, weathering grey and brown, finely-laminated with yellow sand. There are casts of bivalve shells in a band of clay at the bottom. It is based on from 10 to 18 inches of black rounded flint pebbles, often as large as swans' eggs - -	95
		Total thickness - -	653

The fossiliferous beds marked (*b*), (*d*), and (*f*) are very persistent at the various localities where one or another portion of the series is exposed. It is from them that the many splendid collections of fossils have been obtained. Of the well-known shell-beds round the Selsey peninsula, those nearest to Selsey Bill correspond with (*b*) and (*d*). The beds at The Park and Thorney, on the east and west of Selsey, correspond with (*g*), and those of Bracklesham itself with (*k*).

Of the fossiliferous beds near Stubbington, that of Brown Down corresponds with (*d*), and that at Hill Head with (*f*).

Fine collections of fossils, in excellent condition, have also been obtained from the neighbourhood of Brook in the New Forest, from the horizons of (*b*) and (*d*). The large collections obtained from these localities by the late Mr. F. E. Edwards are in the British Museum, and those by the Rev. Osmond Fisher are deposited in the Woodwardian Museum, Cambridge.

More recently (in 1886) clear exposures have enabled Mr. Keeping to fix exactly the junction of the Bracklesham Beds and the Barton Clay.* From the Sandstone or *Tellina* bed (No. XVI. of Mr. Fisher's section) to the *Nummulites elegans* zone the distance is 126 feet. This is about 70 feet less than the distance given by Mr. Fisher and would reduce the total to about 580 feet.

About the same time the measurements given below were made by the Geological Survey of the beds associated with the coal-seam (corresponding with No. VII., VIII., and parts of VI. and IX. of Mr. Fisher).

Section in Whitecliff Bay, measured December 1886.

	Ft.	In.
Brown loam, not well seen.		
Black band of powdery lignite and sand - - -	0	2
Laminated beds of loam, sand, and lignite - -	3	6
Shaly clay, full of slickensides, no fossils observed -	23	0
Worked out [coal, &c.] - - - - -	7	6
Shaly underclay, with roots half an inch thick at the top and dying out below. Some of the roots are casts in clay, some in pyrites; nearly all have a film of lignite on the outside - - - - - - -	7	6
Similar clay with pyritous nodules, no roots observed -	8	0

* *Geol. Mag.*, dec. III., vol. iv. p. 70.

	Ft.	In.
Hidden by talus - - - - - -	24	0
Glauconitic loam with yellow joints and much selenite. Casts of small oysters and other marine shells, and occasional pieces of lignite - - - -	5	0
Blue loamy clay with selenite and badly preserved fossils. *Turritella imbricataria*, fish-scales, &c. - - -	16	6
Clayey loam full of small quartz and flint-pebbles, and crowded with fossils, mostly small. *Ostrea, Cardita, Arca, Solen,* &c. - - - - - -	0	6
Hard loam and clay, full of small fossils - - -	9	6
Clay with beds of *Cardita planicosta* and *Turritella imbricataria*	8	0
Laminated loam, clay, and sand, full of lignite.		

The Beds are perfectly vertical. The above being distances measured along the beach, an allowance must be made for the cliff not cutting the beds at right angles. The true thickness of the measured beds will therefore be 90 feet, instead of 113 feet.

One or two sections where what is perhaps the base of the Bracklesham Beds is exposed have been mentioned in the last chapter, but the only locality yielding fossils is the cutting leading to Ashey Chalk-pit, about three miles south-south-west of Ryde. Here we find, above the London Clay, beds which are full of Bracklesham fossils. It is evident that unless the Bracklesham fauna here extends to the base of the Lower Bagshot Beds and into the London Clay we can only account for the proximity of the Bracklesham Beds to the Reading Beds by a strike fault, which has cut out the greater part of the London Clay, all the Lower Bagshot Beds, and perhaps part of the Bracklesham Beds also.

The section is not perfectly clear, but no fault could be detected, and there being no marked line of division between the two formations it is uncertain how much belongs to the one and how much to the other. Probably if there is really a fault its position will be at the point marked in the subjoined section. Unfortunately the cutting being shallow at its northern end and a good deal overgrown, it was impossible to obtain details of the higher strata. All are nearly vertical. This disturbance will be again referred to in Chapter XIV.

The highest bed which can be traced is a coal or lignite seam, formerly exposed in an old sand pit close to the line. The pit is now overgrown, but the coal was proved by boring. There follow 262 feet of alternations of laminated clay, loam, sand and seams of white clay. These strata cannot be examined, only the lower portion being seen in the northern end of the cutting, which is much overgrown. Then follow the beds with Bracklesham fossils as below :—

Section in the railway cutting south of Ashey.

		Ft.	In.
Bracklesham Beds.	⎧ Light-blue or greenish loamy sand, crowded with Bracklesham fossils (IV. of Fisher ?) -	7	0
	⎨ Dark blue loamy clay with a little lignite -	33	0
	⎩ Blackish shaly clay with a little lignite -	18	0

Probable position of a strike-fault.

London Clay { Clay overgrown - - - - 11 0
            { Sand (Basement Bed of the London Clay) - 6 0
Reading Beds   Red and mottled clay - - - 92 0
            Chalk, nearly vertical.

In the shelly bed 160 feet from the Chalk the following species (determined by Messrs. Sharman and Newton) were obtained, mostly by J. Rhodes (the fossil-collector of the Geological Survey).

B	Arca biangula, *Lam.*	B	Natica acuta, *Sow.*
L B	Cardita planicosta, *Lam.*	B	——— obovata, *Sow.*
L B	Corbula striata, *Lam.*	B	Pleurotoma dentata, *Lam.*
B	Cytherea lucida, *Lam.*	L B	——— denticula, *Bast.*
L B	——— suberycinoides, *Desh.*	L	——— teretrium? *Edw.*
B	——— trigonula, *Desh.*	B	Pseudoliva obtusa, *Sow.*
		B	Rostellaria rimosa, *Sow.*
B	Ancillaria buccinoides, *Lam.*		Solarium, sp.
B	Conus deperditus, *Brong.*	L B	Turritella imbricataria, *Lam.*
B	Fusus longævus, *Lam.*	B	——— sulcata, *Lam.*
L B	——— pyrus. *Brander.*		Voluta, sp. (fragment).

Myliobatis (fragment).

The species marked B (including the whole of the forms determined, with one doubtful exception) are well-known Bracklesham shells; those marked L are found in the London Clay. The *Pleurotoma teretrium* (a somewhat doubtful determination) is the only species elsewhere confined to the London Clay.

Between Ashey and Alum Bay no good sections of the Bracklesham Beds occur. When the strata are again met with, in Alum Bay, their character is so entirely altered that it becomes impossible to correlate the minor divisions, or, as already stated, to be certain whether the upper and lower boundaries have been taken in the same place at opposite ends of the Island.

In the following section the upper limit of the Bracklesham Beds has been taken at the point fixed, on palæontological grounds, by Mr. Fisher, instead of at the pebble bed originally adopted as the junction in the first edition of this Memoir. This increases the thickness of the Bracklesham Beds at this point by 44 feet, making the total 155 feet instead of 111 feet. The details of the fossiliferous beds above the conglomerate are taken from Mr. Fisher's paper,* those of the lower beds are from the first edition of this Memoir.

Section of the Bracklesham Beds in Alum Bay.

	Ft.	In.
BARTON CLAY.—Dark sandy clay with fossils (principally small bivalves).		
Dark sandy clay - - - - - -	15	6
Indurated, dark-greenish, sandy clay, with impressions of fossils - - - - - -	1	0

Fusus undosus?	Cytherea lucida.
Murex asper.	——— suberycinoides.
Pyrula nexilis.	Sanguinolaria Hollowaysii.
Turritella imbricataria.	Modiola, sp.
Natica ambulacrum.	Tellina plagia.

---

* *Quart. Journ. Geol. Soc.*, vol. xviii. p. 85. (1862.)

FT. IN.

Dentalium, sp.	Tellina filosa ?	
Cardium parile.	———. Branderi ?	
Cardita, sp.	———. sp.	
	Arca aviculina.	

Dark sandy clay, containing a bed of septaria - -   11 0

Indurated, greyish, sandy clay, with impressions of fossils   0 7

Fusus undosus ?	Cardita, 2 sp.
Voluta nodosa.	Cytherea obliqua.
Natica, sp.	——— suberycinoides.
Phorus agglutinans.	——— lucida.
Turritella sulcifera.	Tellina tumescens ?
Dentalium, sp.	———, 2 sp.
Teredo, sp.	Sanguinolaria Hollowaysii.
Pecten corneus.	Panopæa corrugata.
Cardium parile.	Leda, sp.
——— sp.	Modiola (or Mytilus) sp.

Dark sandy clay, weathering greenish grey, containing
bands of lignite* - - - - -   16 0

Conglomerate of flint-pebbles, cemented by iron-oxide.
The pebbles are of various sizes, up to a foot in
diameter - - - - - 1 0 to 1 6

Sands (principally white), light tawny-yellow in the upper
part; the lower 3 feet crimson - - - -   45 0

Whitish marly clay - - • - •   25 0

Dark chocolate-coloured marls and carbonaceous clay, ⎫
with much lignite and selenite.

FT. IN.

Clays and marls -	-	-	-	15	3
Lignite band	-	-	-	1	6
Clays and marls -	-	-	-	3	3
Lignite band	-	-	-	1	3
Clays and marls -	-	-	-	6	0
Lignite band	-	-	-	2	3
Clays and marls -	-	-	-	4	3
Lignite band	-	-	-	9 in. to 1	0
Clays and marls -	-	-	-	5	0

39 6

Total thickness of the Bracklesham Beds - 155 0

Whether the lower part of this section really belongs to the
Bracklesham Beds is doubtful. Mr. Fisher takes as the base of
the Bracklesham Beds at Alum Bay the bed of flint pebbles
formerly adopted by the Survey as the base of the Barton Clay.
He therefore places the pebble beds at Whitecliff and Alum Bays
approximately on the same horizon. The pebble bed at Alum
Bay certainly appears to mark the incoming of marine conditions,
after the deposition of the plant-bearing sands and pipe-clays of
the Lower Bagshot Beds. But in the absence of recognizable
fossils throughout the whole of the next 500 feet of strata, it is
possible that we are merely dealing with decalcified equivalents
of the marine beds of Whitecliff Bay and Bracklesham. The
pebble bed at Alum Bay may therefore really belong to the

* This is the lowest bed attributed to the Bracklesham Series in Mr. Fisher's
section.

middle or upper part of the Bracklesham Series, since pebbles occur on various horizons at Bracklesham itself.

Though the Bracklesham Beds of the Isle of Wight have only yielded a small portion of the prolific fauna found at Selsey, yet a considerable number of the most characteristic Bracklesham species occur in both districts. Among the most conspicuous may be mentioned *Nummulites lævigatus, Turritella imbricataria* (Fig. 23), and *Cardita planicosta* (Fig. 22).

Fig. 22.

Cardita planicosta, Lam.

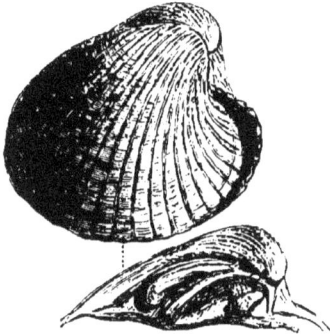

Fig. 23

Turritella imbricataria, Lam.

Specimens of the *Cardita* obtained from the lower portion of the beds at Whitecliff Bay are not only much less in size than those found at Bracklesham, but are pierced by small boring shells; showing that the animals must have perished, and the shells have remained a considerable time at the bottom of the sea before they were covered by the sediment in which they are now imbedded.

The fauna of the Bracklesham Beds of the Isle of Wight appears to show a sub-tropical climate, shoal-water, the proximity of land, and perhaps estuarine conditions. The occurrence of a coal-seam, resting on an ancient vegetable soil, indicates an elevation to a sufficient extent to raise the beds above the sea-level for a portion of the time.

BARTON CLAY.

This group of strata which is displayed in the cliffs at Barton, on the opposite coast of Hampshire, and is so well known to collectors for the richness and abundance of its fossils, is here represented by clays overlying the Bracklesham Beds in Alum and Whitecliff Bays. The nature of these deposits (which are composed of sandy clays, clays, and sands with layers of septaria) is sufficiently shown in the measured sections of Alum Bay, in which locality they attain a thickness of about 250 feet.

Section of the Barton Clay in Alum Bay.*

(Measured in April 1851.)

	FT.	IN.
Ferruginous dark-blue clay, selenite, fragments of univalve shells, numerous fossils - - - - - -	24	0
Pale and ferruginous yellow sandy clay, green in the upper part. Lignite, Corals, *Dentalium*, *Ostrea*, *Corbula*, *Pleurotoma* common and of several species. (The pathway from the chine to the beach cuts through the lower part of these beds) - - - - - - -	69	0
Sands, pale yellowish colour above, green below. (A layer of septaria occurs in this bed about 10 feet from the top, containing pebbles and fragments of wood, and overlying a band of small flint-pebbles) - - - -	35	0
Dark bluish-grey and ferruginous-brown sandy clay, containing much selenite and lignite. *Corbula* abundant. (A layer of septaria, 1 foot thick, occurs 5 feet from the top, 3 feet under which is a band about 2 inches thick of very small pebbles of white quartz, with Shark's teeth. A second layer of septaria occurs at 28 feet; and a third, 5 feet from the bottom of the bed. There is also a band of fossils at 13 feet, and a band of lignite 10 feet from the bottom) - - - - - - -	53	0
Pale grey loamy sand, mottled with yellow, and thinly laminated - - - - - - -	9	0
Dark bluish-green clay, with numerous univalves and other fossils. A ribbed *Dentalium*, *Fusus longævus*, *Voluta spinosa*, *Solarium*, *Cardium*, *Natica* (2 species), *Fusus pyrus*, *Rostellaria*, *Cancellaria*, *Pleurotoma*, *Mitra* (small species) - - - - - - -	65	0
Total - -	255	0

The Rev. O. Fisher gives the following details of the base of the Barton Clay (including 15 feet of beds) at this point :†—

Dark-greenish, coarse, sandy clay. Crowded with *Nummulina Prestwichiana* [now known as *N. elegans*].

Rostellaria ampla.	Strepsidura turgida.
———— rimosa.	Cassidaria ambigua.
Murex asper.	Ancillaria, sp.
Typhis pungens.	Pleurotoma turbida.
Cancellaria, sp.	———— conoides.
Pyrula nexilis.	———— plebeia.
Fusus bulbus.	————, sp.
——— carinella.	Voluta athleta.
——— errans.	——— depauperata.
——— interruptus.	——— maga.
——— longævus.	——— nodosa.
——— Noæ.	Mitra parva.
——— regularis.	Marginella, sp.
——— unicarinatus.	Natica labellata.
——— sp.	Turritella imbricataria.

---

* Another section, differing somewhat in details, will be found in Messrs. Gardner, Keeping, and Monckton's paper. *Quart. Journ. Geol. Soc.*, vol. xliv. p. 600.

† *Quart. Journ. Geol. Soc.*, vol. xviii. p. 84. (1862.)

Phorus agglutinans.
Calyptræa obliqua.
Dentalium, sp.

Corbula pisum.
Pholadomya, sp.

Ostrea flabellula
—— dorsata?
Pecten corneus.
Cardium, sp.

Echinoderm.

Operculina, sp.
Nummulina Prestwichiana.

Lead-coloured clay, with few fossils - - - - 3 0
Rostellaria macroptera.   Corbula pisum.

Dark sandy clay, with fossils (principally small bivalves) - 9 0
Rostellaria ampla.   Arca aviculina.
Fusus regularis?   Leda, sp.
Pleurotoma exorta.   Nucula, sp.
Voluta nodosa.   Cardium parile.
Turritella imbricataria.   Cardita globosa.
Melania?   Cultellus, sp.
Calyptræa, sp.   Corbula pisum.
Solarium plicatum.

These details of the lower beds are given to show how gradual is the upward passage, both lithological and palæontological, from the Bracklesham Series, already described at p. 115, into the overlying Barton Clay.

When the original survey of the Island was made an inland exposure of the Barton Clay was visible at Gunville. This is now overgrown, and no new sections are at present open. The Brick Yard at Gunville showed shelly clay, from which were obtained numerous sharks-teeth and some mollusca. Unfortunately few of these have been preserved, and the new Brick Yard on the west side of the road only shows Pleistocene Brick-earth, resting on the upturned edges of the Lower Bagshot Sands, with perhaps in one place a trace of the base of the Bracklesham Series in some green sandy clay.

At the east end of the Island the Barton Clay is seldom well seen, owing to the accumulation of beach, and to the landslips and mud-streams which constantly obscure this part of the cliff. However in 1886 the sections were exceptionally clear and Mr. Keeping was able to examine this part of the coast and to measure the following section.*

Section of the Barton Clay in Whitecliff Bay.

(Measured by Mr. H. Keeping in 1886.)

FT. IN.

Blue sandy clays, with mottled brown patches of soft earthy ironstone near the base. The upper 15 feet consist of bluish sandy clay, containing - - - - - 50 0

Terebellum sopitum, *Brand.*   Calyptræa trochiformis, *Lam.*
Voluta humerosa, *Edw.*   Ostrea flabellula,*Lam.*
Pyrula nexilis, *Lam.*   Pecten carinatus, *Sow.*
Natica, sp.   ——, sp.

* See Geol. Mag., dec. III. vol. iv. p. 70.

FT. IN.

Lima, sp.	Cypricardia, sp.
Avicula media, *Sow.*	Cardita oblonga, *Sow.*
Arca, sp.	Cytherea tenuistriata, *Sow.*
Pectunculus deletus, *Brand.*	Tellina ambigua ? *Sow.*
Limopsis scalaris, *Sow.*	Corbula ficus ? *Brand.*
Nucula bisulcata, *Sow.*	Panopæa intermedia, *Sow.*
Chama squamosa, *Brand.*	
Cardium porulosum, *Brand.*	Schizaster D'Urbani, *Forbes.*
Lucina gibbosula, *Lam.*	
Crassatella tenuisulcata, *Edw.*	Ditrupa plana, *Sow.*

Imperfect ironstone band, not well seen   -   -   -   3   0

Grey and pale blue clays, with light fawn-coloured bands near
  the base   -   -   -   -   -   -   -   36   0

Stiff laminated clay, with occasionally dark patches. Few or
  no fossils   -   -   -   -   -   -   -   18   0

Pale blue and yellow sandy clays, with very few and badly
  preserved fossils   -   -   -   -   -   -   54   0

Nummulites elegans zone, consisting of rather dark green and
  blue glauconitic sandy clays, much crowded in places with
  Nummulites elegans. Fossils :—   -   -   -   -   1   1

Typhis pungens, *Brand.*	Bulla, sp.
Fusus bulbus, *Brand.*	
Cominella Solandri, *Edw.*	Corbula pisum, *Sow.*
Pleurotoma exorta, *Brand.*	Crassatella sulcata, *Brand.*
Voluta luctatrix, *Brand.*	Cardium semigranulatum, *Sow.*
——— digitalina, *Lam.*	Leda minima, *Sow.*
Mitra parva, *Sow.*	Ostrea flabellula, *Lam.*
Calyptræa trochiformis, *Lam.*	
Dentalium striatum, *Sow.*	Nummulites elegans, *Sow.*

Total   -   162   1

The Barton Clay of the Isle of Wight yields a fauna closely
corresponding to that of the typical locality on the opposite coast
of Hampshire, but at present the list of fossils is much smaller.
This is perhaps partly due to a greater poverty of the fauna, but
in all probability it mainly arises from the difficulty in following
thin fossiliferous seams where the beds are so much hidden by
landslips. Another reason is that the area over which each seam
can be examined is much less in the Isle of Wight than at Barton,
owing to the tilting of the beds and their rapid disappearance
beneath the sea-level.

As in the Bracklesham Beds, the mollusca in the lower part
of the Barton Clay of Alum Bay show a decidedly warm
climate, but the fossils are more exclusively marine, the beds
contain a smaller mixture of lignite, and show altogether less
sign of the proximity of land. Among the more conspicuous
fossils are *Nummulites elegans, Pecten reconditus* (Fig. 35),
Corbula pisum, Crassatella sulcata (Fig. 29), *Pectunculus deletus,
Psammobia compressa* (Fig. 27), *Calyptræa trochiformis* (Fig. 33),
Conus dormitor (Fig. 32), *Fusus longævus* (Fig. 31), *Fusus pyrus*
(Fig. 26), *Murex asper* (Fig. 25), *Phorus agglutinans* (Fig. 24),
Rostellaria rimosa (Fig. 28), *Typhis pungens* (Fig. 34), *Voluta
luctatrix* (Fig. 30), &c.

FIG. 24.
Phorus agglutinans,
Desh.

FIG. 25.
Murex asper,
Brand.

FIG 26.
Fusus pyrus,
Lam.

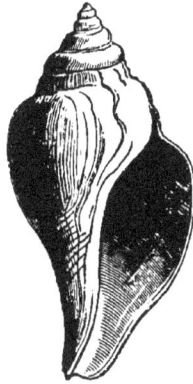

FIG. 27.
Psammobia compressa, Sow.

FIG. 28.
Rostellaria rimosa, Sow.

FIG. 29.
Crassatella sulcata, Sow.

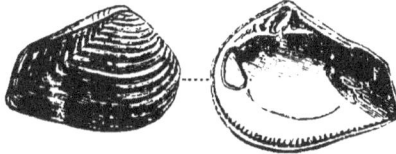

FIG. 31.
Fusus longævus,
Lam.

FIG. 30
Voluta luctatrix, Sow.

FIG. 32.
Conus dormitor, Sow.

FIG. 33.
Calyptræa trochiformis,
Lam.

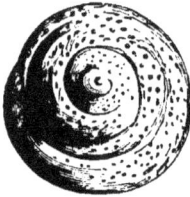

FIG. 34.
Typhis pungens,
Brand.

FIG. 35.
Pecten reconditus,
Brand.

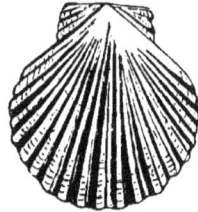

HEADON HILL SANDS.

Between the Barton Clay and the Headon Beds lies a mass of unfossiliferous or sparingly fossiliferous sands. These have been usually called Upper Bagshot Beds, but as they probably belong, as already mentioned, to a higher zone than the Upper Bagshot Series of the London basin, it is better to use for the present the older term "Headon Hill Sands."

The lower part of these strata at Headon Hill consists of about 50 feet of yellow and white sand, succeeded by 60 feet of white sand, with occasional yellow stains caused by the presence of oxide of iron. The total thickness of this group in Alum Bay cannot be determined accurately, in consequence of the disturbed state of the beds there, but probably it ranges from 140 to 200 feet. The Headon Hill Sands are of considerable economic value, their whiteness and purity rendering them particularly suitable for making glass, for which purpose they were extensively worked for many years. Mr. Squire, who rented the cliffs for several years, stated that between 1850 and 1855, 21,984 tons were shipped from Yarmouth, principally to Bristol and London, for the use of the glass-houses there; and a native author, writing in 1795, says,—"Our trade and commerce chiefly is dealing in corn and wool. There are other commodities, such as copperas stones and white shining sand. The former are gathered up in heaps on the sea-shore, and occasionally sent to London, &c. for the purpose of producing the several species of vitriol; the latter is dug out of some very valuable mines, which are the property of David Urry, Esq., near Yarmouth, and from thence sent to London and Bristol for the use of the glass manufactories."

Inland there are at present few or no clear sections of these Sands, but pits, now overgrown, formerly showed the junction with the overlying clays of the Fluvio-marine beds. This junction was formerly seen in a pit about half a mile west of Swainstone, by the side of a road to Fulholding Farm; and, again, further east, under similar circumstances, in the lane a short distance south of Great Park Farm.

South of Gunville about half a mile from Carisbrook in a north-west direction, the Headon Hill Sands and the Barton Clay are thrown up into a vertical position in the brick-pits, where the latter deposit constitutes the brick-earth which was formerly worked there, and, as has been already stated, contained a few fossils.

In East Medina, the Headon Hill Sands showed themselves near Mornhill Farm, and in a pit at the south-east corner of the wood by the side of the road from Arreton Down to Lynn Farm, where they are pure white glass-house sands, together with some of a yellow colour. They are here also vertical, resting with a sharp, well-defined line (marked by a few small rounded flint-pebbles) on green clay—Barton Clay. The age of the strata in this last section is, however, somewhat doubtful, for they are curiously disturbed at this point, and so hidden by gravel that the sands may possibly belong even to the middle division of the Hamstead Beds. Unfortunately this pit being now entirely overgrown cannot be re-examined.

The Headon Hill sands have also been observed in pits at Combley and south of Little Nunwell, as well as on the north side of Bembridge Down, by the side of the road to Bembridge Farm. In Whitecliff Bay the junction between the Headon Hill Sands and the Barton Clay is likewise sharp and well defined, and the former group has a thickness of 184 feet.

Fossil remains are particularly scarce in this member of the Eocene series; though repeatedly searched for during the progress of the survey, no fossils were procured except in Whitecliff Bay, where a few ferruginous casts of bivalve shells were found—chiefly *Tellina*, *Panopæa*, &c.—which, however, could not be preserved on account of their loose and friable condition.

CHAPTER X.

OLIGOCENE.

INTRODUCTION.

THE Fluvio-marine or Oligocene Beds of the Isle of Wight were first described by Webster, who divided them into Lower Freshwater, Upper Marine, and Upper Freshwater, but treated as extensions of the beds in Headon Hill a large series of fluvio-marine beds really lying above the Upper Freshwater.* It was not till the year 1853 that the complete succession was satisfactorily made out, though Prof. Prestwich had already, in 1846,† suggested that the beds seen in Hamstead Cliff were higher than any of the beds at Headon. In 1853 Edward Forbes showed that above Webster's " Upper Freshwater " of Headon Hill, there is found a thick series of beds divisible into several zones characterised by distinct species of fossils.‡ A few years later, in 1856, the observations on which Forbes had been engaged up to the date of his death were published in the Memoirs of the Geological Survey, but the incomplete state in which many of the notes were left rendered it very difficult for Mr. Godwin-Austen, who edited the book, to do full justice to Forbes' work. The divisions and measurements made by Forbes have been adopted with very little alteration in the present Memoir. Later observers have sometimes grouped the beds differently; but this grouping is so much a matter of opinion, and there is such an entire absence of real breaks, that until stronger evidence is brought forward it seems unnecessary to depart from the classification and nomenclature adopted by Edward Forbes.

The following brief summary of the views taken by some of the able geologists who have written on the geology of the strata under notice, may not be out of place here.

Professor Thomas Webster gave the earliest and perhaps the best account of the Fluvio-marine series, founded on observations made in the years 1811–13, and contained in Sir Henry Englefield's work on the Isle of Wight,§ published in 1816. In those letters Professor Webster divided the section at Alum Bay into Lower Freshwater, Upper Marine, and Upper Freshwater

---

* Sir H. C. Englefield. A description of the Principal Picturesque Beauties, Antiquities, and Geological Phenomena of the Isle of Wight. With Additional Observations on the Strata of the Island, &c. by Thos. Webster. (London, 1816), p. 226.

† On the Occurrence of *Cypris* in a part of the Tertiary Freshwater Strata of the Isle of Wight. *Rep. Brit. Assoc.* for 1846, Trans. of Sections, p. 56.

‡ *Quart. Journ. Geol. Soc.*, vol. ix. p. 259.

§ The letters of Professor Webster are illustrated by large copperplate views of cliffs and coast scenery which, for accuracy and spirited execution, have perhaps never been surpassed as drawings illustrating geological phenomena.

Formations; and Headon Hill was considered to comprise a complete section of the whole of the Fluvio-marine series. Although the calcareous strata in the upper part of Headon Hill were noticed, the limestones of other parts of the Island were referred to some of the thick beds of Lower Headon limestone displayed at Headon Hill, and all the marine shells of the Fluvio-marine series to his " Upper Marine " formation, or the Middle Headon beds of Professor Forbes. Hence the limestones of Gurnard Bay, East and West Cowes, and Binstead were referred to the " Lower Freshwater " formations, while the " blocks of calcareous stone containing Limnæa lying on the top, in a detritus of blue clay," seen along the shore eastward of the latter locality, as also the limestones of Dodpits and Bembridge, were considered identical with those of the " Upper Freshwater " formation, or the thick limestones which are displayed in the Upper Headon beds at Headon Hill.

Mr. G. B. Sowerby visited Headon Hill in 1821 and inferred that the Upper Marine formation had been deposited under estuarine rather than under marine conditions, in consequence of observing the occurrence together of shells of marine and freshwater genera.[*]

Professor Sedgwick, in a paper published in May 1822,[†] referred all the strata exposed in the cliffs between Bembridge Ledge and Ryde, between Ryde and Gurnard Bay, and also the argillaceous beds between Yarmouth and Hamstead, to the Lower Freshwater formation of Professor Webster; while the oyster bed and marine marls overlying the Bembridge Limestone, and the upper argillaceous beds of Hamstead, were regarded as the equivalents of the Upper Marine formation of that author.

Professor Prestwich showed,[‡] in 1846, that there were no grounds for the supposition of a want of conformity between the series in Alum Bay and that in Headon Hill, and expressed an opinion that no well-marked divisions could be drawn there, as proposed by Webster,[§] inasmuch as marine shells of the Barton clays re-appear among the overlying freshwater strata in Whitecliff Bay, and that the same freshwater species ranged through nearly the whole thickness of the Headon Hill deposits; the phenomena being such as might be purely local, the result of an accidental irruption of brackish water into a freshwater area.

With respect to the age of the fluvio-marine series of the Isle of Wight, and their synchronism with the deposits of the Paris basin, Mr. Prestwich stated that he felt considerable hesitation in hazarding an opinion ; but, guided by the circumstance that all French and English geologists were agreed in referring the Barton group to the Calcaire grossier, as also by the consideration of the upward range of the Barton species, he was disposed to

---

[*] On the Geological Formations of Headon Hill. . . . *Ann. Phil.*, ser. 2 vol. ii. p. 216.

[†] On the Geology of the Isle of Wight. *Ann. Phil.*, vol. xix. p. 329.

[‡] *Quart. Journ. Geol. Soc.*, vol. ii. pp. 223–259.

[§] Lower freshwater Upper marine, Upper freshwater.

consider the Headon Hill series as the upper portion of the
Barton group, and, as such, to refer the whole to the Calcaire
grossier.

In the autumn of 1846 Prof. Prestwich communicated a paper
" On the occurrence of Cypris in a part of the Tertiary Strata
of the Isle of Wight,"* to the Geological Section at the Meeting
of the British Association at Southampton.

The place from which these fossil Cypridæ were obtained was
the upper part of Hamstead Cliff, near Yarmouth. The author
gives a section of the beds, which will be found to agree most
accurately with the description contained in the subsequent portion of this Memoir, and notes the genera of the included shells,
adding " We have thus in the lower part of the section a deposit
containing essentially freshwater testacea, becoming more mixed,
as we ascend, with shells frequenting estuaries. It is a
singular feature of this group, which I believe to form the upper
beds of the freshwater formation of the Isle of Wight, that a
large portion of the species occurring in it are new ; thus the
two characteristic fossils are a species of *Potamides* and a
Melania, neither of which do I find described. The *Cypris* also
is peculiar to this locality." From the passages here quoted it
will be seen that Professor Prestwich had the clue to the structure
of the Upper Tertiary series of the Isle of Wight, and that time
and opportunity were alone wanting to enable him to work out
details on which the Bembridge and Hamstead groups were shortly
afterwards shown by Forbes to be clearly separable from the
Headon series, with which they had continued to be confounded.

In 1853 Forbes published† an outline of the results of his
work in the Isle of Wight between the years 1848 and 1853. In
this paper he gave a new reading of the succession, and a revised
classification and nomenclature of the beds. This was followed
in 1856 by his posthumous memoir "On the Tertiary Fluvio-
marine Formation of the Isle of Wight,"‡ and in 1862 by the
first edition of the present Memoir.

The only subsequent criticism tending in any way to contra-
dict the work of Forbes was contained in a paper by Prof. Judd.§
This author maintained the correlation of the Headon Beds
at Headon Hill with those of Totland and Colwell Bays to be
erroneous and stated that "the strata exposed at the base of
Headon Hill are not, as supposed by previous observers, a mere
repetition, through an anticlinal fold, of the beds seen in Colwell
and Totland Bays, but are on a distinct and lower horizon than
the latter. These Headon-Hill beds are also found to contain a
different assemblage of fossils from that which characterizes the
Colwell and Totland Bay beds." Prof. Judd also proposed a new
classification of the Oligocene Beds, in which they were divided

* *Report Brit. Assoc.* for 1846, p. 56 (*Candona Forbesii*, T. R. J. in Prof. Prest-
wich's Collection).
† *Quart. Journ. Geol. Soc.*, vol. ix. p. 259.
‡ *Memoirs of the Geological Survey.*
§ *Quart. Journ. Geol. Soc.*, vol. xxxvi. p. 137.   (1880.)

into Headon Group (estuarine), Brockenhurst Series (marine), and Bembridge Group (estuarine).

Subsequently Messrs. Keeping and Tawney maintained that the correlation of the marine beds of Headon Hill and Colwell Bay made by Forbes and the Survey was correct, and that the faunas at the two spots were practically identical, the slight variations being accounted for by the somewhat different conditions under which the beds were deposited.*

Forbes' correlation is followed in this Memoir, for though there are some minor points on which Prof. Judd's criticisms are no doubt just, yet with regard to the main difference the recent re-examination of the Island and mapping of the beds on the scale of 6 inches to the mile have not supported Prof. Judd's contention, but rather shown that Forbes' correlation must still be accepted.

As already observed, the subdivision and grouping of the beds in such a variable series of strata are, in the absence of any real breaks, so entirely a matter of convenience, that without stronger evidence it would be most unadvisable to upset the established nomenclature, and introduce a new mode of grouping, founded on that adopted in other districts. Here also the original nomenclature and grouping used by Forbes have been adopted.

The principal alteration in this new edition of the Memoir is in the use of the term Oligocene for the whole of the Fluvio-marine beds formerly known partly as Upper Eocene and partly as Middle Eocene.† This term is universally adopted on the continent, and the change of conditions at the base of the Fluvio-marine series is so marked in the Isle of Wight, that the division of our Lower Tertiary Beds into two, instead of into three series, and the acceptation of the Headon Beds as the base of the upper group is very convenient. Of course the rarity of fossils in the underlying Headon Hill Sands leaves it still somewhat uncertain to which group they should belong, but the marked change of lithological character at the base of the overlying beds, and the fact, recorded by Forbes, that the Sands contain marine fossils of Barton species, is certainly in favour of their being grouped with the Barton Clay.

TABLE of the OLIGOCENE BEDS of the ISLE OF WIGHT.

				FEET.
Hamstead Series	-	-	-	- about 260
Bembridge Marls	-	-	-	- „ 100
„ Limestone	-	-	-	„ 10
Osborne Series	-	-	-	„ 100
Upper Headon Series	-	-	-	⎫
Middle Headon Series (marine)	-	-	-	⎬ „ 150
Lower Headon Series	-	-	-	⎭
Total	-	-	-	620

* *Quart. Journ. Geol, Soc.*, vol. xxxvii. p. 85. (1881.)
† Lyell referred the highest portion to the Miocene.

Owing to the high dip and absence of any topographical feature, it has been found impossible to separate the Osborne from the Headon Series on the Map. These two series are therefore shown by a single colour, though described separately in this Memoir.

HEADON BEDS.

This series, as a whole, consists of a mass of beds of fresh-water, estuarine, and marine origin, the total thickness of which varies from 147 feet at Headon Hill to 212 feet at Whitecliff Bay. It is only at the western extremity of the Island, between the river Yar and the sea, that the Headon series covers an extensive area, elsewhere it is comprised in a narrow belt of land, between the Headon Hill Sands and the Osborne Series. These beds are best displayed at Headon Hill, in Totland and Colwell Bays, and in Whitecliff Bay. There is also a small section of the upper portion—now almost entirely overgrown or hidded by the sea-wall—on the coast close to Norris Castle and Osborne.

The Fluvio-marine formation, which extends over the northern portion of the Island, forms an undulating tract of country, the scenery of which presents a marked difference to that of the more open district covered by the Cretaceous rocks on the south, owing to the greater abundance of woods with which the surface is in many places covered. The land situated on the limestones is of a more fertile description than that based upon the clays or sands, but over a considerable part of the Island mapped as Fluvio-marine there is a thick deposit of flint gravel spread over the surface, which conceals the underlying strata, and causes the agricultural nature of the soil to bear no relation whatever to the rocks beneath. From the highly inclined position of the beds in the neighbourhood of the Chalk, the lower members of the formation are comprised, for the most part, within comparatively narrow limits, and the chief portion of the superficial area occupied by the Fluvio-marine series consists of the upper members of that group. The thick beds of limestone in this formation thin out towards the north, and nearly disappear in an easterly direction.

The Headon Series was subdivided by Edward Forbes into :—

1. Upper { Uppermost marls, with *Cerithium lapidum*?
{ Upper Headon freshwater and brackish beds.
2. Middle ; Headon intermarine.
3. Lower Headon fresh and brackish-water beds.

The following sections, measured during the original survey of the Island, will give a good idea of the nature and fossils of these beds. It must not be forgotten, however, that each of the minor divisions is extremely variable, and many of them are found to die out or entirely change their character in short distances.

Section of the Headon Series of Headon Hill, measured by Edward Forbes in October 1852 (with a few Corrections made in 1888.)

FT. IN.

Upper Headon Beds, 46 ft. 7 ins.

Blue and yellow clays and marls, passing into grey laminated clays with crushed *Paludina lenta* and *Potamomya gregaria* - - - - -   15 0

Variegated clays with *Potamomya*, especially in the lower part. A 6-inch band of ironstone with *Paludina* occurs in the centre of the bed. *Serpula* -   3 3

Brown and green clays. *Potamomya, Paludina lenta, Melanopsis fusiformis* -   3 4

Limestone, carbonaceous at the top; details :—

Carbonaceous - - 1 0

Sandy, with crushed *Limnæa longiscata* and *Planorbis euomphalus* - - 2 0

Full of fine shells; *Limnæa longiscata, Planorbis euomphalus, P. lens, P. obtusus, P. rotundatus, P. platystomus, Paludina,* &c. - 2 0

Rubbly, with *Planorbis euomphalus* - - - 3 0

         8 0

Bluish and purplish clays, passing into Limestone. *Melanopsis carinata, Limnæa longiscata, Planorbis platystoma, P. obtusus, Bulimus politus* -   5 0

Limestone, compact in places, with many shells and lines of nodular concretions in places. Shells as in the limestone above - - - - -   10 0

Greenish-white compact sands, carbonaceous at the base. *Serpula tenuis* -   2 0

Blue clays and sands, crowded with univalve shells. *Cerithium ventricosum, C. concavum, C. pseudo-cinctum, Cyrena obovata, Ostrea, Natica.* The shells are much broken at the lower part (at 2 feet down) and larger than further northward - - -   3 3

Yellow sand, with bands of lignite and clay. *Cerithium concavum* - -   2 0

Blue-green clay with lignite. Fossils few :—*Cyrena obovata*, scattered *Ostrea*   2 0

Limestone. *Planorbis euomphalus, Limnæa longiscata* - - -   1 0

Middle Headon Beds, 33¼ ft.

VENUS BED.

Blue, green, and brown sandy clay, with oyster-beds at about 5 feet from the top. A few fossils in blue clay above; fossils mostly in the middle and lower part. Occasional flint pebbles. *Ostrea, Cyrena obovata, Cytherea incrassata, Nucula, Natica depressa, Melania, Fusus,* small species. The oysters in this bed are smaller and fewer than at Colwell Bay; the other marine shells are also fewer - - - -   15 0

		Ft.	In.
NERITINA BED.	Sand, clay, and lignite; with bands full of bivalves and scattered univalves. *Cyrena obovata, Cerithium ventricosum, C. concavum, C. pseudo-cinctum, Neritina concava, Melanopsis fusiformis* - - -	10	0
	Cream-coloured limestone in one bed. *Limnæa longiscata, Planorbis euomphalus, P. lens?* This corresponds with the limestone of How Ledge -	3	0
	Sand, clay, and lignite, with seeds. At the bottom 2 feet 9 inches of strong carbonaceous bands with seeds and univalves. *Carpolithes, Melania* -	20	0
	Limestone with shells (much broken) probably brackish water? *Limnæa longiscata, Nematura* - - -	1	6
	Green clays; fossils few or none -	8	0
	Zones of lignite and sand - -	2	0
	Ferruginous bands, alternating with clays full of *Paludina* - -	3	0
	Pale sands with bands of lignite -	4	8
	CYRENA PULCHRA BED.—Green clays, carbonaceous at the base. *Cyrena pulchra, Potamomya, Limnæa* -	0	6
Lower Headon Beds, 61¾ ft.	Limestone, very shelly in the middle, and divided into two beds by a clayey parting. *Limnæa longiscata, L. caudata, Planorbis euomphalus,* fragments of *Paludina* - - -	5	4
	Green clays with purplish streaks (from this clay to the base of the Headon Series the beds vary very much at different places) - - -	1	4
	Sandy limestone, very shelly and ferruginous at the base. Shells crushed -	0	6
	White and yellow sand, with a carbonaceous band at the top - -	0	4
	*Blue clay with shells; becomes sandy below. *Potamomya, Cerithium* -	4	6
	Sandy limestone, passing upwards into sand. *Planorbis euomphalus, Limnæa longiscata* (shells much broken) -	1	6
	Strong band of ironstone 2 inches to -	0	4
	CYRENA CYCLADIFORMIS BED.—Sandy green clays, *Potamomya, Cyrena cycladiformis, Cerithium elegans, C. duplex* - - - -	3	0
	White sands with harder bands -	1	6
	Green clays with a thin ferruginous band 1 inch thick at the base. No fossils? - - - -	6	0
	Total -	146	10
Headon Hill Sands	Bright yellow sands, with white sand, forming lenticular patches in yellow sand - - - - -	11	0

Another Section measured downwards from the beds marked (*), nearer Alum Bay, is slightly different.

		Fᴛ. Iɴ.
Lower Headon Beds.	{Green clays with thick bands of *Potamomya plana. Paludina* in places.	
	Selenite - - - -	3 0
	White sands, without fossils - -	1 6
	Thin band of sandy limestone with *Planorbis*, &c. - - - -	0 6
	White clayey band. No fossils -	0 6
	White sand - - - - -	2 0
	Green marls with lignite bands. Broken *Cyrena* and *Potamomya, Cerithium elegans? C. duplex?* - - -	3 0
	Pink and yellow rather compact sands, with a lignite band at the top -	2 6
	Ferruginous ledge of dark-red sandy beds, with a strong but narrow iron-band at the base. No fossils - -	2 0
	White sands (Headon Hill Sands).	

The Headon Beds vary so much in short distances that other measurements, made only a mile or two from Headon Hill give very different results, though the total thickness is nearly the same. The following were taken about 1852 by E. Forbes and H. W. Bristow :—

Section of the Headon Beds in Colwell and Totland Bays.

		Fᴛ. Iɴ.
Upper Headon Beds, 47½ feet.	{Dark blue clays alternating with ferruginous and septarian bands. *Paludina lenta, P. globuloides, Limnæa longiscata, Serpula, Potamomya gregaria* - - -	6 0
	Red and green marls - - -	1 0
	Sandy beds, greenish clays, and grey shales, with lenticular patches of broken shells and wood. *Paludina lenta* above, *Potamomya? Cyrena obovata*, var. *major*, fragments of *Unio, Melanopsis fusiformis?* and *Melania muricata* - - - - -	8 6
	White, yellowish, and dark sand, with clayey streaks. *Melanopsis fusiformis, M. subcarinata? Cyrena pulchra.* Lenticular patches of dead *Melanopsis* and *Cyrena obovata* in the lower part - - - -	5 0
	Limestone. *Limnæa longiscata, Planorbis* -	1 0
	Greenish clay and sand, crowded in places with univalve and bivalve shells. A ferruginous band at 10 inches from the bottom of the bed. *Potamides trizonatum, Cyrena obovata* - - - - -	10 0
	Argillaceous limestone, passing southward into a bed of sand. A carbonaceous band occurs at the base. *Paludina angulosa, Limnæa longiscata, L. subquadrata?, L. angusta?, L. arenularia?, L. tenuis?, Planorbis euomphalus, P. rotundatus, P. obtusus, P. lens, P. platystoma, Nematura.* (This bed occupies the foreshore at Cliff End Fort) - - - -	3 0
	Bluish-yellow and purplish laminated sands and carbonaceous shales (under the battery, southern end) - - - -	10 0
	Laminated clay and sand, with ferruginous sandy lenticular patches and lines of *Potamomya* in places. *Potamomya* - -	3 0

ɪ 2

		FT.	IN.
	Rather compact pale greenish-yellow sand, without fossils - - - -	2	0
	Verdigris-green clayey beds, abounding in *Cyrena obovata*, *Ostrea*, *Melania muricata*, *Cerithium concavum*, *Natica* - -	3	0
	Band of ferruginous concretions, often calcareous internally. Small *Nematura* or *Hydrobia*, *Neritina* (rare), *Cerithium pseudo-cinctum*, *Melania muricata*, *Cyrena obovata*, *Modiola*, *Ostrea* (rare) 3 inches to	1	0
	Bluish-green clays, often very fossiliferous. *Cyrena* at the top of the bed, and *Ostrea* in lenticular patches in the lower part, which becomes blacker and contains calcareous nodules. *Cyrena obovata*, *Mytilus affinis*, *Cerithium pseudo-cinctum*, *Neritina* 3 feet to	5	0
Middle Headon Beds, 30 feet 4 inches.	Lignite and clay; sand in places. Numerous bands of *Potamomya* near the base. *Cerithium pseudo-cinctum*, *Neritina*, *Melanopsis* - - - - -	2	0
	" VENUS BED." Brownish clay full of marine shells. Bank of oysters varying in thickness in different places. *Ostrea velata*, *Cytherea incrassata*, *Corbula cuspidata*, *Cerithium pseudo-cinctum*, *Fusus*, *Murex*, *Voluta spinosa*, *Cancellaria*, 2 sp., *Pleurotoma*, 2 sp., *Nucula headonensis*, *Arca*, *Natica*, 3 sp., *Bulla*, 2 sp., *Tellina*, 2 sp. (The Oyster Bed rises a little (15 feet) south of Linstone Chine) - - -	9	0
	Very variable alternations of blue and red clays and yellow and white sands, becoming fossiliferous, especially near the base, and with a ferruginous band 4½ inches thick in the centre. *Ostrea*, *Melania muricata*, *Cerithium pseudo-cinctum*, &c. - -	6	0
	" NERITINA BED." Dark-blue sandy clay, with two well-marked bands of *Cyrena*. *Cyrena obovata*, *Cerithium*, 3 sp., *Neritina concava*, *Melanopsis fusiformis*, *Nematura*, *Chara* - - - - -	2	4
	Whitish sandy clay with crushed *Limnæa* -	2	3
	Limestone. *Limnæa longiscata*, *L. pyramidalis*, *L. gibbosula*, *L. minima*, *Planorbis euomphalus*, *P. rotundatus*, *P. obtusus*?, *P. platystoma*, *P. lens*, *Paludina* (rare), *Chara*. (This limestone forms How Ledge) -	2	0
	Whitish and grey calcareous clay, passing into Limestone with thick bands of crushed *Limnæa* and lignite near the top and scattered *Paludina* below. Turtle bones -	4	0
	Blue soft sandy clay, with bands of *Paludina*, small black seeds, and *Unio Solandri* -	4	0
	Purplish-grey carbonaceous laminæ with oblique root like markings. Bands of *Paludina*, *Melania*, *Cyrena*, *Unio*, Seeds -	0	8
	Brown, red, and grey clays and sands, with seams of *Paludina*. Paler sands below -	8	0
	Sand, abounding with small shells above, and with a concretionary band at the base. *Helix labyrinthica*, *Achatina costellata*, *Limnæa pyramidalis*, *L. caudata*, *L.*		

		FT.	IN.
	longiscata, L. mixta?, L. fusiformis, L. tumida?, Planorbis rotundatus, P. lens, P. obtusus, Melanopsis brevis, Melania, Paludina lenta, Cyrena cycladiformis?, C. obovata, Chara - - -	2	6
	Bed partly concretionary, partly sandy with lenticular masses of broken *Potamomya*. (Forms Warden Ledge) - - -	3	6
Lower Headon Beds, 82 feet 4 inches.	Pure white sand with bright yellow stripes. No fossils - - - - -	8	0
	Blackish-grey sands with bands of *Potamomya*. Seeds - - - - -	4	0
	*Carbonaceous sand and clay, with bands of *Potamomya*. A strong band of lignite at the base. Seeds. *Paludina* scarce -	2	6
	Pale-green sandy clays - - -	2	6
	Limestone, with *Potamomya* at the top. *Planorbis euomphalus, Limnæa longiscata, L. pyramdalis, L. sulcata* - - -	1	6
	Greenish and yellowish clay with lignite 2 6 to	3	0
	Imperfect Limnæan limestone - -	0	6
	Pale-green marls, with roots in places and occasional broken *Limnæa* and *Paludina*. *Melanopsis*. (Numerous bands of *Potamomya* near the base) - - -	14	0
	Imperfect Limnæan limestone; very soft, with crushed shells - - - -	1	0
	White and yellowish sand. No fossils -	2	0
	Hard greenish marl. *Melanopsis brevis, Potamomya, Serpula* - - - -	1	0
	Sands - - - - -	1	3
	Greenish marl and sandy clay with bands of *Potamomya* - - - -	6	0
	Limestone with *Limnæa* and *Planorbis*. Ferruginous outside. *Cyrena?* - -	0	8
	Purple calcareous marl, with crushed shells -	2	6
	Strong lignite band - - - -	0	3
	Limnæan limestone - - - -	0	9
	Greenish clay and sand - - -	4	0
	HEADON HILL SANDS (pale grey sand) -	—	
	Total - - 153 2		

A Section measured in Weston Chine, commencing at the bed marked (*) in the foregoing differs somewhat.

	FT.	IN.
Lignite - - - - -	0	3
Green marls, sandy clay, and clay 3 6 or	4	0
Green clay - - - - 3 to	0	6
Hard line of crushed *Potamomya* in bright ochreous sand - - - -	0	2
Limnæan limestone; soft and earthy -	2	0
Greenish tenaceous clay, with carbonaceous matter, especially at the upper part. *Planorbis* and *Limnæa* at the base. Throws out water - - - - -	1	6
Soft earthy Limnæan limestone, impure and thinning away and is then marked by a line of shells - - - - -	0	6
Pale green sandy marl, with *Paludina, Potamomya*, &c. in the lower 3 in. which becomes harder and more marly - -	1	6

FT. IN.

		FT.	IN.
Lower Headon Beds.	Hard irregular band of sandy marl; green and brown and containing ferruginous patches - - - - 2 to	0	4
	Impure limestone, with an undulating irregular surface - - - - -	0	6
	Pale-green marly sand or sandy marl 4 in. to	0	6
	Light-grey sand, with occasional bright ferruginous stains in lines and patches. *Potamomya* at the base - - - -	1	6
	Verdigris-green marls and clays, with occasional *Paludina* and lines of *Potamomya* in the lower 6 inches - - - -	5	0
	Limestone (second ledge of the Chine). *Potamomya* at the top, *Limnæa*, *Planorbis*. 6 inches to	0	8
	Light-grey sands, becoming ferruginous towards the bottom - - - 1 3 to	1	6
	Line of lignite 1 inch. Hard band of variable thickness 1 inch. Imperfect limestone with *Limnæa*, *Planorbis*, *Paludina* (Lignite sometimes disappearing) 3 inches - -	0	9
	Light-green clay weathering brown and becoming harder and concretionary at the base 4½ feet and sands, clays and marls at the upper 3 feet - - - -	7	6

The detailed sections given above will show how thin and variable are the minor divisions which go to make up the Headon Beds at the western end of the Island. This variability largely accounts for the difficulty that is sometimes felt in correlating the beds at Headon Hill with those in Colwell Bay. But if instead of attempting to compare isolated sections, certain marked beds are followed continuously through the cliff, the connexion becomes much clearer.*

So many geologists visit this part of the Island that it will be useful to add a few notes which may assist in the tracing of the beds, and in the identification of the principal fossiliferous zones where the connexion is hidden by landslips.

To obtain a general idea of the structure of the beds, it will be desirable first to examine the cliff from a boat at a distance of half or three-quarters of a mile off Totland, though a very good view may be obtained from the end of Totland Pier. By thus first examining the cliff from a distance, one is enabled to recognise the true structure of the Oligocene Beds, and is not so liable to be misled by changes in the direction of the coast, or by landslips––both fertile sources of error in estimating the relative position or dip of beds in these soft deposits.

Examined this way, the coast section shows that there is a high northerly dip at the west end of Headon Hill, where the cliff runs north and south, but that directly the trend of the coast changes so that the cliff runs parallel to the axis of elevation, the dip apparently

---

* A valuable horizontal section will be found in the paper by Messrs. Keeping and Tawney, "On the Beds at Headon Hill and Colwell Bay." *Quart. Journ. Geol. Soc.*, vol. xxxvii. (1881) p. 85.

disappears. Another curvature of the coast, commencing near Widdick Chine, again shows the true northerly dip, but the angle is much lower, the distance from the line of greatest disturbance being greater. From this point there is a northward dip, till the Headon Beds sink beneath the sea-level a short distance north of the Cliff End Battery. There may be indications of a very slight anticline near the Totland Bay Hotel, but it seems scarcely more than a flattening of the beds for a short distance.

When we attempt to trace the beds on the ground, the landslips at Headon Hill make it impossible to follow most of the horizons continuously. However, the thick limestone which forms so bold a feature all through the hill enables us to identify the beds above and below it.

Commencing with the base, the Headon Hill Sands (the glass sands) can now only be traced for about 5 chains north of the Alum Bay Pier, though formerly they could be seen a short distance further. The extensive working of this sand in old times has much to do with the tumbled and obscure character of this part of the section.

Then for a mile the foreshore is entirely occupied by fallen blocks and landslips and the sands are invisible. It is probable that they have really sunk beneath the sea-level for part of the distance, for the higher beds also apparently sink slightly in the middle of the hill, where the distance from the line of disturbance is rather greater than at either end.

At the east end of the landslip and 8 chains south-west of the boat-house at Widdick Chine, the base of the Headon Beds is again visible. The following section was measured at this point immediately above the beach in May of the present year (1888):—

					FEET.
Lower Headon Beds.	} Clay.				
Headon Hill Sands.	⎧ Black carbonaceous sand and brown sand	-	-	-	9
	⎪ Buff sand and clay	-	-	-	1
	⎨ Buff sand	-	-	-	2
	⎪ Do.	} proved by boring {		-	2
	⎩ Fine white glass sand			-	7
					21

A similar section was seen by Prof. Forbes and H. W. Bristow when the original survey was made, and the junction of the Headon Hill Sands with the Lower Headon Beds was clearly laid open for examination.

As Professor Judd had questioned the accuracy of the correlation of the sands seen at the base of the cliff with the glass sands at the other end of the hill, a boring was made to a depth of 9 feet below the beach level. The buff sand in the upper part might have been referred to either division, for the upper part of the Headon Hill Sands is generally stained for a depth of several feet. But the underlying pure white sands are so unlike anything found in the Headon Beds, that it was not thought necessary to carry the boring deeper, especially as the amount of water

FIG. 36.

Vertical Section of the Beds at the North-East Corner of Headon Hill. (Scale, 8 feet to the inch.)
(Reproduced, by permission, from the *Quart. Journ. Geol. Soc.*, vol. **xxxvii.** (1881), p. 91.)

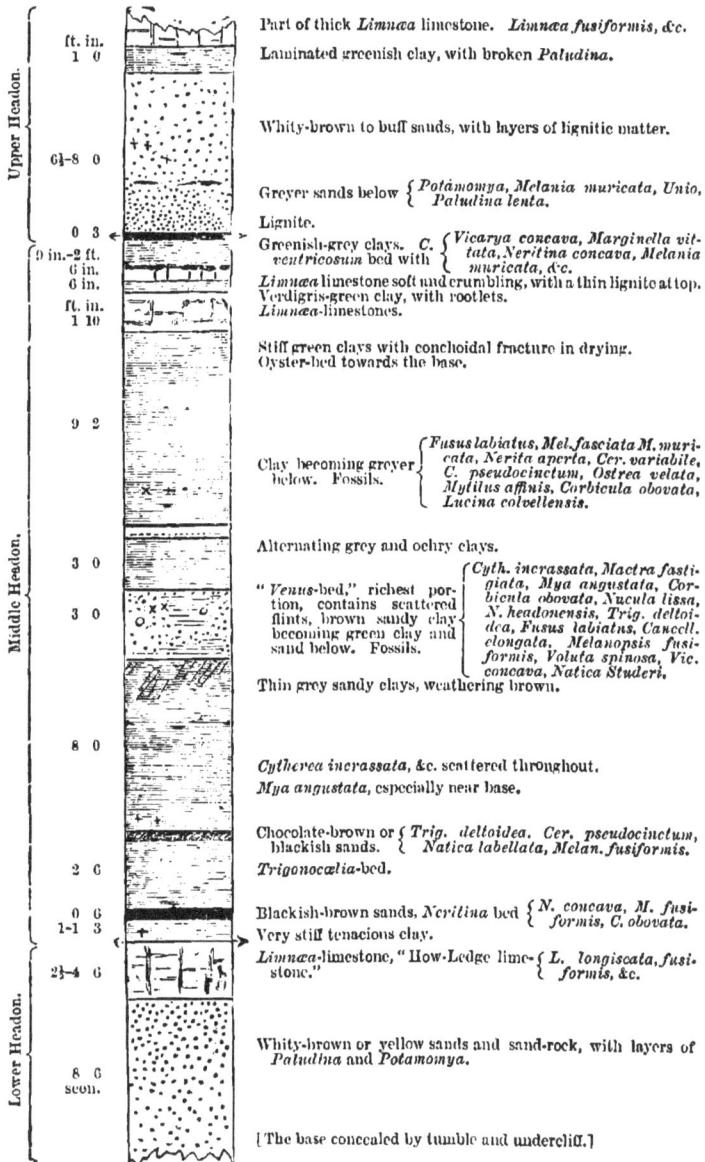

Part of thick *Limnæa* limestone. *Limnæa fusiformis, &c.*

Laminated greenish clay, with broken *Paludina.*

Whity-brown to buff sands, with layers of lignitic matter.

Greyer sands below { *Potamomya, Melania muricata, Unio, Paludina lenta.*

Lignite.

Greenish-grey clays. *C.* { *Vicarya concava, Marginella vitventricosum* bed with { *tata, Neritina concava, Melania muricata, &c.*

Limnæa limestone soft and crumbling, with a thin lignite at top.
Verdigris-green clay, with rootlets.
Limnæa-limestones.

Stiff green clays with conchoidal fracture in drying.
Oyster-bed towards the base.

Clay becoming greyer below. Fossils. { *Fusus labiatus, Mel. fasciata M. muricata, Nerita aperta, Cer. variabile, C. pseudocinctum, Ostrea velata, Mytilus affinis, Corbicula obovata, Lucina colvellensis.*

Alternating grey and ochry clays.

"*Venus*-bed," richest portion, contains scattered flints, brown sandy clay becoming green clay and sand below. Fossils. { *Cyth. incrassata, Mactra fastigiata, Mya angustata, Corbicula obovata, Nucula lissa, N. headonensis, Trig. deltoidea, Fusus labiatus, Cancell. elongata, Melanopsis fusiformis, Voluta spinosa, Vic. concava, Natica Studeri.*

Thin grey sandy clays, weathering brown.

Cytherea incrassata, &c. scattered throughout.
Mya angustata, especially near base.

Chocolate-brown or blackish sands. { *Trig. deltoidea. Cer. pseudocinctum, Natica labellata, Melan. fusiformis.*

Trigonocælia-bed.

Blackish-brown sands, *Neritina* bed { *N. concava, M. fusiformis, C. obovata.*

Very stiff tenacious clay.

Limnæa-limestone, "How-Ledge limestone." { *L. longiscata, fusiformis, &c.*

Whity-brown or yellow sands and sand-rock, with layers of *Paludina* and *Potamomya.*

[The base concealed by tumble and undercliff.]

met with would have necessitated the use of lining tubes if it were to be continued. North of this point the dip quickly carries the base of the Headon Beds below the sea level.

Returning to the western end of Headon Hill we find a thick limestone forming the top of the cliff. The position of this limestone is close to the base of the Upper Headon Beds, and it overlies a series of marine clays and sands full of *Cerithium, Ostrea,* and *Cytherea.* These marine beds belong to the Middle Headon Series, but unfortunately they are not at the present time clearly exposed, except at the two ends of the Hill.

From this point the marine beds are almost entirely hidden by landslips for about a mile but the limestone can be followed, and in a similar position below it at the north-eastern end of the hill the marine beds again occur. Part of these can be well examined at the present time, though they are not easy to find unless one has first identified the thick Limnæan limestone.

Messrs. Keeping and Tawney give a carefully measured section at this point, which is here reproduced, Fig. 36 (*see* page 136).

The base of the thick Upper Headon Limnæan limestone at the point where it leaves the coast is about 120 feet above the sea, and at the north-eastern end of the Headon Hill outlier it has fallen to about 110 feet. Crossing the small valley which divides Headon Hill from a lower hill nearer Middleton, we find the thick limestone at a height of 130 feet. From this point it falls in less than a quarter of a mile to about 110 feet. Then it flattens for another quarter of a mile, and remains at the same level at the northern extremity of the outlier near Amos Hill.

Returning to the coast we find the Oyster Beds in the marine Middle Headon Beds about 95 feet above the sea at the point where the cliff becomes low near Widdick Chine. Half a mile to the north-east there is a small hill on the northern side of Weston Chine which just reaches 100 feet. The upper part of this hill is occupied by a brick-yard, and 7 feet down in the clay, *i.e.,* at about 93 feet, the Oyster Bed is again found. It is full of fossils, but they are not well preserved; the species noted were *Ostrea velata, Cytherea incrassata,* and *Buccinum labiatum.* Thus the same flattening of the beds for a short distance occurs here which we have already noticed in the limestone.

Still further inland, to the north-east, the Oyster Bed is again met with in a large brick-yard near Amos Cottage. Here the height is about 60 feet. In this brick-yard the fossils are all in the state of casts, and only *Ostrea velata* and *Cytherea incrassata* could be determined.

Returning to Totland Bay, we find the dip to become higher and the marine beds again to strike the cliff a few chains north of the Coast Guard station, at a height of about 80 feet. From this point these beds can be followed continuously, except in the parts under Warden Battery, and over short distances where the face of the cliff is obscured by talus. A few yards north of How

Ledge the base of the marine beds falls to the level of the beach, and from this point nearly to Linstone Chine continuous sections are generally exposed, for there is little talus, and the lower part of the cliff is so full of fossils that it presents a vertical face. The thickening of the Oyster Bed, and the way in which it cuts into the underlying clay full of *Cytherea*, are very noticeable in this part of the cliff.

We have now reached the section which all geologists visit, and from which the majority of the marine Headon fossils have been obtained. It may therefore be well to stop for a moment to point out that even this most purely marine portion of the Headon series is full of freshwater shells. A few minutes search is sure to yield several specimens of *Limnæa* and *Cyrena* mixed with the Oysters. The underlying clay full of *Cytherea* is more thoroughly marine, but it also contains a good many valves of *Cyrena*. However there is a decided and essential difference between these marine beds with drifted freshwater shells, and the beds full of *Potamomya, Melania,* and *Potamides,* which lie above and below them. These fossils probably point to deposition in brackish-water lagoons and not in the open sea. Like all accumulations formed in such conditions, therefore they contain abundance of individuals belonging to very few species, instead of a wonderfully varied molluscan fauna like that of the Middle Headon Beds.

The How Ledge limestone, which underlies the marine bed, is another well-marked horizon. This stone is a band, from 3 to 5 feet thick, of freshwater rather tufaceous limestone full of well preserved *Limnæa* and *Planorbis,* belonging to many species. The perfect preservation of the fossils, the softness of the matrix, and the ease with which the bed can be examined, render this the favourite bed from which to obtain these shells. The rock is always visible between How Ledge and Warden Point, and can be traced continuously southward to the Coast Guard Station. Here it passes inland, but Messrs. Keeping and Tawney identify it with the Limnæan limestone at the top of the Lower Headon Beds at the north-eastern end of Headon Hill (see section p. 136). A section of the lower part of the cliff near Colwell Chine, given at p. 242, shows the small reversed or overthrust faults developed in this limestone by lateral pressure connected with the formation of the great uniclinal fold of the Isle of Wight.

A short distance below the How Ledge limestone is a mass of calcareous concretionary sandstone and sand, forming Warden Ledge. This sand is traceable at intervals for about a mile. South of Warden Ledge other thin limestones form a minor ledge on the foreshore. These limestones, full of *Chara* and *Limnæa,* can be traced nearly to Widdick Chine.

The sections of the Headon Beds near Cliff End are, unfortunately, somewhat obscure at present (1889), and the thinning out of the thick Upper Headon limestone renders it difficult to trace the northward limit of the Headon Beds. Messrs. Keeping and Tawney identify the thick limestone of Headon Hill with a bed

1 foot 8 inches thick at Cliff End.* This correlation is probably correct, but it has been found impossible to connect the beds by mapping.

Inland sections of the Headon Beds are rare—at least sections which yield any evidence of definite horizons seldom occur. A very fossiliferous section is exposed in a miniature chine, cut between the north-east corner of Freshwater (All Saints) Church-yard and the marsh. A good deal of gravel has slipped over the beds, which are only clear at the bottom of the channel, so that it was impossible to obtain any measurements. The principal fossiliferous bed consists of a mass of shells in a slightly hardened sandy matrix. The species collected in 1887 were *Planorbis obtusus, Neritina concava, Nematura parvula, Melania muricata, Melanopsis subfusiformis, Limnæa longiscata? Hydrobia Chasteli, Cerithium elegans, Cyrena obovata, Cyrena deperdita, Serpula, Chara.* The specimens of *Neritina* are particularly fine, being unusually large, and with the colour well preserved.

Another manuscript list of fossils from "Wheatlow Brook, near Freshwater Church" (apparently the same locality), gives *Ancillaria buccinoides, Cerithium concavum, C. elegans, C. mutabile, Melanopsis fusiformis, M. carinata, Natica depressa, Nerita aperta, Neritina concava, Paludina lenta, Cyrena obovata.* These fossils were collected about 1852.† In both cases the beds seem to belong to the base of the Middle Headon Beds—the "Neritina Bed" of the coast section.

The well at Golden Hill Fort must have penetrated almost the entire thickness of the Headon Beds, but unfortunately the record of this well has been kept in such a way as to render it almost useless for geological purposes. The section will be found in the Appendix.

Besides those mentioned, there were several temporary sections near Freshwater, showing clays with *Potamomya* and *Paludina.* A well at Poundgreen, 7 chains north-east of the cross-roads, seems to have reached the Headon Hill Sands. It showed :—

Lower Headon Beds.	{ Green clay with *Paludina* and *Potamomya*.
	{ Black clay with crushed *Planorbis*.
	Sand.

The thickness of the beds could not be ascertained.

Crossing the Yar, the old marl pits near the Yarmouth road are in green clay, with *Potamomya*—probably Lower Headon, but no section is now visible. East of these pits the dip becomes high, and there are no exposures for three miles.

Near Little Chessell the beds again flatten somewhat, and sections of the shelly Middle Headon Series can be seen extending for several chains along the stream course about a quarter of a

---

* *Op. cit.,* p. 90.
† I cannot learn definitely who supplied this list or who collected or determined these fossils (though Mr. Bristow thinks it was the late Mr. W. H. Baily), and am unable to find any place named Wheatlow Brook, near Freshwater.—C. R.

mile north-east of the farm.   Here the following species were
collected by J. Rhodes, the fossil collector of the Survey :—

Chara.	Cerithium elegans.
	Hydrobia, sp. (young).
Cyrena obovata.	Melania muricata.
Cytherea incrassata.	Melanopsis subfusiformis.
Tellina, sp.	Natica labellata.
	Nematura parvula.
Ancillaria buccinoides.	Neritina concava.
Buccinum labiatum.	Pleurotoma headonensis.

The *Cerithium* is very abundant, in a shelly sand, and there is
also a bed of clay full of *Cytherea*, but it is difficult to make out
the true succession.

Further north, about 8 chains south of Eades Farm, a ditch
section shows clay full of *Potamomya gregaria*.   On the opposite
side of the stream fossils are ploughed up abundantly in the
fields.   Those collected by J. Rhodes were *Cyrena deperdita*,
C. obovata, *Hydrobia Chasteli*, *Melanopsis curinata*, *Melania
muricata*, *Neritina concava*, *Nematura parvula*, and *Planorbis*.
There is nothing among these to show to what part of the Headon
Series this shelly clay belongs.

From Newbridge eastward to the Medina, the beds are nearly
vertical.   Not a single section of the Headon Series is now
visible there.

At Newport, though the beds cannot be examined at the surface,
the whole thickness of the Upper and Middle Headon strata
seems to have been penetrated in a well at Messrs. Mew and
Company's Brewery (*see* Appendix, p. 305).   It is not easy to fix
the boundary between the Osborne and the Headon Beds, but
taking it as occurring at 259 feet from the surface, we have thickness of 189 feet down to the sand which yielded water.   Of the
189 feet of Headon Beds, at least 82 feet should be referred to
the Upper Headon, and the remainder to the marine Middle
Headon.   Any attempt to correlate the minor subdivisions
would be unsafe, for the samples preserved were small, and the
thickness of the different beds appears to have been greatly
increased by lateral pressure.   Within a few hundred yards of
this well lies the area of sharpest folding.

At West Cowes another well has been sunk to supply the town
(*see* Appendix, p. 313).   Here again the boundary between the
Osborne and the Headon Series is very difficult to fix, but it
seems to lie about 268 feet from the surface.   At 365 feet, *i.e.*,
97 feet below the top of the Headon Series, the shelly " Venus
Bed," commences, and from a sample of clay brought up from that
depth the following species were obtained :—*Cytherea incrassata*,
Cyrena, sp., *Buccinum labiatum*, *Natica labellata*, *Nematura
parvula*, and an otolith of fish.   From 375 feet a sample of green
clay contained *Natica* and indeterminable shell fragments.   From
the spoil heap at the well a considerable number of species were
obtained, and though the exact depth from which they came could
not be fixed, they certainly belong to the clays at about 414 feet.
The species collected were :—

Cardita simplex.
Cytherea incrassata.
Corbula cuspidata.
———— pisum.
Cyrena deperdita.
———— obovata.
Ostrea ventilabrum.

Buccinum labiatum.
Bulla, sp.
Cancellaria elongata.
Cerithium elegans.
Natica labellata.
Pleurotoma plebia.
Rostellaria, sp.
Voluta geminata.

The occurrence of *Cardita simplex* and *Voluta geminata* is interesting, for these are Brockenhurst species previously rare or unrecorded from the Isle of Wight. Both are abundant in this well.

Between 420 and 434 feet grey shelly sand with *Natica*, *Pleurotoma*, *Nematura*, *Potamomya*, *Cyrena*, and *Planorbis* occurs, so the Middle Headon Beds seem to be at least 113 feet thick. This thickness is much greater than at the west end of the island but agrees very well with the Whitecliff Bay section. The increase of thickness of the marine beds is apparently due to the incoming of the Brockenhurst beds, which are absent towards the west. Below the sand the boring penetrated 3 feet into clay, in which no fossils were observed. This clay ought perhaps to be referred to the Lower Headon Series, for the occurrence of *Potamomya* and *Planorbis* in the bed above seems to indicate a change of conditions at this point, but unfortunately the boring was carried no deeper.

Another well, at Woodvale (*see* Appendix, p. 315), a short distance from the last section, penetrates about 13 feet into the Middle Headon Beds, with *Potamomya gregaria*, *Cyrena obovata*, *Ostrea*, *Melania muricata*, *Cerithium concavum*, *C. trizonatum*? The beds seem to correspond with those seen on the foreshore at Osborne.

The Headon Beds reappear for a short distance at the extreme northern point of the Island, brought up by a local undulation connected with the rise of the beds on the north side of the Isle of Wight syncline. During the progress of the first Survey of the Island these beds were well seen at the foot of the cliff near Osborne and Norris Castle. But now the building of the sea walls and the erection of groynes has almost entirely hidden the sections, though abundance of *Cerithium concavum* can still be found on the beach. The following description of the beds is entirely taken from the first edition of this Memoir:—

Due north of East Cowes, a little round the first Point, light-green and red sandy clays, with bands of compressed *Melania costata* and bivalves, forming a shell-marl, have slipped from a higher level on to the shore, and *Paludina lenta*, *Cyrena obovata*, *Potamides* (*Cerithium*) *concavum*, often in a silicified condition, lie scattered in great profusion on the beach.

Immediately under these, apparently, and seen also on the shore, are 1 to 2 feet of greenish-grey clays, with occasional sandy laminæ, and numerous bands of *Potamomya* sparingly mingled with *Paludina lenta*, *Cyrena obovata*, and an occasional *C. pulchra*.

Bands of crushed *Paludina lenta* occur lower down, succeeded by bands of *Melanopsis*, with remains of *Fish* (scales, vertebræ, and teeth). Green sandy clays follow, with thin pyritised bands of shells, a band of *Limnæa longiscata* and smaller subordinate layers of *Potamomya*.

Here the beds undulate, and towards the point above Norris lower beds make their appearance. West of the Point green clays are seen at the base of the cliff 4 inches thick, under a 2-inch band of clay-ironstone. These clays contain *Melania turritissima?* and a black *Cypris*. Upon the clay-ironstone lies a band of *Cyrena pulchra* followed by greenish clay 1 foot thick, full of *Cyrena obovata*, occasionally with the valves in contact, and most numerous towards the upper part. Three feet beneath the ironstone another similar band occurs, separated from the first by green clays, with five or six bands of *Potamomya*. Below the second band of ironstone green clays, with Oysters succeed, associated with *Cyrena pulchra*, *C. obovata*, *Cerithium*, &c.

On the shore, about 50 yards westward from the wall of Norris, pyritiferous bands of *Potamomya* underlie the green clay with oysters, and the section may be there continued as follows:—

	Ft.	In.
Green sandy clays, with an oyster-band 2 inches thick - -	1	6
Grey sands, fossiliferous in the upper part, where they are also laminated, and passing into ferruginous grit - - -	2	6
Light-greyish clayey sands, with 2 inches of *Potamomya* in the upper part - - - - - - -	4	0
Beds not seen - - - - - 3 or	4	0
Greenish sands, with *Melania muricata* and *Potamomya* - -		
Greenish clay, with a few *Potamomya* - - - -	1	0
Consolidated and partly pyritised bands of *Potomomya*, between which are layers of greenish sandy clay full of *Chara*, fish-scales, and *Melania muricata* in patches - - - - -	5	0
Light-green sandy clay, with comminuted Cyrena - - -		

North of Norris, by the sea-wall, the beds on the shore at the Point are crowded with *Cyrena obovata* and *Potamides*; *Cyrena pulchra* and oysters being somewhat scarce.

The shells already noticed as being so plentiful on the beach nearer East Cowes are probably derived from these beds, which are most likely lower than those with consolidated bands described in the preceding section. Opposite the Point they are probably covered by the sea. Hence to the wall separating the Royal grounds from those of Norris the strata are concealed; but on the shore opposite the latter, sands with *Potamides*, *Cyrena*, and Oysters, again appear.

East of Cowes and Newport there are no sections of the Headon Beds till Whitecliff Bay is reached. However the trial borings Nos. 116, 117, and 118, about two miles east of Newport, indicated freshwater beds belonging to the Headon Series, though they yielded no characteristic fossils.

At Whitecliff Bay the Headon Beds are 212 feet thick, and are divisible, as in other parts of the island, into three sections— a middle marine, and an upper and a lower freshwater and estuarine.

The following section is that measured during the original Survey, with some corrections and additions made in 1888:—

Headon Beds in Whitecliff Bay.

		Ft.	In.
Upper Headon Beds, 58 feet.	Grey, reddish, bluish and ash-coloured laminated clays. Layers of *Potamomya gregaria*, with occasional *Paludina lenta*, *Melania* 2 sp., Fish-scales, *Serpula* on the *Paludina* and *Potamomya* - - -	12	0
	Grey laminated clays. *Unio, Cyrena obovata*	5	0
	Sandy clay with calcareous concretions. *Limnæa caudata, Chara Wrightii* - -	1	0
	Ferruginous sands and calcareous hard bands. *Hydrobia*, &c. - - - -	1	0
	Green clay, with *Cyrena obovata* - - ⎫ Brown clay, without fossils - - - ⎬	5	0
	Yellow sand, without fossils - - -	10	0
	Marl and green clay with calcareous concretions. *Cyrena obovata, Limnæa longiscata, Planorbis euomphalus*, pieces of wood -	15	0
	White sand with thin layers of whitish clay -	4	0
	Alternations of carbonaceous clays and greenish sands *Cyrena obovata, Potamides, Chara Wrightii* - - - -	5	0
Middle Headon Beds, 126 feet.	Green sandy loam, with a few casts of marine shells. *Psammobia compressa, Cytherea incrassata, Cyrena* - - - -	12	0
	Blue sandy clay. *Cytherea incrassata* very abundant at the top; *Cerithium pseudocinctum* - - - - -	20	0
	Stiff blue clay, full of fossils. *Cytherea incrassata, Psammobia compressa, Cyrena obovata, Fusus labiatus, Cancellaria elongata, C. muricata, Natica labellata* -	4	0
	Sand or sandy greenish clay weathering brown. Ironstone nodules. Casts of marine shells - - - -	76	0
	Brown sandy clay, often with nodules containing marine shells and fish-remains. *Cardita deltoidea*, &c. - - -	12	0
	Brown clay, containing pieces of the underlying clay and flint-pebbles, and full of marine shells. *Ostrea, Modiola, Cardium, Cardita deltoidea, Cytherea incrassata, Calyptræa*, sp. *Fusus, Voluta spinosa, V. geminata*, &c. (Messrs. Keeping and Tawney record 62 species of mollusca from this bed and compare it with the Brockenhurst zone of the New Forest) -	2	0
Lower Headon Beds, 28½ feet.	Green freshwater marls, with seams of *Potamomya plana, Planorbis, Limnæa*, &c. -	8	0
	Grey sandy clay - - - -	7	0
	Hard ferruginous sandstone - -	0	3
	Pale-green clays, with seams of lignite, and ironstone nodules. *Paludina lenta, Limnæa. Planorbis euomphalus, P. obtusus*, &c. -	8	0
	Carbonaceous clay and lignite - -	1	0
	Green clay, ferruginous at the base. No fossils observed - - - -	4	0
	Total - -	212	3

Here, as at Cowes, there seems to be a tendency in the marine bands to thicken at the expense of the estuarine Lower Headon Beds. These marine bands become more thoroughly marine, losing

to a large extent the admixture of freshwater shells which is so
conspicuous at the west end of the Island. The tufaceous freshwater limestones have all died out, and most of the purely
freshwater beds seem to be largely replaced by beds of estuarine
origin. However, the occurrence of derivative fragments of the
underlying freshwater clays at the base of the marine beds, shows
that the thinning out of the lower series may be due to actual
erosion, and not to a replacement by contemporaneous beds of
marine origin. Messrs. Keeping and Tawney record the occurrence
of a similar line of erosion at the base of the Brockenhurst Beds
in the New Forest.

In Whitecliff Bay two principal horizons in the marine beds
yield most of the fossils. The lowest zone is about 30 feet from
the base of the Headon Series and the greater part of the fossils
are crowded into a seam a few inches thick. The most abundant
species are the *Ostrea, Nucula, Car-
dita acuticosta, Cytherea incrassata*
(Fig. 37). *Pleurotoma*, and *Voluta
spinosa.*

FIG. 37.

Cytherea incrassata, Desh

The other bed is a shaly clay about
90 feet higher. This latter seems to
correspond with the "Venus Bed"
of Colwell Bay, and contains a similar
assemblage of fossils. Among the
common species are *Cytherea incrassata, Corbula deltoidea, Ostrea, San-
guinolaria, Cerithium pseudo-cinctum,
Voluta spinosa,* &c.

A large number of the marine mollusca of the Headon Beds
range downwards into the Barton Clay, but about half are peculiar
to the Oligocene. This apparent break between the Eocene and
the Oligocene will probably disappear when the marine fossils of
the *Lower* Headon Beds and of the Headon Hill Sands are better
known, but at present it is sufficiently marked.

FIG. 38.

Ostrea flabellula, Lam.

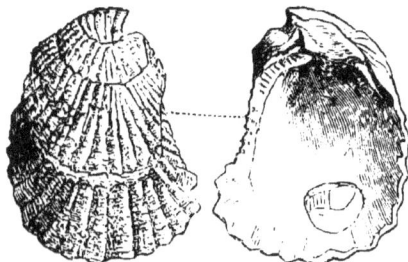

Cytherea incrassata,
though especially abun-
dant in the Middle Hea-
don Series, has a some-
what extended range,
from the Barton Clay to
the Bembridge Beds. It
gives the name to the
well-known "Venus bed"
of collectors, the *Cytherea*
having formerly been
known as *Venus incras-
sata.* Among the other
abundant marine bivalves
may be mentioned the
Ostrea velata, which forms thick banks in Colwell Bay, and the
Ostrea flabellula (Fig. 38), a much scarcer species which ranges

downward into the Barton Clay but does not occur above the Headon Series. *Nucula headonensis* is also very plentiful in Colwell Bay.

The estuarine and freshwater bivalves most commonly met with are species of *Potamomya* (Fig. 39) and *Cyrena*. These occur in

FIG. 39.

Potamomya plana, Sow.

vast numbers in certain beds. *Unios* (Fig. 40) are more rare and are generally confined to thin seams.

FIG. 40.

Unio Solandri, Sow.

The most plentiful univalves in the marine and estuarine beds are several species of *Cerithium*, including *C. concavum* (Fig. 41) and *C. pseudo-cinctum* (Fig. 43), *Melanopsis subfusiformis* (Fig. 42), *Buccinum labiatum, Murex sexdentatus, Nerita aperta,*

FIG. 41.

Cerithium concavum, Sow.

FIG. 42.

Melanopsis subfusiformis, Morris.

FIG. 43.

Cerithium pseudo-cinctum, D'Orb.

Neritina concava, Ancillaria buccinoides, Melania muricata, and several species of *Cancellaria, Natica, Pleurotoma,* and *Voluta.*

E 56786.   K

The mollusca of the freshwater limestones are nearly all Limnæids belonging to the genera *Limnæa* and *Planorbis*, *Limnæa longiscata* (Fig. 45), and *Planorbis euomphalus* (Fig. 44),

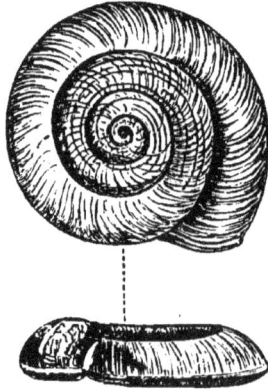

FIG. 44.   FIG. 45.

Planorbis euomphalus, Sow.   *Limnæa longiscata*, Sow.

being perhaps the most abundant and conspicuous species. *Paludina lenta* (Fig. 46) is a very abundant species throughout the Oligocene Beds, especially in the fresh-

FIG. 46.

Paludina lenta, Sow.

water clays and marls. *Nematura parvula* is very plentiful, and more generally distributed than is often thought, for its small size causes it to be overlooked. There is also a considerable number of species of land-shells scattered through the limestones, but these are not so often met with. They however point to the close proximity of the shore.

Of other fossils the most commonly found are valves of *Balanus unguiformis* in the marine beds, and nucules of *Chara*, generally *C. Wrightii* (Fig. 47) in almost any part of the series, but especially in the *Neritina*-bed at the base of the Middle Headon beds. Vertebrate remains are comparatively scarce. Except *Chara*, there are few recognisable plants.

FIG. 47.

Chara Wrightii, Forbes.

Like the other Oligocene beds, the Headon Series seems to be mainly of lagoon or estuarine origin. In the Middle division we have truly marine beds, but these are interbedded with others deposited in brackish water. The Upper and Lower Headon Beds are mainly fresh, or brackish-water deposits, and there seems to be an entire absence in them of purely marine genera, such as *Voluta*, *Ancillaria*, *Pleurotoma*, *Natica* and *Cytherea*.

Every variation in the amount of salt in the water seems to have been marked by a change in the fauna. The purely freshwater beds contain few mollusca except *Limnæa, Planorbis, Paludina, Unio,* and land-shells. The different species of *Potamomya, Cyrena, Cerithium (Potamides), Melania,* and *Melanopsis* appear nearly all to have liked water containing more or less salt. So we have a gradual change to beds containing Oysters, and then to beds with Volutes.

Besides these indications of varying conditions, it is interesting to observe a general tendency in the beds to become more fresh-water towards the south-west, while tufaceous limestones appear in that direction. The land-shells also point to the proximity of land, as do the pebbles of flint.* Unfortunately at the point where the most rapid changes are taking place—at Headon Hill—the beds have been cut off by denudation. We cannot therefore see whether the beds show any tendency to overlap each other, or to overlap the underlying Eocene.

---

* Pebbles of Chalk have been recorded, but they appear to be really white flints. The flint pebbles in the Headon Beds are sometimes weathered to the centre

CHAPTER XI.

OLIGOCENE—*continued.*

OSBORNE BEDS.

Between the Upper Headon beds, containing *Potamomya* and the Bembridge Limestone, intervenes a series of strata to which the name of "St. Helen's Series" was originally applied by Professor Forbes in consequence of the "conspicuous features presented by them between St. Helen's and Ryde." This designation was, however, subsequently changed by Professor Forbes to "the Osborne Series," on account of their being displayed in the cliffs and grounds of the Royal demesne,—a small distance to the east underlying the Bembridge Limestone, and a little to the west in conjunction with the Upper Headon beds, with which they do not appear in connexion at the locality after which they were named in the first instance.

The total thickness of the Osborne Beds varies from about 80 feet at each end of the island to 110 feet at Cowes and Newport.

Commencing at the western end of the island it will be perceived, on comparing the sections of the Osborne beds at Headon Hill with those at Cliff End, that the thick bed of concretionary limestone seen in the former locality altogether disappears in the latter, where it is most probably represented by the mottled clays and marls in which the remains of *Turtle* are found, and by the clays with pale-green nodular concretions containing *Limnæa longiscata, Paludina globuloides,* &c.

Osborne Series at Headon Hill.

	FT.	IN.
Whitish (passing into red and blue) marls, with occasional hard bands, and courses of nodular concretions of light-grey argillaceous limestone in which occur traces of shells and turtle bones. In the concretions are *Limnæa longiscata, Planorbis discus, P. obtusus, P. oligyratus, Paludina,* sp.	40	0
Grey shale, with crushed *Paludina lenta,* fish-vertebræ, &c. Ferruginous and nodular band. Grey shale, *Paludina lenta, Melanopsis carinata, Melania costata.* The FISH and PLANT BEDS	7	0
Yellow, red, and blue sandy clays	3	0
Thick concretionary limestone, with silicious concretions sometimes of large size and used for building. This band almost disappears northward. Fossils scarce. *Limnæa longiscata, Planorbis enomphalus, P. lens, Paludina lenta*	18	0
Greenish-white calcareous clay	4	0
Sandy ferruginous band	2	0
	74	0

Fig. 48.

Diagram of Colwell Bay Cliffs (by Edward Forbes).

A. Cliff End and Sconce Point.
a. Sconce (Bembridge Limestone).

B. Colwell Bay.
b. Osborne series.

C. Warden Point.
c. Upper Headon series.
d. Middle Headon series.
e. Lower Headon series.

The concretionary limestone can be traced inland towards Middleton, forming a bold feature in the hill. At the old limekilns near Greens it contains *Bulimus ellipticus, Limnæa*, and teeth

of *Palæotherium minus*? This rock was formerly referred to the
Bembridge Limestone, but both its lithological character and its
continuity with the concretionary limestone of the coast show
that it ought to be referred to the Osborne Series.

Between Headon Hill and Linstone Chine as will be perceived
by Forbes' sketch (Fig. 48, page 149) the Osborne Series has been
removed by denudation, and the cliffs consist of the subjacent
Headon Beds. At Cliff End it reappears beneath the battery,
and can then be traced at short intervals along the coast nearly to
the river Yar.

The Osborne Beds in this locality were examined by Professor
Forbes and H. W. Bristow in 1852. Forbes revisited them twice
in the spring and autumn of the following year (1853), and in
the present year (1888) they have been re-examined and partly
re-measured. Owing to the constant landslips considerable
difficulty attends the determination of the relative importance
of the several beds. The increased thickness here accepted for
the lower part agrees so well with what has been obtained at
other sections, and was proved so carefully by levelling, that some
of the original measurements must evidently have been taken from
a slipped mass.

Osborne Beds at Cliff End.

	FEET.
Bluish sandy and marly clays. *Cyrena obovata* (this bed is now invisible) - - - - - - - -	About 10
Red and blue marls, with lines of nodular concretions of agillaceous limestone in which fossils occur occasionally - - -	25 to 30
Dark-grey shales, with an ironstone band in the centre. Leaves, Insects, and Fish ; *Candona, Paludina lenta, Melanopsis carinata, Melania costata, Lepidosteus, Alligator.* (Probably the equivalent of the FISH BED.) - - - - - -	7
Reddish and bluish clayey marls, with greenish nodules containing shells ; turtle; *Limnæa longiscata, Hydrobia,* and *Paludina globuloides* - - - - - - -	40
	82 to 92

Following the Osborne Series eastward, we can detect inliers of
mottled clay in the plateau formed by the Bembridge Limestone
south of Wellow, but no measurements can be obtained.

Returning to the coast, we find these beds to be concealed for
four miles by newer formations which occupy the whole of the
cliff. However red and green clays reappear from under the
limestone on the east side of the Newtown River, and can be
examined for a depth of about 30 feet in the cliff and in a
brickyard. No fossils were seen. Half a mile further east the
Osborne Beds again sink beneath the sea-level and are lost for
two and a half miles.

At Gurnard Ledge the mottled clays reappear, but between this
point and Cowes they call for no detailed description, being
almost unfossiliferous and generally much obscured by landslips

The cliffs near Osborne having now been carefully sloped and planted, in this typical locality for the Osborne Series we can only follow Forbes, and the following is his description of the beds.

Osborne Series near Osborne.

" The slips and slopes at the eastern portion of the shore at Osborne* show mottled red and green clays, overlying a limestone composed of broken shells and containing *Melania costata* and *Melanopsis brevis.* On the shore lie flags of

FIG. 49.
Chara Lyellii,
Forbes.

sandstone with fucoidal markings, and blocks of a greenish sandstone containing casts of *Paludina lenta,* often weathered in high relief, *Melania excarata,* and a large-bodied *Limnæa* of considerable size. Among the marls are layers containing entire shells of *Melanopsis carinata,* small *Paludinæ* or *Hydrobiæ,* and *Chara* nucules in abundance. This appears to be an excellent locality for fossils."

" Opposite the lawn that stretches down to the sea in the grounds at Osborne, there are no hard beds or rock masses exposed on shore, but immediately to the west of the landing pier are strata of exceeding interest, for here we see marls and shales belonging to the upper part of the Headon Series. On the shore by the pier outcrops of beds of tenaceous greenish blue clay are exposed, full of *Cyrena obovata,* mingled with *Paludina lenta ;* and in the clay beds in which the foundations of the sea wall are placed are *Cerithia.* At a height of about 20 feet above the shore is a stratum of ragstone, an imperfect limestone, 2 feet or more thick, thickening more westward and thinning out eastward. The ragstone makes but bad lime. Higher up is a sandy limestone, and bands of comminuted shell stone, separated from the rag by marls. In fragments of the limestone I observed numbers of *Paludina lenta,* accompanied by peculiar large-bodied *Limnææ* of considerable size, and occasional lines of *Uniones,* somewhat resembling *U. Solandri* in outline, but a larger shell. The *Paludinæ* were often lying loose in their cavities, and had their shells frequently preserved. I found portions of a large *Planorbis,* apparently *P. euomphalus ;* also *Planorbis obtusus,* and another, *P. platystoma. Melania excarata* and lines of broken *Cyrenæ* occurred in a gritty band. Pale blue and purple shales, about 10 feet thick, capping yellow sands that become white eastwards, surmount the grits, and are succeeded by ferruginous marly and stony bands containing casts of *Paludina lenta,* hollow and having their cavities lined with crystals of calc-spar, *Limnæa* and *Planorbis.* Dark shales, with partings of *Cyrena obovata,* form the highest portions of the broken cliff. The details of this important section are obscured by land slips and cultivation, but it is evident that here the ground to the surface is occupied by

---

* The old name of Osborne, according to Worsley, was Austerborne.

typical beds of the Osborne Series, those on the western side of
the lawn belonging to the lower or Nettlestone division, whilst
eastwards we find the members of the higher or St. Helen's group.
The Osborne section is peculiarly interesting for the link that it
affords between the very different aspects of these beds at Cliff
End as compared with those at St. Helen's."

A section in red and mottled clays of the Osborne Series is seen
in the East Cowes Park Brick-yard. Here J. Rhodes obtained
Chara, impressions of plants, casts of *Limnæa*, Fish vertebræ,
scales of *Lepidosteus*, *Chelone*, *Trionyx*, Crocodile, and the
astragalus of a small mammal.

During May of the present year (1888) the Osborne Beds near
Ryde were re-examined, under the guidance of Mr. Colenutt, who
has paid special attention to this division. The principal point of
interest was the occurrence of a bed of clay in which are multi-
tudes of small fish (*Clupea vectensis*), evidently suddenly killed
and buried before they had time to decay. The thin seam in
which these occur is difficult to find, but such has been the
minuteness of Mr. Colenutt's examination that he has been able to
trace it from King's Quay, near Osborne, to Sea View.[*]

The first locality at which these fish were discovered was near
Ryde House, but during this visit the section was obscured at that
point, though another one was measured close to King's Quay.
Here the cliff is so obscured by landslips and so much overgrown,
that the exact position of the Bembridge Limestone cannot be fixed,
and only the beds on the foreshore can be well seen. Though the
measurements are only approximate, the changes of character and
colour of the different clays are sufficiently marked to enable the
different beds to be recognised. The fish-bed is generally just
below the level of high-water, and being slightly harder than the
other clays it often projects through the beach.

Section east of King's Quay (measured with the assistance of
Mr. Colenutt).

	FEET.
Bembridge Limestone.	
Red and mottled clay (only seen in landslips) - - -	About 40
Green clay, with scattered fish bones. Scales and vertebræ of *Lepidosteus* abundant, *Alligator*, *Emys*, *Trionyx*, and *Chelone*, *Theridomys* and snake vertebra - - - - -	About 4
Hard grey shaly clay, full of fish bones, and whole fish (*Clupea vectensis*)	2
- - - - - - - -	2
Similar clay with grass-like leaves and lenticular masses of cement stones - - - - - - - -	3
Blue clay, with abundance of mollusca. *Paludina lenta, Melanopsis carinata*, &c. - - - - - - -	6
Unfossiliferous green clay, to low water.	
	55

* See *Geol. Mag.*, dec. III., vol. v. p. 358. (Aug. 1888.) The fish, which is new
to science, has recently been described by E. T. Newton under the name *Clupea
vectensis*. See *Quart. Journ. Geol. Soc.*, vol. xlv. p. 112. (Feb. 1889.)

West of Binstead Point, thirty feet of red and green marls are displayed at the base of the cliff, supporting hard light-green marl with small white concretions ; above this succeeds a thin band of decayed shells (forming a soft shelly limestone, the greater portion of which is composed of fragments of bivalve shells), with a sort of laminated appearance. The calcareous band contains comminuted *Cyrena*, *Limnæa longiscata*, *Unio*, *Melania excavata*, *Melanopsis*, *Planorbis discus*, &c., with two feet of interstratified sands and sandstones and grits above it, which are probably the equivalents of the silicious beds beneath the Bembridge Limestone at the Binstead quarries. Two feet of soft sand complete the section.

FIG. 50.

Section at Binstead.

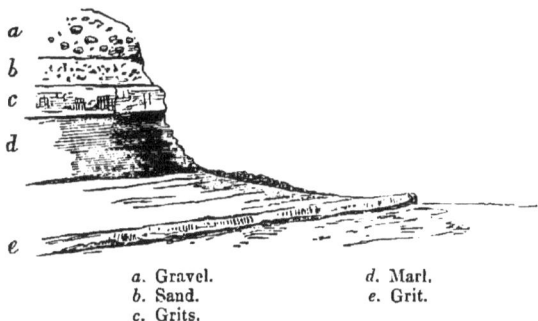

a. Gravel.   d. Marl.
b. Sand.   e. Grit.
c. Grits.

At Ryde House a ripple-marked flaggy sandstone (probably bed *e* in the above woodcut) immediately overlies the fish bed.

At Binstead Point the upper calcareous portion of the thick bed at Nettlestone comes to the shore, capped with green marls, and assumes the character of a hard and compact white limestone with *Melania excavata*. Westward of the Point it forms a ledge on the shore, which strikes nearly due west in the direction of Osborne. About a quarter of a mile east of the Point, sandstone appears, dipping 10° W. of S. at 5°. Gravel and the enclosed nature of the ground now conceal the strata for a considerable distance ; but a few scattered blocks of grit lie under the sea-wall opposite the first houses west of the town of Ryde, and again midway between Ryde Pier and Apley.

At the west corner of Apley Wood a bed of calcareous sandstone, about four feet thick (full, in places, of casts of *Paludina*, associated with numerous large *Unio*, *Limnæa*, *Planorbis*, and occasional bones of Turtle), appears on the shore beneath the sea-wall. The shells, which are as much crowded as in Sussex marble, are sometimes filled with a greenish marl, the rock itself being somewhat ferruginous, and of a pale ochreous colour. It rests upon ragstone similar to that at Nettlestone, ten feet or more thick, under which sandstone, in layers eighteen inches thick, continues to a depth of ten or eleven feet. Under all lies a strong greenish-blue clay for thirty feet more, which contained, apparently, crushed *Paludina*. Much of the stone used in the construction of the sea-wall has been obtained from the shore here, opposite the wood. Red and white clays are based upon the upper bed of stone; they are seen in the cliffs for a considerable distance, and have furnished the earth manufactured at the brick-pits inside Little Apley Wood.

The strata begin to arch from about this place, and in so doing disclose a good section of the Osborne Series, especially between Nettlestone and St Helen's, as far as Watch House Point, where the Bembridge Limestone rapidly descends to the shore. The centre of the arch is somewhere near the old Salterns, but among the fossils found, or the strata brought into view, there is no evidence of any portion of the Headon series being brought to the surface.

From the semicircular projection halfway along the bay, to the notch in the coast near the eastern termination of the wood, hard beds with *Chara* appear at intervals on the shore and beneath the sea-wall, dipping W.S.W. 2°. Opposite Puckpool Farm, and between the Point further east and Nettlestone, there is a broad expanse of bright green marl, which, although dry at low water, and free from blocks of stone, is generally concealed from observation by a thin layer of sand. Two hundred yards west of Nettlestone Point, thick beds of hard sandstone containing *Limnæa* and large and small *Paludina*, and calcareous bands, sometimes formed of comminuted shells, which are the same beds as those seen further onwards beneath Priory (Summer-house) Point, appear on the shore forming a cliff, and support the pathway in front of the Crown Inn. Under the Flagstaff, the shelly limestone which constitutes the upper five feet of the bed is almost entirely made up of comminuted *Melania excavata*, with bands of *Paludina lenta* the whole resting on flaggy siliceous grits containing ripple-marks. The rocks at Nettlestone Point are thick-bedded concretionary limestones, in some places soft and composed of comminuted *Paludina lenta*, in others passing into hard siliceous grit. They constitute large blocks on the shore, eight feet thick, which weather very unequally into irregular cavities, and contain a few small rounded pebbles of flint, larger fragments of sub-angular flint, Turtle bones, and fossils with the shells preserved. The lower four feet become more indurated and cavernous (honey-combed) and pass into hard grit; while in the freestone, about two feet six inches from the top, there is a well-defined band of *Limnæa*, six inches in thickness. Green sand, with large flat lenticular concretions of a yellow colour, which have an irregular surface and resemble *septaria*, overlies the limestone.

Round the Point, the upper part of the thick grit becomes an indurated marl of an ochreous colour, with greenish-grey, argillo-calcareous concretions; while further east, a short distance west of the boat-house, it becomes a limestone (containing *Chara* and *Limnæa longiscata*), which has been quarried on the shore for building stone. This change of mineral character apparently escaped the notice of Professor Forbes, who has described the bed, both under its normal and altered aspect, in his section of the Nettlestone Grit, at pages 74 and 75 of his memoir on the Fluvio-marine formation of the Isle of Wight, as two distinct and separate strata, Nos. 9 and 10.

The following is Forbes' detailed section of the beds in the centre of this anticline :—

(1. ST. HELEN'S SANDS.)

1. Immediately under the lowest bed of the Bembridge Lime-stone (here divided into three bands) occurs a band of dark greenish carb naceous clay, breaking with a sub-conchoidal fracture, and forming a truncated stratum in the cliff; 1 ft. 6 in.

2. Pale greenish white and yellowish marls, with patches of calcareous sand and comminuted shells; also argillo-calcareous nodules of various sizes. In this bed a characteristic fossil, *Melania excavata*, occurs in abundance, and has the shell preserved. 8 ft.

3. Pale green, yellowish, and white sands, hardening into sand-stones, with large lenticular siliceous concretions and spongoid bodies. *Melania excavata* is plentiful here and there, and occasionally occurs crowded. A small *Hydrobia* is also present; and from a mass of loose sand I extracted a *Helix* with the shell entire, apparently *Helix omphalus*, but unfortunately destroyed the specimen. 14 ft.

4. Greenish-yellow irregular and concretionary sandstone, with siphonoid or fucoidal bodies; 3 ft.

5. Yellowish and whitish sands, with a line of purple (manga-nese?) nodules and siliceous concretions below; 9 ft.

6. Laminated white sands, indurated into quartzose flags above and below; the upper surface exhibiting strong current marks. This band is remarkable for its contents, including *Limnæa longiscata*, a shorter species of *Limnæa*, resembling *L. peregra*, *Planorbis obtusus*, and *Melania excavata*, all in the condition of casts. The fossiliferous portion is in the lower part. 3 ft.

7. White sandy clay, with a band of broken *Cyrenæ*; 2 ft.

8. Greenish-blue clay, seen on shore at low-water, containing *Cypridæ* and traces of *Melania* and *Cyrenæ* (*C. obovata?*). The thickness may be estimated at 8 ft. [This apparently contains the fish-bed discovered by Mr. Colenutt.]

(2. NETTLESTONE GRITS.)

9. Imperfect softish bright yellow limestone, riddled by minute confervoidal cavities, hardening into a building stone by exposure

to the weather. Not very fossiliferous, but contained *Limnæa longiscata*, a large full-bodied species, *Hydrobiæ*, and *Chara nucules* (*Chara Lyellii*). This limestone may be seen opposite the boathouse near Nettlestone, but as it is much carried away is not evident except at a low water. It is the equivalent of the band in the slope at Whitecliff Bay.   2 ft.

10. Bright yellow and white marly clays, with patches of greenish sand, filled with argillo-calcareous nodules of various sizes. In these nodules the *Melania excavata* abounds. These clays do not appear to exceed a thickness of 4 ft.

11. Freestone or rag, with siliceous concretions passing into a grit. A great part of this bed is made of comminuted univalves, the fragments smaller and finer below. In the middle portion occur bands of unbroken *Paludina lenta*. This is the bed of which portions are thrown up in the line of the fault below Summerhouse Point, where it is very conglomeratic and includes pebbles of flint. Similar pebbles are seen here and there in it at Seafield. It is used for a building stone there, and for making the groins on the shore east of Ryde. In these beds the casts of *Melania excavata* occur in myriads, also *Paludina lenta*, *Hydrobiæ*, a short *Melanopsis* apparently *M. brevis*, *Melanopsis carinata*, *Planorbis rotundatus* (scarce), *Limnæa longiscata*, and the short-spired species, vertebræ of fish, and fragments of turtle.   8 ft.

In a block in a neighbouring wall I observed impressions of a small and peculiar *Cerithium*, and remains of a large shell, apparently *Achatina costellata*.

12. Softer and whiter sandstone, with frequent calcareous concretionary bands, containing *Limnæa longiscata*, and separated by a thin layer of compact sandstone with impressions of *Unio*, form a compact flagstone with fucoidal impressions.   4 ft.

13. Shelly sandstones, often studded with angular flints; 6 in.

14. Soft calcareous stone, with *Paludina lenta* ; 6 in.

15. Flags of sandstone, with large ripple marks; 6 in."

At Sea-View the fish-bed occurs at the base of the cliff a short distance east of the Pier, and as the Nettlestone Grits sink beneath the sea-level close to the Pier, it is probable that the fish-bed is in the clay at the base of Forbes' higher division, or St. Helen's Sands. At this locality, as near Ryde House, ripple-marked flags are found immediatly above it.

At Priory (or Horestone) Point, thick-bedded sandstone (No. 11 of Professor Forbes' Nettlestone section) forms the base of the cliff, containing in some parts bands of small rounded flint pebbles; in others, layers of partially decomposed angular flints. The upper part is full of broken shells, and patches of comminuted shells occur about two feet from the top, which is calcareous, and less hard than the lower portion of the bed. There are also occasional fucoidal markings and large irregular concretions, which, weathering unequally, cause the rock to assume a honey-combed cavernous appearance.

A fault at the Point, running in a direction 30° E. of S., skirts the shore and brings up the Nettlestone division of the Osborne Beds, in a manner that at first sight appears to be very puzzling.

Nothing more is seen of the Osborne strata between Watch House Point and Whitecliff Bay.

The strata composing the Osborne series were better displayed at Whitecliff Bay in the summer of 1856 than at the time of Professor Forbes's visit, when they were concealed by landslips, or in grass-covered undercliffs. The following is a list of the beds then observed :—

Section of the Osborne Beds in Whitecliff Bay.

	FEET.
Dark bituminous clay, with *Limnæa* in patches - -	2
Grit - - - - - - - -	1
Dark olive-green clayey sand - - - - -	3
Red and green mottled clays, with 1 to 2 inches of clay iron-stone on the top of the bed - - - - -	18 or 20
Green clays - - - - - - -	3 or 4
Dark grey sandy clays - - - - -	3
Shelly band, large *Paludina*, *Melanopsis carinata* - -	4½
Dark green marls - - - - - -	8
Olive-green clay, *Melanopsis carinata*, *Paludina lenta* - -	15 to 18
Fine cream-yellow limestone, running out to sea in a direction 10° N. of E. No fossils observed - - - -	1
Green clays; *Paludina*, *Melanopsis* - - -	About 15
Total thickness of Osborne beds - -	79½

The foregoing sections will show how uncertain and difficult to fix is the boundary between the Headon and the Osborne Series. When one examines the fossils also, not a single mollusc can be found that is confined to the Osborne Beds, and the only peculiar fossils are small and delicate fish and prawns, the preservation of which is due to exceptional circumstances In fact, so little is yet known of the fauna of the Osborne Series, that it still remains doubtful whether these beds ought or ought not to be separated from the Headon.

The paucity of species seems to be mainly due to the conditions under which the beds were deposited. There is an absence of truly marine beds, though a few marine shells occur. Purely freshwater strata are also rare. The mass of the clays seems to have been deposited in lagoons, varying in saltness, in which could live brackish-water molluscs like *Melania* and *Potamomya*, and a few of the more hardy freshwater and marine species. Lagoons of this character are at the present day favourite places for turtles and alligators, like those so abundant in this deposit.

No doubt the Osborne Beds have been undeservedly neglected, owing to their proximity to the much more interesting Headon and Bembridge Series. But the fish-bed, especially, is well worth further examination and tracing into other parts of the Island.

Not only is this horizon noticeable for the occurrence in it of shoals of small fish and prawns, but the abundance of scales and vertebræ of the ganoid *Lepidosteus* is of great interest. A bed which yields such well-preserved fish and prawns is likely also to contain plant-remains and insects. A few plants have already been obtained from it near Ryde. During a recent visit to Cliff End numerous well-preserved plants were discovered on this horizon (by Clement Reid and Henry Keeping). No attempt was then made at systematic collecting, but during an hour or two's search grass or sedge-like leaves of several genera, palm ?, fern, and fragments of several peculiar reticulated leaves were found. This locality would repay more minute examination, as scarcely anything is yet known about the botany of the Osborne period.

Bembridge Limestone.

Of the Fluvio-marine strata of the Isle of Wight, the Bembridge Group is by far the most constant in lithological characters, and the changes exhibited by its component strata throughout their range are for the most part slight and unimportant. It is consequently everywhere easily recognizable by mineral composition, and, as might be expected, its most characteristic fossil contents are, in the main, very uniformly distributed. Its lower portion is most calcareous, and everywhere in the Island exhibits more or less compact limestones, occasionally separated by shales, and accompanied by marly beds.

These limestones in the first edition of the Map and Memoir were treated merely as part of the Bembridge Series. But it has been found easy to separate them on the more accurate topographical map now available, for they form the most marked feature to be seen in any bed above the Chalk in the Island. There is also in places a distinct line of erosion between them and the overlying marls, and everywhere proof may be found of a sudden break and change in the conditions of sedimentation, from an almost purely calcareous freshwater deposit, to a marine clay or sand.

As there is an equally sharp line at the base of the limestone, where it rests on the mottled clays of the Osborne Series, the Bembridge Limestone is here treated as a separate subdivision, not necessarily differing greatly in age from the older or newer deposits, but showing a marked change of physical conditions at the time of its formation.

The Bembridge Limestone includes the uppermost limestones of Headon Hill and Sconce, and the well-known limestones of Hamstead and Gurnard Ledges, Cowes, and Binstead. On the same horizon lies the rock which, owing to a dip slope, spreads over so wide an area near Wellow and Newbridge.

Headon Hill.—This important member of the Isle of Wight Tertiary series plays but an inconspicuous part in the Headon section. Among the grassy slopes beneath the gravels that crown the summit of the hill, white and yellowish sandy marls appear here and there in the broken ground, occasionally varied by containing hard white compact limestone nodules that break with a sharp-edged, splintering fracture. A little to the north of the summit these beds, dipping northward, become rather more developed, passing into concretionary and travertinous limestones. The bodies regarded by Mr. Edwards as turtle's eggs occur among them in regular lines. The fossils found in the concretions are almost invariably terrestrial, and consist of *Helix D'Urbani, H. omphalus, H. occlusa, H. headonensis? Bulimus ellipticus, Pupa perdentata,* and *Cyclotus cinctus.*

FIG. 51.

Bulimus ellipticus, Sow.

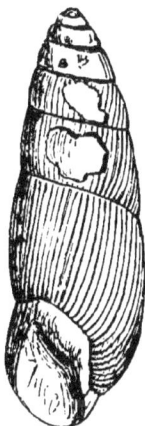

FIG. 52.

Helix globosa, Sow.

Bulimus ellipticus (Fig. 51), *Helix globosa* (Fig. 52), *Planorbis discus* (Fig. 53), &c., have been obtained from these beds by the fossil collectors of the Geological Survey, mostly in the condition of casts, but the shell is sometimes replaced by calc-spar, which also occurs in a crystalline form lining and filling small cavities in the stone. As a general rule, the Bembridge Limestone may be distinguished from the thick Upper Headon Limestones, as well as from those in the lower groups, by its greater whiteness and its peculiar brecciated or tufaceous character, as well as by the fossils either being casts, or having their shells replaced by calc-spar. The Headon Limestones, on the contrary, are of a somewhat darker cream-colour, more earthy and soft in composition, and have the shells of the *Limnææ* and other fossils preserved.

The total thickness of this limestone at Headon Hill is from fifteen to sixteen feet. It is surmounted by a greenish-grey marl with *Cyrena obovata* having both valves in contact, which passes upwards into a soft, unctuous, earthy limestone, containing *Planorbis* and a large *Limnæa*, which again merges upward into very tenacious grey clay, weathering brown and black, and carbonaceous on the top. In thickness these deposits are variable, even within short distances, the limestone being sometimes as much as three feet, while the clay resting upon it varies from three to fourteen inches. In one place, however, where the three deposits formed but a single bed, the aggregate thickness was three feet six inches; viz., clay six inches; limestone, one foot ten inches; and green marl, one foot two inches.   Above the carbonaceous clay is a soft cream-coloured earthy limestone, also containing *Limnæa* and *Planorbis.*   The thickness of this upper limestone, which has apparently a denuded surface, varies considerably, but from 5 to 8 feet of it appear from beneath the white sands which form the lowest member of the gravel series constituting the summit of Headon Hill.

In a section pointed out by Mr. Keeping, further north, the *Bulimus* limestone, uneven and irregular, is covered in places with brown and black carbonaceous clay, filling irregularities in its surface.   The green clay with *Cyrena* above the thick limestone (here from one foot nine inches to four feet thick) contains a layer of *Cyrena* fifteen inches from the bottom of the bed, while the limestone, which (in addition to *Limnæa* and *Planorbis*) also contains *Cyrena* in the lower three inches, is only one foot thick. The clays above are irregular, and of variable thickness, but average about two feet, the lower six to nine inches of which is brown clay, becoming occasionally dark and carbonaceous towards the bottom, and dark grey carbonaceous clay six to fifteen inches, the upper six to nine inches of which frequently consist of lignite ; two or three inches of sand, with carbonaceous laminæ, succeeded by green marl, complete the section.   Hard thick beds are quarried at the eastern extremity of this outlier.

Another outlier, over three-quarters of a mile long, covers the high ground upon which Hill Farm is built.   A pit has been opened in it at the end of the lane running in a north-westerly direction from the farm.   In the road to More Green casts of *Limnæa, Planorbis,* and small *Helix* have been found.   A short distance further north the limestone is overlain by green clay containing comminuted fragments of *Cyrena.*   At its northern extremity, the limestone based on red clay is cream-coloured, soft and earthy (somewhat similar to dried mortar), becoming, however, occasionally harder in places, and assuming a kind of tufaceous character.   Another inconsiderable patch of limestone similar to that last noticed, occurs half way between it and Norton.

Sconce.—For years this locality has yielded many of the most interesting fossil shells found in the Isle of Wight Tertiary Series, especially species of terrestrial origin.   Not a few of the rarer and more curious pulmoniferous molluscs, so well figured and described in Mr. Frederick Edwards's excellent monograph, were discovered at Sconce.   At present (1888) the section being much overgrown, the following details are taken from Forbes' Memoir.

" The Bembridge limestone at Sconce, a mass of limestone and marls, is from 16 to 20 feet in thickness.   It rises with the slope of the hill opposite Yarmouth, and forms the partly mural crest cropping out at Cliff End.   The entire thickness is composed of calcareous beds passing into each other, very concretionary,

variable within short distances, and of a highly travertinous
character.   Indeed, very much of the limestone in this locality is
a true travertine, or calcareous-tufa.   Much of it has a peculiarly
brecciated appearance not presented by the Headon limestones,
and the porosity dependent on the presence of irregular confervoid
tubular cavities, so characteristic of the Bembridge limestone in all
its localities, and so strikingly comparable with a like appearance
exhibited by the travertines of the Paris basin, is very manifest in
the rock at Sconce.   The cause of this structure, first noticed
by Von Buch, and afterwards laid stress upon by Cuvier and
Alex. Brongniart, has been frequently discussed by French geologists, who are inclined to refer it to the effect of the disengage-
ment of gaseous vapours.   I am inclined to refer some of these
appearances to the ancient and now obliterated presence of vegetable bodies, such as chara stems and algæ.   The distinctive
palæontological feature of the Sconce locality for this limestone
is the remarkable abundance of land shells in it.   These occur for
the most part in the upper half of the beds, freshwater shells
being more frequent in the lower, but much of the strata here
seems entirely unfossiliferous.   In some places the mass of land
shells seems to lie in irregular tufaceous bands between harder
strata, the latter abounding in *Limnæa longiscata, Planorbis discus,*

Fig. 53.

Planorbis discus, Edw.

P. obtusus, and *P. oligyratus,* mostly
in the condition of casts, but never-
theless exceedingly well preserved
and easily extracted.   Great blocks
of grey sandy limestone lie along the
shore, fallen from the hill crest, full
of *Planorbides* and *Limnææ,* mingled
with occasional *Helices* (*H. occlusa,
H. D'Urbani,* and *H. rectensis* being
most common), and the fine *Paludina
orbicularis.*   These blocks are broken
up by the native collectors, who seek
especially for the last-named shell, and
for *Bulimus ellipticus, Achatina cos-*
tellata, and *Helix globosa,* all species of great size and beauty, that
find a ready sale among visitors.   In a thin white band beneath a
belt of *Limnæa longiscata* I find here the little *Paludina
globuloides* occupy the same horizon as at Bembridge and Cowes,
and remarkable for its constancy of place.   The most concretionary and brecciated portion of these beds consists of a white
band from 6 inches to a foot thick not far from the uppermost
layer, and evidently comparable with the cap of the limestone at
Bembridge.   Just below the top, every here and there, a hard
band of silex, often nodular, reminds us of the cherty layers near
the summit of this limestone at St. Helen's.   Four or five inches
of soft calcareous marls, with small limestone pebbles (or possibly
concretions), form the very uppermost portion.   In the line of the

tufaceous concretionary portion is a curious layer or old surface, in which lie the remarkable bodies regarded by Mr. Edwards as turtle's eggs."

Besides the fossils mentioned, *Helix omphalus* and *H. tropifera*, *Pupa perdentata* and *P. oryza*, *Clausilia striatula*, *Cyclotus cinctus*, and *Succinea Edwardsii*, were all collected by Prof. Forbes and Mr. Gibbs in this prolific locality, the *Clausilia* and *Cyclotus* being by no means uncommon. Although diligently searching for many days these observers met with no remains of vertebrata.

The following list of shells procured by Mr. William Cotton of Freshwater, during the course of a single morning, will show the variety and abundance of the fossils contained in the limestone here :—

Fossils from the Bembridge Limestone of Sconce.

	No. of Specimens.
Achatina costellata (Fig. 54) - - - - -	1
Helix globosa ? - - - - - -	3
„ vectensis, var. depressa - - - - -	8
„ D'Urbani - - - - -	12
„ occlusa - - - - - -	4
„ tropifera - - - - - -	1
„ (or Paludina) carinata, [probably Paludina angulosa]	5
Clausilia striatula ? (young) - - - - -	2
Planorbis obtusus - - - - - -	3
„ discus	
„ oligyratus (young) - - - -	25
Limnæa longiscata, var.	
„ slender var. small.	
„ ? large bodied var.	
Cyclotus cinctus - - - - -	6
„ nudus - - - - - -	1

Bulimus ellipticus, Achatina costellata, and *Helix globosa,* are all large conspicuous species. *Paludina angulosa* and *Achatina costellata* (Fig. 54) are the shells especially sought

FIG. 54.

Achatina costellata, Sow.

for by the native collectors; but good specimens with the shell preserved are rare. The blocks which have fallen from the crest of the hill are crowded with specimens of *Planorbis* and *Limnæa,* and occasionally *Helix,* the most common being *Helix D'Urbani, H. occlusa,* and *H. vectensis.*

The Bembridge Limestone of Sconce descends below the 50-foot contour at its eastern end, and the small outlier further east nearly touches the 25-foot line. Continuing the dip shown by these outliers, we observe that the limestone ought to plunge beneath the sea within a short distance. We accordingly find an isolated rock at a quarter of a mile from the shore off Norton. This is known as Black Rock. It is only visible at extremely low spring-tides, and we have not been able to examine it, but have been told that it consists of a hard freestone.

The depth of the old channel of the Yar prevents the Limestone from being traced continuously to the east side. But near Yarmouth Gas Works it reappears on the foreshore, and was also well seen in the railway cutting close by. Crossing Thorley Brook it gradually spreads out, so as to occupy an extensive dip slope, such as one scarcely expects from so thin and soft a bed.

In the neighbourhood of Wellow, Shalcombe, and Newbridge an area of nearly 3 square miles is covered by the Limestone, which forms a bold escarpment rising to a height of about 270 feet near Shalcombe. A dip of about 2° to the north-north-east causes the Limestone to pass beneath the Bembridge Marls near the Yarmouth and Newbridge high road.

Notwithstanding this large spread not many sections are now open, for brick has taken the place of limestone as a building material, and chalk is preferred for agricultural purposes. One would have thought, however, that this limestone, with its greater quantity of phosphoric acid, would have made a better manure; we have not been able to learn the reason for the substitution of chalk, even on farms where the Bembridge Limestone would be cheaper. The stone was formerly extensively dug in pits near the escarpment, but these are all overgrown, the only remaining sections being near Newclose Farm, in Thorley Street, near Marshfield, in Wellow, and near Bank Cottage, Newbridge, where the outcrop becomes more narrow. None of these pits are of much interest, or show the upper or lower surface of the stone.

Other sections are seen in the old pits between Newbridge and Fullholding, and for nearly a mile the road runs along a ridge formed by the Limestone. From Fullholding eastward the bedding becomes vertical. The limestone, therefore, occupies a very small area at the surface. There seems also to be a tendency for it, like other thin limestones, entirely to disappear for a depth of several feet from the surface, where exposed to the solvent action of rain water. For these reasons it is often difficult to follow the outcrop; but limestone has been seen south of North Park Farm; north of Swainstone; at Great Park; for nearly three-quarters of a mile west of Gunville; and in an old quarry half-a-mile east of Gunville.

Returning to the coast, we find the Bembridge Limestone to sink beneath the sea at Yarmouth,* to reappear on the northern side of the syncline with a west-north-west strike. The limestone of Hamstead Ledge consists of three beds, with other softer bands between, and contains numerous specimens of *Limnæa longiscata, Planorbis, Chara,* &c. It can be traced nearly as far as the Newtown river, making a conspicuous feature, though the old cliff is now much overgrown.

---

* In ancient charters it is called Eremuth (Worsley).

On the east side of the Newtown river it appears above the Osborne Beds at the Brick Yard, but sinks when traced in a south-easterly direction, and is lost beneath the marsh of Spur Lake, to reappear in the bed of the stream near Porchfield for a quarter-of-a-mile. Continuing eastward along the coast, the Limestone in the cliff gradually falls till it spreads out on the shore, forming two ledges with an expanse of dark green marl between. Near Thorness Wood the stone is lost, and does not rise again for about a mile and a half.

The section in the cliffs near Burnt Wood is of great interest, for it is almost the only place in the Island where the Bembridge Limestone contains perfectly preserved shells and not merely casts. It also shows a distinct line of erosion between the Limestone and the overlying marine base of the Bembridge Marls. (*See* Fig. 55.)

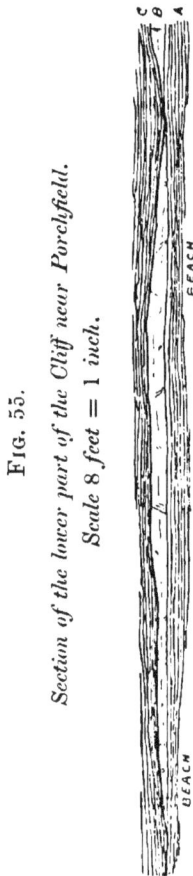

FIG. 55.

Section of the lower part of the Cliff near Porchfield.

Scale 8 feet = 1 inch.

C. Black Clay, with *Ostrea, Modiola, Cytherea, Cyrena, Mya, Cerithium, Melania.*
B. Bembridge Limestone—upper Block—with well-preserved shells.
A. Green marl, with *Paludina globuloides.*

The bottom block of Limestone (not seen in the cliff at this point, but exposed on the foreshore opposite) calls for no remark. It is merely a freshwater limestone of the usual character, with casts of *Limnæa*. Above it comes a mass of dark green somewhat mottled marl, the upper part of which is crowded with perfect specimens of the minute *Paludina globuloides*. On this lies the top block of Limestone; a soft earthy stone, easily cut when first dug out, but hardening by exposure. This stone is full of uninjured specimens of *Limnæa pyramidalis, L. mixta*, and *Planorbis obtusus*, but only for a short distance. The preservation of the shells here is due to the stone being sealed up in a mass of impervious clay. The upper surface of the limestone is much broken up and eroded, and in the cracks are found marine shells, *Panopæa* (or *Mya*) *minor* having the valves united. In some places the erosion has cut entirely through the upper block of the Limestone, so that the base of the Bembridge Marls rests directly on the green marl with *Paludina globuloides*.

In Thorness Bay the Limestone rises again, showing the same three divisions. The bottom block forms Gurnard Ledge, and the thin upper block makes a minor ledge nearly opposite Sticelett Farm. From Gurnard Ledge the Limestone runs as a marked feature in the cliff as far as Gurnard Bridge, but on the east side of the marsh the sections are obscure and hidden by talus, though abundance of fallen blocks can be examined as far as Egypt Point. From this Point eastward through West Cowes another marked feature, now overgrown or hidden by buildings, shows the outcrop of the Limestone, which was formerly seen in the foundations of several of the houses. Near the West Cowes Gas Works the same rock is again met with, and from this point to Bottom Copse, where it sinks beneath the Medina, there is no difficulty in following its characteristic feature.

Crossing the Medina, the Limestone is seen on the foreshore exactly opposite the point on the west bank where it was lost, thus proving that here the beds are continuous across the river and are not displaced by any fault.

On the feature that marks the outcrop towards East Cowes a large abandoned quarry may be seen in Little Shambler's Copse. The stone has also been quarried near East Cowes Park, in places now occupied by houses, and it is again seen at Elm Cottage, close to the south-western corner of the grounds belonging to Norris Castle. Here, at a height of about 120 feet, it is lost under the Plateau Gravel.

At Newport the Limestone, though masked by Drift and rainwash, has been proved in several wells (*see* Appendix). Unfortunately the well at Mew's Brewery—the only one that passed through the stone—was bored many years since, and the samples that have been preserved do not show the thickness of this bed.

East of Newport the stone was formerly quarried about 200 yards north-east of Great Pan Farm; and again nearly due north of Little Pan Farm. It was also touched in a trial boring at Durton Farm. From this point it is lost for about a mile, owing to a covering of Gravel and wash from the Downs.

Close to Combley Farm it re-appears, and can then be traced continuously, either by feature or by blocks ploughed up, as far as Little Duxmore, where it is vertical. East of the last locality the Limestone cannot now be seen for about 3 miles, though blocks were formerly ploughed up near Ashey. During the original survey, a section was also seen south of Little Nunwell, in a ditch under a newly-made fence.

At Brading, where the dip becomes lower, the Limestone forms a more marked feature which passes under the Church. Wall Lane is also carried along the ridge; the stone having formerly been dug close to the road on the south side, there is now a vertical wall of rock running parallel with the lane. At the Cement Works the dip in the quarry is 5° at the northern boundary, but it increases to 10° close to the road, and to about

20° on the south side of the road. The flexure is as sharp as
in Whitecliff Bay.

East of the Yar and Brading Harbour, the Limestone reappears
at two spots at the edge of the marsh, and from Peacock Hill
eastward to Whitecliff Bay it forms a marked ridge.

At Osborne, the Limestone, which is lost under the Plateau
Gravel, ought to reappear in the upper part of the Pier Wood,
but the grounds are so well planted, and the features so obscured
by rainwash, that no trace of it is met with till King's Quay is
reached. Here, though the beds cannot be measured, part can
be seen on the foreshore, and fallen blocks are abundant. From
King's Quay to Wootton Creek and Binstead, there is no difficulty
in following the limestone-feature through the woods and tumbled
ground, but there are now no open sections, even at Binstead, for
the celebrated stone quarries are all worked out or abandoned.
The Binstead quarries are so celebrated that the following notes,
taken from the first edition of this Memoir, may be acceptable,
though the sections cannot now be examined.

" In a quarry in the wood west of Binstead Church, and opening
to the sea, the upper part consists of thick-bedded, nodular, shelly
limestone, with *Bulimus ellipticus*, *Limnæa*, *Planorbis* (like *rotundatus*), *Cyrena*, or *Cyclas*, resting on soft sandstones, and hard,
calcareous, flaggy beds, sometimes well-laminated, and containing
teeth of *Anoplotherium*, claws of Lobster, *Paludina orbicularis*,
P. (small sp.), *Limnæa*, and a small *Planorbis*. The upper part of
the quarry is made up of green marls, and an irregular surface of
Limnæan limestone, which is covered with from one to four feet
of ferruginous loam, almost free from flints. There are, however,
a few small scattered flints in the loam, generally in the lower
part, which is clayey, while in the upper half are lines of small
fragments of limestone, with an occasional pebble. Under the
rubbish, in the quarries between this and the road to Ryde, concretionary shelly limestone rests on sandy beds, with layers of clay,
beneath which are four feet and a half of grey, flaggy sandstone,
forming the bottom of the quarry. The Binstead limestone was
formerly highly esteemed as a building stone, and has been used
in the construction of several churches in Sussex, the interior of
Winchester Cathedral, Lewes Priory, Yarmouth Castle and Quarr
Abbey (I. W.), an old Saxon ruin at Southampton, noticed by
Webster, &c., &c."*

In Ryde, according to Mr. Barrow, the Bembridge Limestone
was met with in laying down some drains in George Street. It

* The quarries near Quarr Abbey were in estimation for many centuries. They
furnished some of the stone for building Winchester Cathedral, as appears by a grant
made by the Conqueror (and confirmed by William Rufus) to Bishop Walkelyne,
and by two precepts from Henry I. to Richard de Redvers, Lord of the Island, for
stone to be dug there for the Cathedral at Winchester ; and subsequently to Stigand,
when he transferred his See from Selsey to Chichester. The registers of Winchester
record that William of Wykeham used this stone in building the body of Winchester
Cathedral.

is now visible near St. John's Road Station, at a height of about 15 feet above the sea, but it soon sinks beneath the marsh level, and is altogether lost half a mile further south. The dip at Ryde is southward, but the amount is only about half a degree.

At the west corner of Apley Wood, about 200 yards south of the sea-wall, an earthy limestone of the ordinary Bembridge type has been quarried beneath the site of some unfinished houses. This was probably the lowest bed of the Bembridge Limestone, but the place is now covered with underwood. The blocks were from fifteen to eighteen inches thick, and contained *Limnœa*, *Chara*, &c. From this point the Limestone is invisible for more than a mile, reappearing in the road, and in a small pit about a quarter of a mile south of Sea View.

At Horestone Point the Limestone again makes a distinct feature, traceable through the tumbled cliff as far as Watch House, or Node's Point, where we again meet with clear sections. The dip is south-south-west. On the south side of Watch House Point the following section was measured :—

Bembridge Limestone at Watch House Point:

	Ft.	In.
Limestone, irregular, marly, and most compact in the lower half of the bed, which is, also, the least fossiliferous. Full of *Chara*, with a few *Limnœa* and *Paludina globuloides*. The upper 2 feet more ferruginous and less indurated, and is frequently marked by the abundance of *Limnœa*	4	0
Dark laminated clay ; the lower part of a lighter colour, and more sandy	1	3
Compact greenish clay (slightly bituminous), with fragments of *Cyrena*, and now and then a perfect valve	0	9
Earthy limestone ; the upper part soft and of variable thickness. *Planorbis discus* in the upper part, *Limnœa* throughout	1 6 to 2	0
Hard green marl, with concretions in the lower part	2	6

At St. Helen's the Bembridge Limestone passes into the sea close to the old church tower, and reappears at Bembridge Point. The upper bed has an uneven, undulating surface, and is covered with a cap, of variable thickness, containing Oysters throughout its entire depth.

From Bembridge Point to the Foreland the Limestone becomes nearly horizontal, spreading out to form extensive ledges on the foreshore, but not rising above high-water level till Whitecliff Bay is reached. Between Foreland Point and the margin of the bay it forms in great part the floor of the shore, with a hollow and slightly basin-shaped curve, dipping inwards and landwards on the east and south-east. The extension of the broken margin of this shallow trough constitutes the reef of rocks known as Bembridge Ledge, and formerly quarried at low water for building stone. Rolled fragments of the Limestone strew the bay, and mingle with the flint gravel of the drift to form the shingle. At a distance it is conspicuous among the neighbouring strata, owing to its general creamy-white hue, and the angular fracture of its beds. When closely inspected it is found to consist of a number of distinct strata varying somewhat in thickness in different parts of the bay, and yielding different measurements to observers in different years, owing to the occasional swelling out of the individual beds. Their mutual relations and distinctions seem, however, to be tolerably constant at this locality.

In the cliff, not far from the hotel, the Limestone rises from the shore with a rapid and sudden curve; its uppermost portion inclining at a high angle. The best point for examination will be found where the great curve of the limestones first reaches the shore, and where these strata are exhibited in their entirety with perfect clearness. Here this division of the Bembridge group is composed of the following elements:—

Bembridge Limestone at Whitecliff Bay (Measured in 1856 by Professors Ramsay and Morris and H. W. Bristow).

	FEET.
Hard white crumbly marl, with a few concretions and scattered shells, and becoming harder and more shelly for the lower 6 inches. Throws out water at the top. *Planorbis discus, Limnæa* in places. Passes gradually into the bed below. This is No. 6. of Professor Forbes' section (see below) -	2½
Hard, compact, very shelly limestone, sometimes forming two beds, with a harder and darker-coloured parting between. *Chara tuberculata* and *Ch. sp.*—very abundant. *Paludina orbicularis* at 2 feet from the top. *Limnæa, Planorbis discus, Planorbis*	5
Hard bed of compact sandy limestone, weathering white; *plant-like markings. Limnæa* (a few); *Paludina* (sm. sp.) - -	1
Dark grey and carbonaceous clays, laminated with sand in the lower part; light green in the upper 2 feet, where they are compact and marly, and separated from the lower 12 inches by a band of *Cyrena obtusa* with both valves joined - -	3
Cream-coloured cavernous limestone, with a hard brecciated concretionary cap, 6 to 9 inches thick, on the top of the bed, which weathers to a very irregular surface. *Limnæa*, numerous *Taxites* and *Planorbis* (sm. sp.), *Chara tuberculata*, especially 2 feet from the top. Emits a bituminous odour when struck - -	4 to 6
Soft, white, earthy limestone, with a few casts of shells; *Planorbis, Limnæa, Fish* - - - - - - - -	2
Concretionary cream-coloured limestone, with an uneven surface above and below; weathering irregularly, and emitting a bituminous odour when struck. *Chara, Limnæa longiscata* - -	4 or 5

Another section measured in 1853, near the same spot, by Professor Forbes and Mr. Bristow is interesting for comparison with the above, as it shows how the strata vary.

Bembridge Limestone in Whitecliff Bay (1853).

6. Crumbly white marl, with small globular concretions. *Chara tuberculata* has its uppermost limit apparently in this bed. *Planorbis obtusus* is common in it, but, like all other shells in the Bembridge limestones, is almost always in the condition of a cast. 2 ft. 7 in.

5. Greenish white limestone, very concretionary and fossiliferous. Small patches of a white mineral are highly distinctive of this band. *Limnœa longiscata* is the most abundant fossil. Of other shells I find in this locality *Planorbis discus, P. rotundatus, P. Sowerbii,* and *P. obtusus,* a new *Paludina* (identical with that in No. 1), *Helix occlusa, H. labyrinthica,* and two other species. The uppermost 6 inches are very conglomeratic. This cap weathers pebbly, and contains freshwater shells; when removed by the action of the waters the stone below weathers with a rough and pinnacled surface, speckled by the white mineral and very shelly. The substance of the bed is much less shelly below. The thickness at the margin of the bay is 4 ft. 3 in.

4. Pale, often white marly limestone, in some places becoming very compact; remarkable for abounding in myriads of a small, rather globose *Paludina* (*P. globuloides*); containing also *Limnœa longiscata,* a small *Hydrobia,* and, more rarely, *Cyclostoma mumia.* When this bed is much exposed superficially it forms a flat white platform, with an undulated and much cracked surface, the cracks extending throughout its thickness. In its uppermost part is a paleish carbonaceous strip abounding in comminuted shells of *Cyrenœ.* The *Chara tuberculata* occurs in it. 3 ft.

3. Compact creamy yellow limestone, abounding in casts of *Limnœa longiscata,* of which parts of it seem almost entirely made up; also *Planorbis oligyratus?* The nucules of *Chara tuberculata* occur in this bed, but not so plentifully as in No. 1. The uppermost portion of it is conglomeratic. 5 ft. 6 in. This is the bed most sought after here for building, yielding blocks of considerable dimensions.

2. Greenish grey marly clay, with an irregular and crumbling fracture; it contains crushed shells of *Limnœa longiscata* and *Planorbides.* 4 ft. 6 in.

1. Yellowish compact limestone, weathering rather darker, exhibiting in the fracture minute confervoid ramifying cavities. This bed is very full of casts *Limnœa longiscata* and nucules of *Chara tuberculata* are scattered abundantly through its substance. A small *Paludina,* a *Hydrobia,* and a *Planorbis* (*oligyratus*) occur occasionally. The average thickness is 3 ft. 6 in.

Total thickness at Whitecliff Bay, as exposed in November 1853, 24 ft. 3 in. When measured near the same spot by Captain Ibbetson and Professor Forbes in 1854, it was made 27 feet. Professor Prestwich, in his section, states the thickness as 26 feet.

The fauna of the Bembridge Limestone has been very carefully collected. As a rule it consists entirely of freshwater mollusca. In a few places, however, abundance of land shells have also been obtained, and in others, as at Headon Hill and Binstead, mammalian remains are not uncommon. The land shells comprise tropical-looking gigantic species of *Bulimus* and *Achatina* (see pp. 159, 162). Among the mammals *Anoplotherium, Chœropotamus, Hyopotamus,* and *Palœotherium* are the most abundant. Very little is yet known about the associated plants, for though nucules of *Chara* (Fig. 56), are abundant, the limestone seldom yields determinable leaves or fruit of the higher orders. Near Foreland Point the palm leaf (*Palmacites*) figured by Dr. Mantell in the "Geological Excursions round the Isle of Wight," 1854, p. 311, is said to have been found in one of the beds of the Bembridge limestone, but the specimen is in an ironstone nodule.

FIG. 56.

Chara tuberculata, Lyell.

BEMBRIDGE MARLS.

Above the Bembridge Limestone lies a series of freshwater, estuarine, and marine clays and marls. These attain a thickness of about 120 feet at the east end of the island, but thin away to about 70 feet towards the west. The Bembridge Marls were divided by Forbes into the Oyster Bed, Lower Marls, and Upper Marls. But there is no break or definite boundary between the Upper and Lower Marls, and no marked palæontological change anywhere in the series, except that the marine shells are confined to the base. The Marls are therefore here treated as one subdivision, in which certain marked beds can be traced for considerable distances, but which it is not necessary or practicable to separate into smaller sub-groups, except locally.

At the east end of the Island, where the beds are thickest, the following section was measured by Forbes :—

Bembridge Marls in Whitecliff Bay.

	Ft.	In.
Variegated yellow and brown clay (occasionally sandy) containing lines of nodular concretions, but no fossils	8	0
Pale shaly clays, the lower part with a band of septarian concretions, containing *Paludina lenta* and other shells	3	0
Lead-coloured clays, laminated above, paler below. *Paludina lenta, Melanopsis, Melania turritissima*, occasional *Cyrena*, and remains of Fish	20	0
Pale bluish sands and sandy clay. *Melania turritissima, Melanopsis fusiformis, Paludina lenta*, Fish, Seeds	8	0
Sandy grey limestone occasionally passing into marl ; sometimes very fossiliferous, often concretionary with few fossils. *Bulimus ellipticus, Achatina costellata, Limnæa longiscata, Melania costata, Paludina lenta, Cyrena transversa, Unio, Chara Wrightii*	4	0
Red marl, without fossils	5	0
Pale blue laminated sandy clay, containing a few pebbles of limestone and flint. Traces of Fish	3	0
Variegated red and green marls. *Cyrena ;* fragments of *Trionyx incrassatus*	24	0
Clays with whitish streaks. *Melanopsis fusiformis, Paludina lenta*	2	0
Seam of *Serpula.*		
Clay, with *Cyrena semistriata, C. pulchra, C. obovata, C. obtusa, Cerithium mutabile* and *Melania costata*	4	0
Hard unfossiliferous bluish septarian stone [probably the equivalent of the Insect Limestone further west]	0	6
Dark shaly clay. *Cyrena semistriata*	2	6
Green sandy beds. *Ostrea vectensis, Cytherea incrassata, Mytilus affinis, Nucula similis*, &c.	2	0
Whitish sands interstratified with fine stripes of clay ; occasional pebbles. Lines of *Cyrena semistriata* and occasional *Cerithium*	2	0
Greenish marls, with lines of white nodules in the lower part	3	0
BEMBRIDGE LIMESTONE.		
	91	0

Another measurement of the Marls, made near the same place in 1888, gave a total visible thickness of 93 feet; but about 15 feet of the upper part of the cliff are overgrown and hidden. Possibly there may be an outlier of Hamstead Beds here, but if not, the Bembridge Marls must be at least 106 feet thick, with the top not reached. This computation agrees with the thickness proved at St. Helen's.

The marine base of the Bembridge Marls is so variable that the following detailed notes of the beds seen on the coast will be useful, especially as portions are often entirely hidden by beach sand or talus. The account is that given by Forbes, with some additions from notes made in 1888.

The blue septarian limestone strikingly resembles in mineral character the harder insect-bearing limestones of the Purbeck beds. It is thickest (about 1 foot) and finest about half way between Whitecliff Point and the Foreland, where its upper surface forms part of the floor of the shore. Everywhere it preserves the same peculiar mineral character. Near the same place the finest display of the oyster bed is seen, the surface of which also, for some distance, forms the floor of the shore. There it is underlain by a pale concretionary blue marl, containing occasional pebbles, and abounding in casts of shells, especially of *Cerithium* (probably *C. mutabile*), occasionally mingled with casts of freshwater shells (*Limnæa longiscata*, *Planorbis discus* and *P. obtusus*), *Cyrenæ* of more species than one, a small angulated *Corbula*, *Murex Forbesii*, a curious pupa-like *Bulimus*?, occasional *Mytili*, *Hydrobiæ*, a Tellinoid bivalve, occasional examples of *Melania muricata*, and traces of fish. Between this blue marl and the oyster band is a thin sandy bed, filled with comminuted shells, and on this rest numerous individuals of *Cytherea incrassata*, with their valves closed, but the shells are in so exceedingly decayed a condition that, after many trials, Forbes was unable to remove any entire. The internal casts, however, are fine and transportable. Then come the oysters, mostly, but not all, single valves, here and there mingled with good double specimens. They are thinly distributed, but occasionally occur in clusters of considerable number, bristling the surface of the shore. Individuals vary much in shape even in the same cluster. With them are *Mytili* (*M. affinis*), *Nucula similis*, a *Solecurtus*-like bivalve, and Forbes once met with a *Natica*. The *Mytili* and *Nuculæ* retain the substance of their shells perfectly. Occasional pebbles are mingled with the oysters.

In a few places interesting indications can be found that marine conditions lingered for some time. *Cliona*-bored oysters occur, on and in which *Serpula* and *Balanus* have grown, and the dead *Serpulæ* and *Balani* have been subsequently covered by a growth of *Polyzoa*. The best preserved marine shells will be found at about half-tide level, a short distance south of the Foreland Inn, where even the *Cytherea* may occasionally be obtained in a perfect condition, though fragile. *Nucula similis* is abundant

here, and uninjured specimens are sometimes washed up by
the sea.

At the foot of the cliff, half way between Whitecliff Point and
Foreland Point, just beyond the place where the oyster bed is
best displayed on the shore, the strata immediately surmounting
the septarian stone-band are well exhibited.   Dark blue clays,
with scattered shells (double) of *Cyrena obtusa* and *C. obovata*
first appear.   Then come darker and more friable shaly clays,
including a strongly marked band of *Cyrenæ*, the species being

FIG. 57.

Cyrena pulchra, Sow.

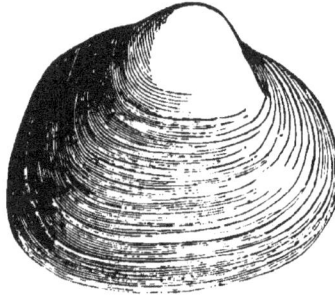

C. pulchra (Fig. 57) *obtusa,* and *obovata,* mingled with occasional
large examples of *Cerithium mutabile* (Fig. 58) of which now and

FIG. 58.

Cerithium mutabile, Lam.

then a specimen may be found with a *Balanus* attached.   After
some pale laminated clays, containing the same shells, succeeded
by greenish marls, crowded with little knots of *Serpulæ,* clays and

shaly strata follow, including a thick band composed almost entirely of *Melania muricata*, asso-

FIG. 59.

Cyrena semistriata, Desh.

ciated with *Cyrena semistriata*, which latter shell forms also a band of its own. The specimens of all these shells are beautifully preserved. In the clays and mottled marls that follow, shells are scarce or wanting, but fragments of turtle occur, and Forbes had the good fortune to find *in situ* the greater part of the carapace of a *Trionyx incrassatus*.

North of Foreland Point the cliff becomes low and the Bembridge Marls are almost entirely hidden by slipped gravel. Crossing Brading Harbour these Marls re-appear in the cliffs near St. Helen's Church. Owing to the destruction of the sea-wall good sections of the lower parts of the group are now visible between here and Horestone Point.

The greater portion of the Marls is exactly similar to the corresponding part of the Whitecliff Bay section; but with some slight though interesting differences in the region of the oyster beds, worthy of detailed notice.

The top of the Bembridge limestone in this locality, as mentioned in the account of that rock, presents a surface somewhat irregular, and including oysters, *Cyrenæ*, and casts of *Cerithium*. This is immediately succeeded by half a foot of greenish clay containing oysters. Then come six inches of brown clay charged with *Cyrenæ*; a coarse greenish clay, 1 foot thick, succeeds, having a crumbly and angular fracture, and including *Melanopsis fusiformis*, *Cyrenæ*, and a small *Melania*. The next overlying dark shaly clay, 1 foot thick, contains *Cyrenæ*, and is surmounted by some four or five inches of pale-lilac, compact, septarian stone, weathering white. Nearly two feet of dark laminated clays and marls succeed, containing in their upper part a band filled with *Cyrena semistriata*, accompanied by *Cyrena obovata* and *Cerithium mutabile*. Then come greenish and variegated marls.

A peculiarity in this section is the presence, in the brown clay above the oyster clay, of some shells of marine origin not noticed elsewhere; these are a pretty little *Arca (A. Websteri)*, and a *Modiola*.

FIG. 60.

Arca Websteri, Forbes.

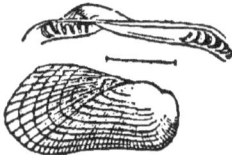

The above is the account given by Edward Forbes of this section, but it may be interesting to add a fuller list of the mollusca, from specimens collected in 1888 by Mr. Henry Keeping and Clement Reid.

Mollusca of the Lower Bembridge Marls at St. Helen's.

Arca Websteri.	Mya minor.
Cyrena obovata.	Mytilus affinis.
,, obtusa.	Ostrea vectensis.

Cerithium elegans.
    ,,   mutabile.
    ,,   plicatum.
Fusus Forbesii.
Melania Forbesii.
    ,,   muricata.

Melania turritissima.
Melanopsis carinata.
    ,,   fusiformis.
Paludina lenta.
Balanus.

The occurrence of *Melania turritissima* so low in the Bembridge Marls breaks down the palæontological line drawn by Forbes between the Upper and Lower Marls. Like many other Oligocene fossils, this species is commonly confined to certain thin beds, but reappears on widely separated horizons. Even in the Headon Series, a scarcely distinguishable variety is met with under the name of *Melania peracuminata*. The specimens of *Cerithium plicatum* are not perfect, but there seems little doubt that they are correctly determined, and that the occurrence of this species so far from its principal horizon is another case of the same kind.

North of the Priory the beds rise, and the Marls are lost in the overgrown part of the cliff, or pass inland. The only section near Sea View is in an old pit on the road to Fairy Hill, where a bed with Oysters overlies the Limestone, but no measurements can be obtained.

The inland sections near Bembridge and St. Helen's are few and unimportant, the most interesting being a small exposure of the Oyster bed on the Limestone at the edge of Brading Harbour north of Woolverton ; and a brick pit, showing mottled clay with bones of turtle, on the northern border of the harbour, near Carpenters.

A well-section, communicated by Mr. Parsons, shows that close to St. Helen's Station the Bembridge Limestone has sunk considerably beneath the sea-level, rock being reached at 28 feet from the surface, which is about 5 feet above high water. In the soil heap were found the ordinary fossils belonging to the base of the Bembridge Marl, including *Ostrea vectensis*. (*See* Appendix, p. 309.)

Another well on the top of the hill about a quarter of a mile north of St. Helen's reached the limestone at 133½ feet. Perhaps 15 feet of this depth belong to the Hamstead Beds, leaving 118½ feet as the total thickness of the Bembridge Marls. Scarcely any determinable fossils were found in the samples, but *Serpula* occurs about 7 feet above the Limestone.

Hamstead Beds or Gravels hide much of the Bembridge Marls west of Brading Harbour. As there are also few pits and no clear cliff-sections, little can be said about the changes this series undergoes between Sea View and the Medina. At the time of writing there are no good sections near Ryde.

The upper part of the Bembridge Marls can be examined in Ashlake Brick Yard, near Wootton Bridge. Here the following section was observed :—

Ashlake Brick-yard.

			Ft.	In.
Drift or Rainwash.	Stony clay	- - - -	3	6
Hamstead Beds -	Weathered clay - - - -		4	0
	Black clay full of lignite (the BLACK BAND) - - -		0	5

		FT.	IN.

Bembridge Marls —
{ Green clay, weathered in the upper part.
  Melania turritissima in pyrites - - 3 0
  Seam of marl with Melania muricata,
  Melanopsis and Hydrobia Chasteli.
  Green clay - - - - 1 0
  Line of ironstone nodules.
  Green clay (2¼ feet seen) - dug for 12 6

24 5

FIG. 61.
Hydrobia Chasteli, Nyst.

The occurrence of a thin shelly seam full of *Hydrobia Chasteli* three feet below the Black Band is noticeable. This *Hydrobia* was formerly considered to be characteristic of the Hamstead Beds, but now we find that wherever there is a clear section of the upper part of the Bembridge Marls this thin seam—never more than two inches thick—is found at from three to eight feet below the Black Band. It is well seen on the foreshore at Hamstead and near Yarmouth.

On the east bank of the Medina, near Whippingham, one of the trial-borings made by the Survey reached clay full of *Serpula*, apparently belonging to the lower part of the Bembridge Marls.

Crossing the Medina good sections were exposed in the Zerena slip-way, near Shambler's Copse. At the time of the re-survey the section was obscured, but it appears to have cut through the marine beds and the underlying Limestone. The following fossils were found in the spoil heap:—

Cyrena obovata.
„ obtusa.
„ semistriata.
Cerithium elegans.

Cerithium mutabile.
Melania muricata.
Melanopsis carinata.
Lamna (tooth).

Similar beds were well seen in a deep ditch by the side of the railway cutting a quarter of a mile further south, close to Bolton Copse. Here the base of the Marl is crowded with *Melania muricata*, so that the heaps looked quite white after rain. The other species obtained were *Serpula tenuis, Cyrena obovata, C. obtusa* and *Cerithium mutabile*. In places, a seam of white marl hardens into a shell-limestone containing *Cyrena semistriata, Cerithium mutabile,* and *Neritina concava*. In this *Cyrena* limestone J. Rhodes found a new Cyprid (Fig. 62).

FIG. 62.
Pseudocythere Bristovii, Jones & Sherborn.*

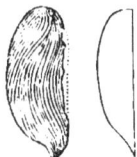

a. b.
a. Right valve (slightly broken along the ventral edge).
b. Edge view. Magnified 20 diam.

A mile further south, at Werror Brick Yard, J. Rhodes obtained Plant-remains, Fish-bones, *Paleryx,* and a phalanx of a Bird. These were found immediately below the Hamstead Beds, which are also shown in the same pit. At this point the Bembridge Marls are lost beneath the marsh level.

A series of wells at Cowes, the West

* Supp. Monogr. Tert. Entom. *Pal. Soc.,* 1889.

Medina Cement Works, and Newport will be found in the Appendix. Unfortunately the samples preserved are not sufficient to prove the exact position of the base of the Hamstead Beds, or to show the palæontological character of the different parts of the Bembridge Marls. However, they show that the Marls are about 120 feet thick, and that they consist of variously coloured clays, as in other parts of the Island.

The cliffs between Cowes and Gurnard are now much overgrown and obscured by landslips, but the marine beds overlying the limestone seem to have been better exposed when Forbes visited this part. He observes that : " At Gurnard Bay, whitish marls, separated by a carbonaceous band, immediately surmount the limestone, and then succeeds about a foot thick of blue clay and shelly stone full of *Cyrenæ*. This is surmounted by nearly three feet of dark shaly clays containing oysters, *Cyrena pulchra* and *oborata*, and *Cerithium mutabile*, a shell here much more plentiful than I have observed it elsewhere. A well-marked band of pale blue septarian stone succeeds ; then come some 10 feet of shales and clays, with *Cyrena obtusa* and *oborata*, *Melania muricata*, and the *Cerithium*, which fossils re-occur in clays and shales occasionally forming compact bands to the summit of the cliff. At the point where this section was noted the upper beds of the Bembridge limestone only are above the shore."

A short distance north of Gurnard Ledge, the upper part of the Marls can be examined, for a small outlier of the Hamstead Beds caps the hill. Here the shelly seam full of *Hydrobia Chasteli*, *Melania muricata*, and *Melanopsis carinata* is found 8 feet below the Black Band. Further south, near Sticclett, the same seam is again met with in the upper part of the cliff.

The lower portion of the Bembridge Marls in Gurnard and Thorness Bays is of great interest, for it contains a thin seam of insect-limestone, which adds very largely to our knowledge of the land fauna of this period. This limestone was discovered by Mr. E. J. A'Court Smith nearly thirty years ago, but no account of it appears to have been published till Dr. Henry Woodward, recognising the great interest of the fauna, read notes on it before the British Association and Geological Society in 1877.* Unfortunately a misunderstanding of the relation of the beds led to the " insect limestone " being referred at first to the Osborne Series and subsequently to the Bembridge Limestone. Its true position, however, is in the lower part of the Bembridge Marls, above the oyster bed.

This part of the series was re-examined in May 1888 (by Clement Reid) in company with Mr. Smith, who pointed out the exact position of the insect limestone and showed a number of the fossils which he had obtained. A short distance west of Gurnard Ledge the section of the lower part of the cliff was :—

	Ft.	In.
Blue clay · · · · · · ·	1	0
Fine-grained blue-hearted limestone, like lithographic stone. Many insect remains, and occasional leaves and fresh-water shells. This bed does not appear to be perfectly continuous, but forms large thin cakes dying out for a few feet and coming on again at the same horizon. One portion, a little further west, thickened to 2 feet, and was full of insect remains, but is now entirely destroyed · ·	0	3
Blue clay · · · · · · ·	0	3
Sandy bed, full of *Cerithium mutabile* · · ·	0	3
Blue clay, with *Cyrena obovata, Melania muricata, Melanopsis carinata,* and *Paludina globuloides* - · · ·	2	6
Ferruginous loam, with *Ostrea vectensis, Cytherea incrassata, Cyrena, Cerithium mutabile,* &c. · · · ·	0	10
Bembridge Limestone.		

Mr. Smith has traced the Insect Bed from West Cowes nearly to the Newtown river. He states that it varies in thickness from 2 inches to about 2 feet, though the extreme measurement of 2 feet is quite exceptional. Its distance above the Bembridge Limestone also varies slightly, sometimes being as much as 9 feet.

The fossils of the Insect Bed have been collected during many years by Mr. A'Court Smith, to whose industry we owe the whole of our knowledge of this interesting fauna. Among the forms contained in his collection are numerous beetles, flies, locusts, and even spiders and caterpillars. These have been as yet only partially studied, but Mr. Frederick Smith gave the following list of genera* :—

I. Coleoptera.

1. Staphylinus.
2. Dorcus (Lucanidæ).
3. Anobium.
4. Curculio.

II. Hymenoptera.

5. Wings of.
6. Formica.
7. Myrmica.
8. Camponotus.

III. Lepidoptera.

9. Lithosia.

IV. Diptera.

10. Wings of.
11. Tipulidæ.

V. Neuroptera.

12. Phryganea.
13. Termes ?
14. Hemerobius.
15. Perla.
16. Agrion.
17. Wings of Libellula.

VI. Orthoptera.

18. Gryllotalpa.
19. Acridiidæ.

VII. Hemiptera.

20. Wing of ?
21. Triecphora sanguinolenta.

Arachnida.

1. Spider.

---

* In Dr. Woodward's paper.

To this list Dr. H. Woodward adds two new crustacea: a Phyllopod, *Branchipodites rectensis*, and an Isopod, *Eosphæroma fluviatile*. A second species of Isopod, *Eosphæroma Smithii*, was discovered by Mr. Smith in a " fine yellow marl or pipe-clay, full of rootlets of aquatic plants " somewhat higher in the series. Ostracoda also occur, and in the last volume of the Palæontographical Society's monographs, Messrs. Jones and Sherborn describe a new species of *Potamocypris* (Fig. 63).

FIG. 63.

Potamocypris Brodiei, J. & S.

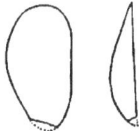

a.   b.

a. Right valve (slightly broken at the posterior margin).
b. Edge view. Magnified 20 diam.

The determinable plant-remains in the Insect Bed, though not abundant, are also interesting, but until Mr. Gardner has finished his monograph on the Oligocene flora not much can be said about them.

More to the west, at Thorness Point, Forbes measured a good section of the middle beds of the marls, exhibiting the following succession in descending order:—

	FT.	IN.
Green clays, with plentiful specimens of *Melanopsis carinata*, and, less abundantly, *Paludina lenta*, *Melania turritissima*, and *Cyrena obovata* - - - - -	6	0
Band of comminuted *Melaniæ* - - -	0	2
Dark-green shaly marls, with ferruginous concretions, and numerous specimens of *Melania muricata* and *Melanopsis carinata*, a belt of which shell forms the base of this bed -	2	2
Green marls, with *Paludina lenta* - - -	2	3
Pale-yellow stony band, composed of comminuted shells, and becoming a limestone. Broken *Cyrenæ* and *Melania muricata* form the mass of it - - - -	1	2
Green clays, with lines of broken *Melania muricata* - -	0	2½
Band of comminuted *Melania muricata* - - -	0	2
Green marly stone, with a well-preserved band of *Melania muricata* - - - - - -	0	4½
Band of comminuted *Cyrenæ* - - - -	0	4½
Grey septarian stone band, capped by a thin layer of greenish stone, with fucoidal markings - - - -	0	5
Greenish marls, with bands of finely preserved *Cyrena obovata*, very abundant, patches of *Melania muricata*, and scattered shells of *Paludina lenta* - - - - -	2	6
Band of septarian stone - - - - -	0	2
Green clays, full of *Melania muricata*, constituting the last bed exposed upon the shore.		

In this neighbourhood the thickness of the Bembridge Marls has apparently decreased to about 90 feet, but the Hamstead Beds are so overgrown that it is difficult to obtain exact measurements.

Crossing the Newtown River, we find, at Hamstead, the only locality where a section of the entire thickness of the Bembridge Marls is displayed. Here the whole of the beds which compose the subdivision, from the Bembridge Limestone forming Hamstead ledge to the "*Black Band*" which constitutes the base of the Hamstead Series, may be examined in detail at low water without a break, as they successively crop out on the shore, and their beautifully preserved fossils may there be collected.

The following section taken along the shore, at low water, in 1856, by Sir A. Ramsay, then Director of the Geological Survey, Professor John Morris, and H. W. Bristow, is more complete than that of Professor Forbes, in consequence of its supplying the thicknesses of the several beds, which are omitted in his section.

Bembridge Marls of Hamstead

Measured along the Shore at Low Water.

	Ft.	In.
Black Band (base of Hamstead Series) - - -	1	9
Green clays, with large bands of *Paludina lenta* - -	4	6
Ironstone - - - - - - -	0	9
Clay - - - - - - - -	4	0
Clay with *Paludina* - - - - - -	4	6
Concretionary ironstone - - - - -	0	3
Clay with *Paludina* - - - - - -	4	6
Clay with two or three small black bands - - -	2	6
Ferruginous brown sandy clay, with *Paludina* at base, thickness variable - - - - - 0 to	0	6
Thin bituminous bands, with reed-like plants, and a layer of *Paludina lenta* below filled with green clay	0	1
Grey clay, with short zones of *Melania Forbesii* and nodules containing *Paludina lenta* - - - -	3	0
Band of scattered nodules of iron pyrites, overlying verdigris-green clays, with bands of *Paludina lenta* (occasionally of very large size). *Melania Forbesii* - - -	5	0
Dark-grey clays, with *Paludina* and numerous oval seed-vessels, and containing thin carbonaceous sandy bands, with (reed-like) plant impressions, *Cypris*, *Paludina*, *Planorbis* immediately overlying a layer of large *Limnæa*. This is altogether a freshwater deposit - - - -	3	0
Bands of *Melania turritissima*, *Planorbis*, and *Paludina* -	0	1
Greenish shaly clay, with concretions of indurated marl, and containing near the base a band of *Melania turritissima*, *M. costata*, *Melanopsis carinata*, *Paludina*, &c.	4	0
Hard shelly band chiefly made up of *Melania turritissima*, a few *Melanopsis*, and fragments of *Fish* - -	0	3
Pale-grey clay, with bands of compressed shells, chiefly *Paludina* - - - - - - -	2	0
Sandy band, full of *Cyrena obtusa* (with both valves), *Cerithium* and *Melanopsis fusiformis* - - -	0	9
Pale greenish shaly clay, with a thin band of *Melania turritissima* 6 inches from the top, and a 1-inch bituminous band at 3 feet. Compressed *Carpolithes*, *Melania turritissima*, *Melanopsis carinata* - - - -	6	0

FT. IN.

Sandy clay, with *Melania muricata*, *M. turritissima*, *Melanopsis*, 3 inches thick, resting on sandy clay, almost entirely composed of *Melania muricata*, with a few broken *Cyrena*, and some *Melanopsis* - - - - 0 4 to 0	7
Bluish irregular shaly clay, with selenite. A band of *Melania turritissima* and *Melanopsis carinata*, 2 inches from the top, *Paludina* - - - - - - - 3	0
Indurated, greenish marly clay, with bands of *Paludina lenta* - - - - - - - 4	6
Greenish clay, with 2 bands of broken *Cyrena obtusa*, and *C. semistriata*, and on the top a bed of *Melania muricata*, with scattered *Melanopsis*, and occasional *Cerithium* - 0	9
Green clays, *Melania turritissima*, with scattered *Melanopsis* and *Melania muricata*, mixed with patches of *Cyrena semistriata*, *C. obovata*, Fish remains, &c., about the middle 4 inches 1	0
Green clays, with *Melania muricata*, *M. turritissima*, and numerous *Melanopsis carinata* and *Paludina lenta* - - 1	6
Verdigris-green clay, with *Cyrena semistriata* - - 1	0
Bright-green clays, with *Cyrena semistriata* on the top - 2	0
Bluish-green clays, with bands of *Melania muricata* and *Cyrena obovata* - - - - - - 1	6
Hard, sandy green marl, with scattered *Cyrena semistriata* - 0	6
Verdigris-green clays, with 5 bands of *Cyrena semistriata* and *Melania* - - - - - - - 1	6
Greenish clay, with 2 marked bands 1 and 2 inches thick, full of *Melania muricata* (small var.), occasional *Cyrena pulchra*, and a few *Cyrena semistriata* - - - 1	0
Green clays, with *Cyrena semistriata* (finely preserved) *C. pulchra* and *Melania* - - - - - - 0	9
Dark clay, with soft green sandy concretions, 6 inches; greenish clay 1 foot. Scattered *Cyrena obovata* - - 1	6
Blue clay, with small *Cyrena (obovata?)*, *Melania muricata*, and *Melanopsis* - - - - - - 1	0
BEMBRIDGE LIMESTONE in 3 beds, with softer beds between, forming a ledge (Hamstead Ledge), out at sea, in the direction of the Buoy, and containing numerous *Linnæa longiscata*, *Chara*, &c.	

Total of Bembridge Marls - • - 69	6

Forbes gives the total as nearly 75 feet, but it is difficult to obtain exact measurements in these soft beds. A recent measurement in the cliff above Hamstead Ledge gave 82 feet, so probably Forbes was nearly correct.

Though not mentioned in the above section, the thin seam full of *Hydrobia Chasteli* will be found about 5 feet under the Black Band. The marine bed with *Ostrea* at the base of the series, was recorded by Forbes; but marine fossils are not abundant, and like most of the subdivisions in the Oligocene series, this bed also tends to become more estuarine towards the west.

On the southern side of the syncline, the Bembridge Marls rise from below the Hamstead Beds due north of the point where

the high-road strikes the coast. Here exceptionally good exposures were visible during 1887 and 1888, for though the cliff is
low and overgrown, a continuous foreshore of clay was laid bare
as far as the first houses in Yarmouth. The strata are so like
those on the north side of the synclinal that it is unnecessary to
give detailed measurements. The seam with *Hydrobia Chastch*
again occurs a few feet under the Black
Band, and the section below is continued
down to the *Melania turritissima* (Fig. 64)
beds, which lie 10 or 15 feet above the
Limestone. The base of the Marls cannot
be examined here, but the Limestone outcrops on the other side of Yarmouth, at the
Gas Works and Station.

FIG. 64.

*Melania
turritissima,*
Forbes.

A good deal of drift wood occurs in the
Bembridge Marls between Hamstead and
Yarmouth, and thin seams rendered quite
black by the number of seeds they contain
are often conspicuous on the shore or in the
washed base of the cliff. The drift wood does not occur as rafts,
but generally as isolated trunks and branches, often of considerable
size. One of these trunks, examined by Mr. Keeping and Clement
Reid, was cleared for 18 feet without reaching the end. It
measured $3\frac{1}{2}$ inches thick at the broken smaller end, only increasing to 5 inches 13 feet below. The thickness of the overlying clay prevented us from following the tree further, but its
straightness and slenderness showed that it had probably grown
in a forest—not in open ground. In the Marls near Yarmouth
Toll Gate we also obtained portion of the bones of a large teleostean
fish.

Inland sections on the southern side of the syncline are few,
and do not expose any of the more characteristic beds. In the
railway cuttings near Shalfleet, *Cyrena obtusa* (Fig. 65) is common,
but no other species were noticed.

FIG. 65.

Cyrena obtusa,
Forbes.

On the west side of the Yar there are
three outliers of the Bembridge Marls.
The first caps the long ridge between Sconce
and Cliff End for about half a mile. It
exhibits no clear sections, and all that can be
made out is that clays with *Cyrena obovata,
C. semistriata, Melania muricata, Melanopsis
fusiformis,* and *Serpula* overlie the limestone. The thickness cannot be great;
probably it is under 20 feet. The second
outlier, of still smaller extent and thickness,
occurs at Hill Cross, south of Norton.
Shelly marl is found in the road cutting, but there is no section.
The third outlier underlies the gravel capping Headon Hill.

No section is visible, and the surface is much obscured by washed gravel. The Bembridge Marls here consist of grey and white clays with *Cyrena*.

The fauna of the Bembridge Marls is not a prolific one. Leaving out the mammals, which are little known, though apparently the same as those of the Limestone, the vertebrates are turtles, crocodiles, and fish, such as occur throughout the Oligocene Beds.

The assemblage of mollusca is poor, consisting of abundance of individuals belonging to comparatively few species, the common genera being *Cyrena*, *Melania*, *Melanopsis* (Fig. 66), and *Paludina*. However, though the species are few, the shells are fine and remarkable for their beautiful state of preservation. The species of *Cyrena* often retain their colour-markings. In the marine beds, fossils are not usually well preserved, but this horizon is especially worthy of careful examination, for any correlation with other districts must be founded mainly on the marine mollusca. Though few species of these mollusca have yet been obtained, it is important to note that a considerable proportion of them is confined to this horizon. Among these are *Arca Websteri* (Fig. 60, p. 173), and *Ostrea vectensis* (Fig. 67).

Fig. 66.

Melanopsis carinata, Sow.

Fig. 67.

Ostrea vectensis, Forbes.

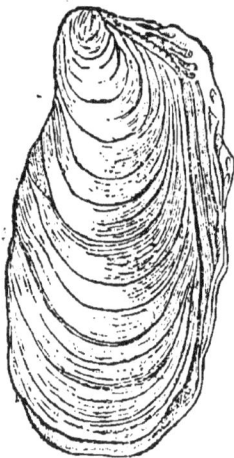

Certain thin seams in the Bembridge Marls contain only freshwater forms; but the usual character of the deposits and their included fauna points to an estuarine origin. Red-mottled lagoon clays, with nothing but remains of turtles and crocodiles, are comparatively rare in these beds, though they appear again and again on different horizons throughout the Oligocene and Eocene formations.

Drift-wood, seeds, and fruit are common in the Bembridge Marls, especially near Hamstead and Yarmouth, but few plants have yet been determined. The only good leaf-bed yet observed seems to be the insect-limestone, where, however, leaves are not abundant.

The following list of the Bembridge plants has been revised by Mr. Gardner; but the whole flora being under examination, the list can only be regarded as provisional:—

Chara Lyellii, *Forbes.*
—— medicaginula, *Brong.*
—— tuberculata, *Lyell.*
Chrysodium lanzæanum, *Visiani.*
Gleichenia, sp.
Arthrotaxis (Sequoia) Couttsiæ, *Heer.*
Carpolithes (Folliculites) Websteri,
  Brong.
Cinnamomum lanceolatum, *Unger.*
—————— polymorphum, *Unger.*

Doliostrobus Sternbergii, *Goepp.*
Engelhardtia, sp.
Ficus, sp.
Flabellaria Lamanonis, *Sternb.*
Lygodium, sp.
Myrica, sp.
Pinus Dixoni, *Bowerb.*
Rhus, sp.
Sabal major, *Heer.*
Viburnum, sp.
Zizyphus Ungeri, *Ett.*

CHAPTER XII.

OLIGOCENE—*continued.*

HAMSTEAD BEDS.

In 1853 Forbes pointed out that a thick series of beds overlies the Bembridge Marls, and yields in its upper part a marine fauna, which includes a large number of characteristic species. These beds attracted great interest, for they were the highest of our older Tertiary series, and were separated by many writers from the rest of the Fluvio-Marine Beds and were referred to the Miocene period. Though no longer classed as Miocene the interest of these deposits has not decreased, for nearly all the recent additions to the fauna are characteristic of the upper part of the Hamstead group.

Not only has our knowledge of the fauna increased since Forbes' time, but the deposits prove to be both thicker and more extensive than was originally imagined. When Forbes' Memoir was published the only known strata of this age occurred in Hamstead and Bouldnor cliffs, with a doubtful outlier in Parkhurst Forest. Now it has been ascertained that they cover a much larger area, for they extend over about half the Tertiary basin in the Isle of Wight, stretching continuously from Yarmouth to Brading, and occupying the greater part of the wide trough of the Isle of Wight syncline.

In thickness also Forbes under-estimated the importance of this group. Instead of only reaching about 170 feet in Hamstead Cliff, new measurements prove that the Hamstead Beds are there 256 feet thick (probably rather more), and that at Parkhurst Forest, and at Wootton, in the East Medina, they also exceed 200 feet.

Forbes divided the Hamstead Series into:—
  Corbula beds.
  Upper freshwater and estuary marls (full of *Cerithium plicatum*).
  Middle freshwater and estuary marls (full of *Melania fasciata*), with the "White Band" at the base.
  Lower freshwater and estuary marls (with *Melania muricata* and *Melanopsis carinata*).
  "Black Band."

FIG. 68.

Cerithium plicatum,
Lam.

FIG. 69.

Corbula pisum,
Sow.

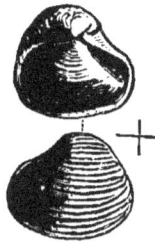

FIG. 70.

Cerithium elegans,
Desh.

FIG. 71.

Corbula rectensis, Forbes.

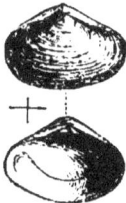

This classification might with advantage be considerably simplified. The *Corbula* Beds and *Cerithium plicatum* Beds pass imperceptibly into each other, and form one marine division, with *Corbula* becoming scarcer below, and *Cerithium* dying out above. In fact, these strata become more truly marine upwards, though *Corbula rectensis* (Fig. 71) extends downwards even to the base of the marine bands.

The line between the Middle and Lower Freshwater and Estuary Marls is a very indefinite one, and proves to be only of local value, for the White Band, which Forbes took as the junction, soon dies out and there is no palæontological evidence on which to separate the two horizons.*

The Hamstead Beds may therefore be divided into :—

	FEET.
Marine Beds, with *Corbula, Cytherea, Ostrea callifera, Cuma, Voluta, Natica, Cerithium,* and *Melania* - - - - -	31
Freshwater, estuarine, and lagoon beds, with *Unio, Cyrena, Cyclas, Paludina, Hydrobia, Melania, Planorbis, Cerithium* (rare), Turtles, Crocodiles, Mammals, Leaves, and Seeds -	225
	256

---

* *Melania fasciata* seems to be only a stunted form of *M. inflata*. Selected specimens are sufficiently different, but certain beds contain forms that might be referred to either.

As the beds were evidently considerably thicker than had been thought, they were re-measured during the summer of 1888. The following section is the one then made :—

Section of the Hamstead Beds at Hamstead.

(Measured by Clement Reid and Henry Keeping.)

		FEET.
Drift	{ Irregular clayey gravel of subangular flint, flint and quartz pebbles, &c.	0 to 5
	Pale bluish-green clay (much weathered), with seams of *Ostrea callifera* bored by *Lithodomus* and overgrown by *Balanus*	11
	Carbonaceous and ferruginous clays, full of broken and waterworn *Cyrena semistriata. Cuma Charlesworthii* and *Voluta Rathieri* also occur occasionally	1
	Stiff blue clay, full of *Corbula pisum, Cerithium plicatum, C. elegans, Voluta Rathieri,* and *Strebloceras cornuoides.* A layer of flat septaria about the middle	7
	Black clay, full of *Corbula vectensis.* Also *Cytherea Lyellii, Cyrena semistriata, Hydrobia Chasteli, Cerithium plicatum, C. elegans, Melania fasciata, M. inflata*	½
Upper Hamstead Beds (marine) 31¼ feet.	Shaly clays, with *Cyrena semistriata, Cerithium plicatum, Hydrobia Chasteli,* seam of *Mya minor.* Occasional *Paludina lenta* and *Unio* towards the base	4½
	Shell-bed, full of *Cerithium plicatum, C. elegans, Hydrobia Chasteli, Melania inflata,* and *Cyrena semistriata*	¾
	Shaly clays. Seams of *Paludina lenta* and *Unio*	4
	Stiff blue clay, carbonaceous at the base. Scattered *Cyrena semistriata.* At base seams of *Mya minor,* with *Cerithium plicatum, C. elegans, Hydrobia Chasteli, Melania inflata, Corbula vectensis, Cyrena semistriata,* and *Balanus*	1½
	Laminated carbonaceous clay, and sand-partings with *Cyclas*	1
	Green clay, with *Paludina*	3½
	Carbonaceous clay, with *Chara*	½
	Mottled red and green clay. *Unio, Paludina,* seeds, &c., occasionally	11
	Obscure (mottled clays ?)	18
	Carbonaceous seams, with *Carpolithes ovulum, Unio,* and *Paludina lenta*	2
	Carbonaceous clays, with seams of *Melania inflata,* var. *lævis, Unio, Paludina, Planorbis, Hydrobia Chasteli,* and Seeds	15
	Bluish loam	1
	Clay, with *Paludina,* Seeds, &c.	3
	Obscure	4
	Clays, with occasional *Paludina* and *Melania,* and seams of *Carpolithes ovulum.* Fossils rare	20
	Obscure	5
	Laminated carbonaceous clay, with Seeds, Palm-leaves, leaves of Water-lily, *Unio, Paludina lenta, Melania,* and *Candona*	1¾
	Green and red marls	8
	Obscure (? sparingly fossiliferous)	60

		FEET.
Lower Hamstead Beds (lacustrine and estuarine) 224½ feet.	WHITE BAND—Green clays and white shell-marls. *Melania fasciata, Cerithium inornatum, C. Sedgwickii, Mya minor*, &c.	6
	Green clay, with lines of ironstone nodules at the base - - - - -	7
	Obscure - - - - -	28
	Black or slate-coloured carbonaceous clay, full of *Cyrena semistriata* and *Nematura pupa*. Also *Bythinia conica, Cerithium* 2 *sp.*, and *Cyclas Bristovii* - - -	3
	Green clay - - - - -	1½
	Black laminated clay, full of *Planorbis obtusus, P.* sp., and *Cyclas Bristovii* - -	½
	Green clays - - - - -	4
	Mixed black and green clay, full of *Melania muricata, Hydrobia Chasteli, Limnæa, Planorbis,* &c. - - - -	¼
	Green clays with *Paludina, Nematura,* sp. (in the upper part), *Melanopsis carinata* -	20
	BLACK BAND, full of *Paludina lenta; Unio* at the base - - - - -	1½
	Total - -	255¾

The constant landslips from Hamstead cliff render it impossible to measure the whole of the beds at any one time. It will therefore be necessary, as far as possible, to fill up the gaps in the above section from notes taken during the original survey of the Island. Unfortunately the 60 feet of beds above the White Band do not appear ever to have been measured in detail, and the following section only gives the lower part:—

Section of the lower part of the middle freshwater and estuary marls and of the white band, measured in a low cliff on the shore, at the base of a great founder. (*By Professors Ramsay and Morris.*)

	FT.	IN.
Rubble.		
Small bands of clay, with apparently nodules of ironstone -	0	4
Band of crushed *Cyrena* 1 inch - - - -		
Shaley clay 1 inch - - - -	0	3
Band of crushed *Cyrena* and *Melania fasciata* in great abundance, 1 inch - - - - -		
Brown clay - - - - -	0	1½
Ferruginous band: *Melania fasciata, Cyrena, Paludina* separated by a very thin layer (sometimes passing into it) from another similar layer - - - - -	0	1½
Ferruginous band, containing the same shells; with *Cerithium* -	0	2
Laminated clay, with bands of compressed *Melania* and *Cyrena* about	0	9
Greenish tenacious clay, with bands of *Paludina* and Fish remains	3	0
Ferruginous concretions, covering a layer of compressed *Cyrena* -	0	2
Tenacious green clay; a layer of *Melania* (and bones of turtle) 1 inch from the bottom, with concretions of clay-ironstone -	3	0
Thin layer of vegetable matter, with reed-like plants and long seeds (*Folliculites*), patches of *Paludina* - - -	0	0½
Greenish sandy clay forming the shore, with large concretions; patches of shells, *Paludina* - - - -	3	0
WHITE BAND. { Verdigris-green clay, with two bands of *Cyrena semistriata*, more or less perfect; in the lower half of the bed (between which is a thin layer of ironstone with *Panopæa Gibbsii?*) *Cerithium, Melania,* Fish, and *Folliculites* in the lowest band of *Cyrena* - - - - -	1	0

Further east, about 100 yards west of some stakes driven into the beach, where the *White Band* comes to the base of the cliff, it contains occasional large white nodular concretions 10 inches thick and a yard long, while above it are two bands of clay-ironstone, each one inch thick.

Section measured eastward of the preceding section, where the white band appears at the base of the cliff.

	Ft.	In.
Dark clay	2	0
Tenacious dark-greenish clay ; at two inches below the top of the bed are two 1-inch bands of tabular ironstone, containing *Paludina,* and separated by two inches of clay also with *Paludina*	1	0

		In.		
	Very shelly clay, with *Cyrena, Melania,* and *Cerithium*	3		
	Clay	0½		
	Shelly clay, as before	2		
	Clay with compressed *Cyrena*	1		
	Shelly bands, as before; sometimes separated by a thin layer of clay	3 to 4		
WHITE BAND.	Clay, with bands of *Cyrena, Cerithium,* and *Melania*	4	4	1½
	Green clay, mostly filled with broken *Cyrena* and *Cerithium.* The shells in this band are much blackened, and occasionally at the bottom are *Panopæa* in upright positions, partly sinking into the clay below, *Melania fasciata, Cerithium Sedgwickii*	2		

	Ft.	In.
Greenish clay about two feet from the shore, forming the base of the cliff, and containing bands of crushed *Paludina*	3	0

Further east of the place where the above section was measured, all the bands forming the *White Band* unite, and are well seen in the cliff, forming a distinctly marked white line at its base, with about twelve feet of dark clay weathering brown above.

The next three sections help to fill up the gaps in the beds between the White Band and the Black Band.

About S.E. of the buoy the following section was measured in a projecting point of the undercliff.

	Ft.	In.
Laminated tenacious clay, with shelly bands, mostly made up of broken *Cyrena*	15	0
Laminated clays, with *Hydrobia Chasteli* and *Cypridæ*	0	9
Band of broken *Cyrena*	0	2
Lenticular patches of white marl, containing fragments of lignite and disseminated vegetable matter, with reed-like stems	0	9
Tenacious blue clay	0	3
Fossil band ; *Melania muricata, Melanopsis, Cyrena* or *Cyclas, Nematura pupa, Candona; Planorbis* on surface of bed	0	6
Tenacious greenish clay, with layers of *Paludina* and seeds towards the middle	3	0
Ochreous clay, passing into ironstone	1	0
Stiff lead-coloured clays with several bands of *Paludina* towards the upper part; more sandy and ferruginous towards the base, where the beds sometimes become very finely laminated	20	0
Probable position of the *Black Band.*		

Section measured further west, in the broken ground a few feet above the shore, about 29° E. of S. from the buoy on Hamstead Ledge.

	Ft.	In.
Traces of *White Band* on the top of broken ground - -	0	0
Ground not seen - - - - - - about	5	0
Clay, weathering brown, with traces of *Cyrena* bands on the weathered surfaces - - - - - about	20	0
Ferruginous band, with fragments of shells - - -	0	1
Laminated clays, unfossiliferous ? - - - -	4	0
Clays, with laminæ of *Cyrena semistriata* - - - -	1	0
Dark tenacious clay, with two bands of *Cyrena*, the upper containing numerous *Nematura pupa* and valves of *Cyrena*, often perfect and united at the hinge - - - -	5	0
Green clay, with *Cypridæ* - - - - - -	0	4
Dark fossil-band, *Planorbis* on top of bed, *Cypridæ* throughout, associated with *Melania muricata*, *M. fasciata*, a smooth *Melania*, *Melanopsis fusiformis*, *Limnæa*, and *Cyrena* or *Cyclas*	0	5
Clay, with band of *Cypridæ* and occasional *Melania* at the base -	0	6
Clay - - - - - - - - -	0	7
Clay, with compressed *Paludina* and seed vessels - -	1	6

Measured (by pacing) along the shore at low water, under Hamstead Hill.

	Ft.	In.
Shelly band with *Melania, Hydrobia, Limnæa,* and *Planorbis* on top of bed - - - - - - - - -	0	6
Green clays, with *Paludina* and seeds - - - -	4	6
White marl, with *Paludina lenta* - - - -	0	3
Green clays, with bands of *Paludina lenta* and *Melanopsis carinata*	18	0
Black Band, with reed-like plants, and *Unio* and *Paludina* at base	1	9

As the Hamstead Series, though of considerable thickness, presents no breaks and no marked lithological changes, special attention may with advantage be called to its different fossiliferous horizons that can be identified inland. These, with their approximate distances above the Black Band, are as follows :—

	FEET.
Corbula and *Cerithium plicatum* beds -	224 to 256
Water-lily and leaf beds - - -	140
White Band - - - -	65
Nematura beds - - - -	30
Black Band - - - -	0

The Black Band was taken by Forbes as the base of the Hamstead Series " for several reasons, and foremost, because it is apparently the first bed that succeeds to those which terminate the Bembridge marl at Whitecliff and elsewhere in the Isle of Wight. This circumstance, combined with those of the beginning of a new series of fossils, of which the *Rissoa Chasteli* (Fig. 61, p. 175) is the first conspicuous representative, of the disappearance of others, and the probable indications of a terrestrial surface indicated in some of the features, both of this bed itself and the bed below it, may fairly warrant the choice of so well marked an horizon."

FIG. 72.
Cyclas Bristovii, Forbes.

FIG. 73.
Unio Gibbsii, Forbes.

Forbes thus describes the Black Band :—
"It consists of nearly two feet of firm carbonaceous laminated clays, abounding in fossils. These are *Paludina lenta*, very numerous; *Hydrobia Chasteli major*, scarce; *Melanopsis carinata; Limnæa; Planorbis obtusus*, of large size; a peculiar small *Cyclas*, (*C. Bristovii*, Fig. 72), which I have not met with elsewhere; and fish vertebræ. Impressions of the linear leaves of gramineous plants, occasionally large seed vessels, and *Gyrogonites* are found in it, and lumps of lignite. At its base is found a seam of *Unio* (*U. Gibbsii*, Fig. 73) containing well-preserved specimens."

"The Black band rests in perfect conformity on a bed, three feet in thickness, of dark green marls, becoming paler below, and separated by an irregular seam of broken univalves (*Paludina lenta*) from greenish blue pale marly clays, with lenticular seams of crushed *Paludina*. In the dark green marls are scattered fine specimens of *Paludina lenta* and *Melanopsis*, also numerous fossil bones. There are, moreover, in this bed, curious vertical or slanting tubular concretions, with hollow cavities, as if formed round the roots of plants."

This weathering of the surface of the underlying Bembridge Marls is very noticeable. It is a character still more marked inland, where repeatedly after boring through unweathered Hamstead Beds we penetrated a carbonaceous soil (the Black Band), and then again entered weathered clays full of roots, like the surface soil many feet above.

Though this thin bed, however, can be traced nearly throughout the Island, there seems to be no evidence of any real break. Fossil species die out upwards one by one, and are replaced by other species. Even the species which Forbes considered to be most characteristic of the Hamstead Beds—*Hydrobia Chasteli*—we have shown in the last chapter not to be confined to this Series, but to appear several feet down in the Bembridge Marls. Similarly *Nematura pupa* comes in somewhat higher : and so on with others. Probably if the beds were now for the first time to be sub-divided, we should class the the Bembridge Marls and the greater part of the Hamstead Beds together, and separate the marine beds as the commencement of a new series formed under different conditions.

But though no palæontological break occurs at the Black Band, it was so necessary to sub-divide the thick mass of clay above the Bembridge Limestone, that some marked and easily recognisable bed had to be traced. The Black Band proved to be the only horizon that could be followed, and that would give a satisfactory line from which to calculate dips and thicknesses.

Borings were therefore made, and the Black Band traced inland; with the result that this horizon has been identified in many places, and over a wider area than any other part of the Hamstead Series.

Forbes' description of the Black Band at Hamstead is excellent, and will also apply to the inland sections. Unfortunately the *Nematura pupa* bed at Hamstead has occasionally been confounded with the Black Band, and it is now probable that in a few of the well-sections and trial borings the bed at first thought to be a modified representative of the Black Band is really the *Nematura pupa* bed about 30 feet higher.

In the trial-borings the difference between the Black Band and the clays lying below and above it is even more marked than on the coast. The Black Band is generally a brown clay or loam, turning a sooty black after a few seconds exposure, in which abundant seeds and fish-bones are found, but few shells except *Paludina lenta, Melanopsis carinata*, and *Unio*. In it occasionally occur small angular fragments of flint.

The next marked zone is the *Nematura pupa* bed, nearly 30 feet above the Black Band. This is a bed of laminated slate-coloured carbonaceous clay crowded with *Cyrena semistriata, Nematura pupa, Bythinia conica*, and Cyprids, and more rarely yielding other species. On the coast it is the first bed that yields *Nematura pupa*, but inland this species perhaps ranges down into the Black Band, though it is not always possible to distinguish these horizons in borings. Though recognised at many localities this bed seems to be more variable than the Black Band. A few feet underneath it there is generally a line of *Melania muricata*.

The *Nematura pupa* bed indicates slightly estuarine conditions of deposit, yielding *Modiola* and *Cerithium* at several localities. It is perhaps the best horizon in the Lower Hamstead Beds for fossils, for not only are these exceptionally well preserved but the fauna is also more varied than is usual in these freshwater beds. The following is a list of the species obtained, but no doubt it could be considerably increased :—

Cyclas Bristovii.	Melania muricata.
Cyrena semistriata.	Melanopsis carinata.
Modiola Prestwichii.	———— subcarinata.
	———— subulata.
Bythinia conica.	Neritina tristis.
Cerithium elegans.	Nematura pupa.
———— sp. (like C. plicatum).	Paludina lenta.
Hydrobia Chasteli.	Planorbis (small sp.).
Melania Forbesii.	

About 36 feet above the *Nematura pupa* bed and 65 feet above the Black Band occurs the White Band. This consists of green clay in which are seams of white shell-marl. Though so conspicuous at the base of Hamstead and Bouldnor Cliffs, it is not persistent, being only traceable as far as Parkhurst Forest. In the East Medina it is apparently represented by a seam of fine sand which, commencing near Newport, expands eastward till it reaches a thickness of about 40 feet and forms a marked topographical feature.

The White Band is characterised by two species of *Cerithium* (*C. inornatum* and *C. Sedgwickii*). It also contains abundance of

FIG. 74.
Melania fasciata, Sow.

Melania fasciata (Fig. 74) and *Mya* (*Panopæa*) *minor*, (Fig. 75), but the fossils are so much decayed and so fragile that no determinable specimens were obtained from any of the borings in Parkhurst Forest.

FIG. 75.
Panopæa minor, Forbes.

Above the White Band there is a gap of 70 or 80 feet before another marked fossiliferous horizon is met with. The intervening beds are generally much obscured by mud-streams and landslips, but they appear to be very sparingly fossiliferous. None of the inland borings or well-sections yielded much of interest in this part of the Hamstead Beds. As the series of borings in Parkhurst Forest penetrated the whole without meeting with any conspicuous shell beds, it is probable that such are absent.

About 140 feet above the Black Band, and 120 feet below the marine beds lies a bed of compact laminated clay full of a peculiar creeping root, and containing leaves of Palm and Water-lily (*Nelumbium*), &c.* This horizon forms a ledge or low cliff, over which the softer overlying beds slip. It has not yet been recognised inland; but as there are no open sections on this horizon, and the plants would not be preserved in the small cores obtained by boring, the leaf-bed may cover a considerable area.

The marine beds commence about 224 feet above the Black Band, and range upwards to the highest point reached (see sketch by Edward Forbes, Fig. 76, p. 193). Unfortunately they are confined to a small outlier of a few acres on Hamstead and Bouldnor Cliffs, and another about half a mile long at Wootton in the East Medina.

A reference to the table on p. 189 will show the approximate position of the beds mentioned in the description of the inland sections, for though, as in all the Oligocene Beds, a considerable amount of lateral change may be remarked, yet certain marked beds extend persistently over the whole of the area.

The notes made in the course of the re-survey were so voluminous that it has been necessary greatly to condense them; but all the well-sections will be found in the Appendix, and the position of each of the trial-borings is marked on the 6-inch maps deposited in the Office of the Geological Survey. As the number of the trial-borings in Hamstead Beds amounted to nearly three hundred, it has not been thought advisable to print so bulky a record, but wherever fossiliferous strata of marked character were met with, the occurrence will be found recorded in the text.

* J. S. Gardner. Report of the Committee for . . . . exploring the Higher Eocene Beds of the Isle of Wight. Report Brit. Assoc. for 1887, p. 414.

FIG. 76.

Sketch of the upper part of Hamstead Cliff
(By Edward Forbes).

a. Gravel. *b. d.* Marine Corbula beds. *c. e.* Lower Cyrena band.
f. Shaly clays.

In Parkhurst Forest, where a note by Mr. Godwin-Austen led us to expect an outlier of the marine beds,* the survey had to be entirely made by boring, for there are no open sections. Commencing at the highest part of the Forest, we made radiating series of borings and continued them southward and northward, till the Black Band was reached. These excavations, and those made by Mr. Keeping, lead us to conclude that the note of the occurrence

---

* Mr. Godwin-Austen mentions the occurrence of *Ostrea callifera* in the Forest, but does not state by whom it was found. Forbes, " On the Tertiary Fluvio-marine Formation," p. 37, footnote.

of *Ostrea callifera* (Fig. 77) is founded on some mistake.   There seem to be no strata in the Forest so high as the base of the *Cerithium plicatum* bed.

FIG. 77.

Ostrea callifera, Lam.

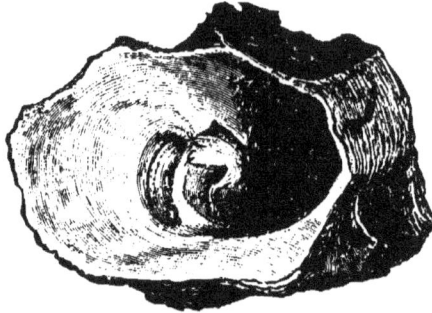

The highest strata in the Forest are found immediately west of the Signal House.   Here an old gravel pit has been dug, only a foot and a half below the top of the hill, and at the bottom of it a boring was made to a depth of 24 feet (B. H. 11).   The surface at this point is 273 feet above the sea, and the base of the gravel lies at 266 feet.   The strata passed through are red and mottled clays, like those immediately beneath the marine beds at Hamstead and Wootton.   *Paludina* occurred at 15 feet from the surface, but no other recognisable fossils were met with.

About 8 chains south-west of the Signal House another old gravel pit lies at a height of 254 feet.   In this a boring (B. H. 10), commencing 10 feet from the surface, was carried to a depth of 33 feet, in alternations of red and carbonaceous clays, with *Melania, Paludina, Unio,* and *Chara* in the lower part.   This boring is important, as the upper part seems to show strata that are too much obscured to be measured at Hamstead.   In this upper portion—probably corresponding with some of the beds marked " obscure," about 25 feet below the marine beds*—a tooth of *Theridomys* was found at 11 feet, and another small mammalian bone at 15 feet.   Mammalian bones are of rare occurrence in the Hamstead Series, and the finding of two specimens in a small boring makes it probable that this horizon might turn out to be exceptionally fossiliferous, if it could be examined in Hamstead Cliff.

Other borings continue the section in a southerly direction (B.H. 12, 33, 34, 35, 36, 37, 38, 39, 40, 41, 42, 43) into lower beds, but nothing of interest is met with till we descend to 170 feet.   Here shell marls commence (B. H. 35), but it is difficult to

---

* *See* section, p. 186.

say whether they represent the White Band, or whether this band should be identified with other white marls at about 120 feet (B. II. 40). Probably the lower bed is more nearly equivalent to the conspicuous seam in the Hamstead Cliff.

At the southern border of the Forest we meet with a bed that probably corresponds with the *Nematura pupa* bed of the coast. This is a stratum full of *Cyrena semistriata* and Entomostraca (B. H. 43). It is conspicuous over a considerable area, and its position and fossils correspond so well with those of the *Nematura pupa* bed, that the local absence of the *Nematura* is counter-balanced by the other evidence. No other band of the sort occurs over the same area, and a bed, apparently on the same horizon, is full of *Nematura pupa* at Newport.

From Forest Side to Gunville the succession is carried on by other borings (B. H. 44 to 52). The first of these is in the same beds as B. II. 43, but as it commences at a higher level there must be a northerly dip of less than 1° between these points.

About 15 or 20 feet below the *Cyrena* bed lies a seam of shell marl crowded with *Melania muricata* and *Hydrobia Chasteli* This seems to correspond with the similar seam below the *Nematura pupa* bed on the coast.

About 26 feet lower lies the Black Band, first met with a few yards north of Gunville Bridge (B. H. 49), and again a quarter of a mile further south, in the village (B. H. 51, 52). The section is interesting from its exact correspondence with that seen at Hamstead :—

Section 1 chain north of Gunville Bridge (B. H. 49).

		FEET.
	Soil - - - - - - - - 1	
Hamstead Beds.	Blue and gray loam. Nodule, with casts of small univalves at 4 feet - - - - - 7	
	Lead-coloured clay. Abundant shell fragments (*Paludina lenta*) and small angular flints between 11 and 11¾ feet	3¾
	Hard black laminated clay, with shells, pieces of lignite, pyrites, and small angular flints - - -	1¾
Bembridge Marls.	Green marly clay, with much 'race' (concretions of carbonate of lime commonly found in weathered marls) and carbonaceous remains like small roots. Crushed *Paludina*	¾
	Hard green clay - - - - - - -	¼
		14½

On looking through the series of borings already referred to, it will be seen that the levels and distances have been so arranged that each boring slightly overlaps the preceding one. By this means the whole succession of strata has been penetrated, and we can construct a section of the Hamstead Series comparable with that seen in the cliff at Hamstead.

The total thickness of the Hamstead Series on the south side of Parkhurst Forest appears to be 220 feet. This calculation was made before the re-measurement of the typical locality, and it is interesting to find that it agrees thoroughly with the corrected thickness. The highest strata in Parkhurst Forest are extremely

N 2

like those immediately below the marine beds—a comparison of the measurements shows that they ought to be within 4 feet of the *Cerithium plicatum* bed.

We will take next the rest of the sections in the West Medina, none of which show continuous exposures of any great thickness of beds. Commencing with the cliffs, we encounter Hamstead Beds for more than two miles, from near Yarmouth Toll Gate to above Hamstead Ledge. Then travelling eastward we pass over a gap of two and a half miles, to Thorness Cliffs, where the Black Band again strikes the coast, much overgrown and hidden by landslips. From Thorness to Sticclett the cliff sinks, but in the higher cliff near the latter place the Black Band is well seen, overlying Bembridge Marls with the usual seam of *Melania muricata* and *Hydrobia Chasteli*. Still further to the north it may again be examined in a small outlier cut through by the cliff. In both sections the Hamstead Beds are much weathered, only the lower part being exposed.

Taking next the inland sections of the Black Band, we will give the evidence on which the division has been made on the map between the Hamstead and the Bembridge Series, commencing with the north side of the syncline.

Above Hamstead ledge the Black Band strikes inland in a south-south-easterly direction. There was formerly a large brick-yard at Lower Hamstead, but this is now overgrown. However, a boring (B.H. 276) was made in the pit near the cottage. This proved hard brown and bluish-green clays, like those 50 or 60 feet up in the Hamstead Beds. So another boring (B.H. 277) was put down on the northern shore of the creek immediately north of the brickyard. This proved beds crowded with *Cyrena semistriata* and Entomostraca. Another boring (B.H. 278) close to the shed at the Saltworks was in tough blue clay, in the upper part full of *Paludina* and *Nematura*. Unfortunately the specimens were destroyed, and it is uncertain whether the *Nematura* is the typical *N. pupa* or the other form which occurs lower down and near the Black Band. At any rate the outcrop of the Black Band lies only a few yards further north.

Eastward the base of the Hamstead Beds disappears under the wide alluvial flat north of Newtown. A boring (B.H. 275) north of Newtown Coastguard Station proved hard clays like those found at Lower Hamstead. Another boring (B.H. 274) on the southern margin of Clamerkin Lake showed the bed with *Cyrena* and Entomostraca, as in B.H. 277. A third boring (B.H. 271), further east and near Clamerkin, proved the Black Band. Beyond this point the strike changes and gradually curves to the north round Porchfield.

At Locksgreen, close to the Smithy, the *Cyrena* bed was met with (B.H. 270). A quarter of a mile south of Porchfield the Black Band was reached (B.H. 268), and two other borings (B.H. 269 and 267) also pierced the lower part of the Hamstead Series. Half a mile east of Porchfield two borings (B.H. 264 and 263) were perhaps sunk in the lower part of the same Series, but the evidence was not quite satisfactory. All the borings between Porchfield and Burnt Wood leave the age of the beds somewhat doubtful, and it is still uncertain whether the thin carbonaceous seam met with (B.II. 264, 262) represents the Black Band.

A similar uncertainty affects most of the borings near Great Thorness, but the Black Band occurs again 9 feet below the surface at the cross roads north of Whitehouse Farm (B.H. 242). A boring immediately south of Little Thorness (B.H. 228) was put down into the beds above the Black Band. In the high road near the junction of the road to Little Thorness a seam of white marl, perhaps representing the White Band, was met with (B.H. 212). It lies at about the right distance above the base and is full of *Cyrena* and *Melania fasciata.*

A quarter of a mile further east along the high road, clays with *Melania muricata* and *Melanopsis* were found in the spoil heap of a well, similar beds occurring at about the same level near Hillis Farm (B.H. 208). South of Hillis Farm a boring (B.H. 209) in the valley reached the Black Band at 16 feet, another boring (B.II. 211), at Rolls Bridge, disclosed clays with *Paludina, Melania turritissima,* and *Folliculites thalictroides.* This last may be in Bembridge Marls.

Of three borings near Whippance the highest (B.II. 220, 221) seems to have been sunk in or near the *Nematura pupa* bed, and the lowest (B.II. 222) reached the Black Band, in which again occur small angular flints.

Near Sticcelett either the strata undulate, or, as is more likely, they are slipping downward towards the sea. The highest boring (B.H. 213) commenced at 92 feet, but others at lower levels seemed still to be in Hamstead Beds. Much of the upper part of this hill is covered with gravel, through which it would be difficult to bore. At the junction of Tinker's Lane with the Gurnard Road the Black Band was again met with (B.II. 207), though the whole of the beds to a depth of 14 feet were much altered and full of selenite.

Skinners Grove Tile Works show clays with seams of *Cyrena semistriata* and *Cytheridea Mülleri.* J. Rhodes also obtained bones of Turtle and Crocodile. These beds lie probably 30 or 40 feet up in the Hamstead Series, for in the valley a quarter of a mile to the east-south-east clays like those immediately above the Black Band were reached (B.H. 200).

Two borings near Pallance, one north of the Farm (B.II. 205), and one south (B.H. 201), both reached the Black Band, but the fossils are very much decayed. North of Pallance the junction of the Hamstead and Bembridge Series soon becomes much obscured by wash from the gravel plateau, but a well near Upper Cockleton showed shelly clay, full of *Cyrena semistriata,* beneath the gravel. North of this point the gravel descends and overlaps the junction of the Hamstead and Bembridge Beds.

In the middle of the plateau, shaly clay with *Paludina angulosa* and *Melanopsis carinata* has been dug at Place Brickyard; but though from its position this clay must belong to the Hamstead

Series, there is nothing characteristic among its fossils. Another section of the beds beneath the gravel was exposed in the new well at the West Cowes waterworks, which apparently penetrated about 30 feet of Hamstead Beds, including a shaly carbonaceous clay like the Black Band ; but unfortunately few samples were preserved from the upper part of this well.

Descending towards the Medina we find another Brickyard at Werror. In this the junction of the Hamstead and Bembridge Beds is apparently shown. Above a black seam were found *Melania turritissima*, *M. Forbesii*, *Melanopsis*, *Paludina lenta*, Fish bones, and *Folliculites thalictroides* with other seeds, but the strata are so weathered that it is not easy to obtain details of the section, and it is possible that this black seam may be somewhat higher than the Black Band.

Beyond this point the base of the Hamstead Series sinks beneath the sea level, and a boring (B.H. 97) a few hundred yards farther south showed carbonaceous clays full of *Nematura pupa*, *Paludina*, and *Melanopsis*, probably the *Nematura pupa* bed.

At the West Medina Cement Works the *Nematura pupa* bed re-appears at the sea-level near Dickson's Copse, but in the pit close to the Kilns it is nearly 10 feet lower. A good section may be seen at the latter place ; and by means of boring (B.H. 93) it was carried 10 feet below the bottom of the pit and 8 feet below high-water level. It shows :—

	FEET.
Blue and yellow clay, with faint red mottling in the upper part; no mollusca observed. Turtle bones. (Seen in the pit)	25
Soft greenish clay (in B. H.)	3
Soft light-blue and yellow loams, with sand partings and selenite. Decayed *Cyrena* at 6 feet ; Entomostraca from 7 to 7½ feet	7
Lead-coloured, dark-grey, or black laminated clay, full of shells between 11½ and 12¼ feet. *Nematura pupa*, *Hydrobia Chasteli*, *Neritina tristis*, *Melania muricata*, *Cyrena semistriata*	2¼
Green loam, with sandy partings	¾
	38

The carbonaceous clays were at first taken to represent the Black Band, but there is now little doubt that they are really 20 or 30 feet higher in the series.

South of Medina Cottage clays with *Melania muricata* and *Melanopsis carinata* are seen at several places in the river bank. They are the beds immediately above the lower *Nematura* bed, for a boring (B.H. 95) at the western end of the Mill Pond shows the succession :—

	FT.	IN.
Soil	1	0
Free-cutting loam, with much selenite	4	0
Darker blue stiff clays. *Melania muricata*, *Melanopsis subulata*, *Paludina*	5	0
Seam of black clay	0	2
Green and blue clay, with sandy partings and some carbonaceous matter. *Nematura pupa*, inflated var. at 11 feet	2	10
Blacker shaly clay	0	2
Green clay, with sand partings	2	10
	16	0

At the Reservoir about 50 feet above this level there seems to be some representative of the White Band, for J. Rhodes obtained from the spoil-heap *Mya minor, Cerithium, Cyrena semistriata, Cytheridea Mülleri* and a new species of *Cytheridea* (Fig. 78.) However, no white marl like that of Hamstead Cliff is visible at this spot.

FIG. 78.
Cytheridea montosa,
Jones and
Sherborn.*

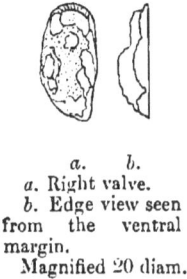

a.   b.
a. Right valve.
b. Edge view seen from the ventral margin.
Magnified 20 diam.

Having traced the base of the Hamstead Series till it has now passed out of reach beneath the sea-level, we will follow the southern margin of the syncline from Yarmouth to Newport, taking afterwards the higher beds met with here and there in the West Medina. It will be remembered that the Black Band was traced on the foreshore to within a quarter of a mile of Yarmouth Turnpike (see p. 196). In the overgrown cliff it was again found 200 yards further west, and a boring by the side of the high road (B. H. 355) reached it at a depth of 13 feet, showing that the Hamstead Beds must extend westward along this ridge to within 130 yards of Yarmouth Turnpike.

Half-way between Bouldnor and West Bouldnor, and also a quarter of a mile south-east of Bouldnor, the Black Band is again met with (B.H. 352, 348). Then the strike suddenly curves, and the Hamstead Beds extend southward in a tongue corresponding with the similar feature in the Bembridge Limestone. This curve is proved by a boring (B.H. 334) in the lane north of Lee Farm, and by another (B.H. 332) a quarter of a mile north of Freeplace, but as these only show the usual character of the Black Band there is no need to give the details.

Near Ningwood the position of the base of the Hamstead Series is exactly fixed by a series of borings, all reaching the Black Band (B.H. 325, 321, 319, 317, 314, 313). South-east of Shalfleet the boundary makes another sudden bend to the south, this time approaching the Chalk so closely that the beds come within the influence of the more violent flexure and have a high northerly dip. It is therefore often difficult to strike the exact base in a boring, though its place can be fixed within a chain of its true position.

Two borings at Stonesteps (B.H. 296 and 297), within a chain of each other, show, the one Bembridge Marls, the other free-cutting loams some distance up in the Hamstead Series. A boring (B.H. 295) on the road to Fullholding happened to strike the Black Band at 4 feet below the surface, while another (B.H. 292) close to Fullholding reached it at 16 feet, though this latter commenced at a level 30 feet lower. There must be an average northerly dip of about 3° between these points, probably the dip is much higher at the first boring and rapidly decreases

---

* Suppl. Monogr., Tert. Entom. *Pal. Soc.*, 1889.

near Fullholding. Other borings (B.H. 290 and 291), a quarter of
a mile further north, showed grey clays, with *Cyrena semistriata,
Melania muricata,* and *Cytheridea Muelleri*—probably the *Nematura*
bed—and similar beds occur in the railway cutting near North
Park, and again north-east of Great Park.

Near Alvington Farm the Hamstead Beds approach nearer to
the Chalk than anywhere else. A boring (B.H. 245), a quarter of
a mile north-west of the Farm, descended into the *Nematura* beds ;
so that the Black Band cannot be more than 27 chains from the
Downs, and the dip must be high. Due north of the farm a boring
(B.H. 244) seemed to reach the beds immediately above the Black
Band, while an adjoining one (B.H. 243) showed the green clays of
the Bembridge Series.

This brings us to the series of borings at Gunville already
described (p. 195). Passing these, the Black Band can be traced
towards Newport in several borings (B.II. 64, 66 ?, 69 ?, 70, 75),
the first of which showed small angular flints in the carbonaceous
mud. The bed of *Cyrena semistriata* and *Cytheridea Muelleri*
occurred in two borings (B.H. 60, 62) near Little Kitbridge, and
probably crops out also in the road-cutting between Newport and
Hunny Hill.

In Newport itself there are no clear sections, but the *Nematura*
bed was well represented in a boring (B.H. 92) in the siding between the Station and the river. The fossils in this boring were
exceptionally numerous and well preserved, and seem to prove
that the strata containing them lie some distance up in the Hamstead Series and are equivalent to those found at the Cement
Works. Lithologically the black clay resembles the Black Band,
and like that bed, rests on a green clay with ' race' and root-like
markings. The well at Mew's brewery (*see* Appendix, p. 305) must
also have penetrated the lower part of the Hamstead Series, but
no samples of the beds above the Bembridge Limestone were
preserved.

The only inland sections of the higher portion of Hamstead
Beds in the West Medina are borings; there are no open pits,
and no samples have been preserved of the beds passed through in
wells. During 1887 Mr. Keeping sank a pit for the British
Association Committee in Parkhurst Forest, on the hill near
Marks Corner, but only found clays that probably lie about 25
feet below the marine beds. They yielded *Paludina,* fish, and small
globose fruits.

Another pit on the Signal Hill showed mottled green clay, with
Paludina, Planorbis, Unio, Chara, and a fragment of *Emys.*
This Mr. Keeping took to correspond with the mottled bed about
15 feet below the *Corbula* beds.* As this pit is somewhat lower
than the highest boring made by the Survey (B.H. 11), which
seemed to be sunk in the clays immediately below the marine beds,
this correlation is probably right.

On the southern end of the ridge above Northwood a trial
boring (B.H. 91) below Noke Farm showed beds that seemed to .

---

* *Report Brit. Assoc.* for 1887, p. 414–423.

correspond with those on the Signal Hill (B.H. 11). It is therefore not improbable that an outlier of the marine beds may be
found higher up near the Farm, where the land is 20 feet higher.
But the exact position of the synclinal axis has not been fixed;
if it lies south of this boring the ridge will be a dip slope and
there will be no outlier.

East of the Medina the beds continue with the same character,
except that in the middle portion is developed a bed of sand.
Tracing first the lower beds, we made a series of borings (B.H.
1 to 9) between Newport and Whippingham. These show that
the strata on opposite sides of Medina exactly correspond, and that
there can be no fault down the valley. The *Nematura* beds are
found opposite the Cement Works at exactly the same level as at
the Works. Lower down the river the Black Band occurs. There
is no necessity to repeat the details of the borings.

At Whippingham the Black Band rises quickly, so that it must
cross the 100-foot contour near the village. A short distance
further north characteristic fossils of the *Nematura* beds were
found by Dr. Wilkins in a well at the Keeper's Cottage at
Osborne.* From this point eastward the base of the Hamstead
beds cannot be traced till Palmer's Brook is passed. But between
the Brook and Palmer's Farm four borings (B.H. 175, 176, 177,
178) seemed to have been sunk in the lower part of the Hamstead Series, one of them reaching the Black Band.

Half a mile to the south-west Alverstone Brick and Tile Works
deserve notice as one of the few localities where the Hamstead
Beds can be examined in an open section. The strata there
visible belong perhaps to that part of the series which overlies the
Nematura beds, but the fossils are not sufficient to settle this
point, though a boring was carried 17 feet below the bottom of
the pit. The following is the section obtained:—

Alverstone Brick Yard.

	FEET.
Blue and yellow clay, with a thin seam of shelly marl. *Paludina angulosa*, *Hydrobia Chasteli*, *Melania muricata*, *M. Forbesii*, *Melanopsis subulata*, Fish bones, and *Folliculites thalictroides*	6
Ferruginous clay and ironstone - - - - -	$\frac{3}{4}$
Laminated clay, with sand partings. *Folliculites thalictroides*, *Sequoia*, and other plant remains, *Trionyx*? - - -	5
Blue and yellow laminated clay, with selenite, becoming stiffer below. *Paludina* and *Melanopsis carinata* at 20 feet from surface - - - - - - - -	17
	28¾

The whole of the beds, except those reached in the lower part
of the boring, are much weathered. Then the Black Band is again
lost, though wells show that the Lower Hamstead beds are well

* On a newly-discovered Outlier of the Hamstead Strata, on the Osborne
Estate, Isle of Wight. *Proc. Geol. Assoc.*, vol. 1, p. 194. (1861.)

represented at Wootton. East of Wootton Creek the Black Band
re-appears in Ashlake Brickyard, the section showing the usual
weathered soil underneath it, and also the thin seam of *Melania
muricata* and *Hydrobia Chasteli*. The base of the Hamstead
Series lies unexpectedly low in this pit, and various indications
appear to show that its position is largely due to a landslip of
ancient date.

Borings in Firestone Copse did not yield any definite results,
but one about the middle of the wood (B.H. 172), and two beyond
the southern end (B.H. 168, 167) seemed to traverse the lower
parts of the Series.

East of Ashlake the boundary is again much obscured by gravel,
but about a quarter of a mile south of Binstead Lodge the
Nematura beds were well shown in a boring (B.H. 180). As this
is the most easterly point to which the *Nematura* beds have yet
been traced, it may be interesting to note that the beds remain
unaltered and contain the same assemblage of fossils as at Hamstead
Cliff. The section is :—

		FEET.
	Free-cutting loam, full of ' race ' - -	3
	Stiff dark-blue and brown clay, rather carbona-	
	ceous and with small pieces of lignite - -	8
Nematura Beds.	Bluer clay, not so carbonaceous. *Nematura pupa* and *Cyrena semistriata* - - -	3
	Blacker clay, *Nematura pupa*, *Hydrobia Chasteli*, *Neritina tristis*, *Cerithium elegans*, *Cyrena semistriata*, *Modiola Prestwichii*, *Cytheridea Muelleri*, and otolith and bones of Fish - - -	2
	Green carbonaceous clay - - -	1
		17

Similar beds, or perhaps beds a few feet higher or lower, were
found in another boring (B.H. 199) by the side of the high-road a
quarter of a mile west of Stroud Wood. Between Binstead and
Brading the Black Band has not been found, though the Hamstead
Series undoubtedly extends as far as Brading, and the *Nematura*
beds were reached in a boring (B.H. 199) at Hardingshute. No
sufficient evidence of the occurrence of Hamstead Beds has yet
been obtained at St. Helen's, but from the height of the hill
there may be an outlier of considerable size under the gravel.
Returning to Newport, we will now follow the southern margin
of the basin towards Brading. The first section of the Black
Band met with was found in a boring (B.H. 99) at the angle of
the road north of Great Pan, but the dip is there so high that
other borings a few yards away pierced quite different beds. Near
Little Pan the Black Band is again met with, and a series of
borings (B.H. 108 to 103) showed the change upwards into red
and mottled clays and then into fine sands. None of these borings
were markedly fossiliferous, but there seems to be a gap in the
series of borings just where the *Nematura* beds ought to occur.

North of Durton Farm a boring (B.H. 115) showed carbonaceous
clays belonging to the *Nematura* beds. The Black Band has not
been reached in this neighbourhood, and the dip is so high and

variable that it would need a large number of borings a few yards apart to follow it. However, the boundary on the map is correct within a chain or two.

North-west of Duxmore Farm a number of borings (B.H. 126 to 140) was made, but though most of them evidently cut the lower part of the Hamstead Series, none happened to yield characteristic fossils. Still further north a boring (B.H. 143) in the bed of the stream reached the Black Band after passing through clays with *Paludina*. A fragment of a dicotyledonous leaf was brought up by the auger from this boring. Close to Little Duxmore similar beds were found (B.H. 146), but the dip is evidently high. Strata apparently of the same age as those just mentioned occur near West Ashey (B.H. 149) and East Ashey (B.H. 150), but the only fossils obtained were *Paludina*. At the junction of the road to Nunwell with the road to Brading the Black Band was again found (B.H. 154). This brings us to the point where the dip decreases and the boundary curves to the north.

There now only remain to be described the higher portions of the Hamstead Series in the East Medina. It has already been pointed out that the White Band seems to die out east of Parkhurst Forest, and that on or about the same horizon a bed of fine sand appears in the East Medina. This sand is so useful as fixing a definite horizon in a mass of clay, and also as a water-bearing bed, that wherever it could be traced it has been laid down on the map. It seems to form an obscure feature above Cross Lane (about half a mile north-east of Newport), but is apparently thin at that place. As this feature is traced to the south-east it becomes bolder, and the springs given out along its course make a belt of wet land near Heathfield and Buckbury, but no section is visible. Between Buckbury and Little Pan the sand seems suddenly to have expanded to a thickness of about 40 feet, for three borings (B.H. 112, 113, 114) at different levels were all in this bed, and another lower down (B.H. 103) also showed trace of it. A pit at Staplers Brickyard affords the only open section of these beds in the neighbourhood. It shows alternations of loam, fine sand, and shaly clay, the only fossils being casts of freshwater shells, principally *Paludina* and *Limnæa*, and also some casts of cyprids.

The same sand bed can be traced along Long Lane, till at Longlane Shute it approaches closely to the Downs. It is evident that at this spot the sharp monoclinal curve affects all the strata up to the middle part of the Hamstead Series. The high dip, however, dies away so suddenly that the beds flatten immediately and the sand can be traced for a long distance northward with only a gentle dip. Near Blackland the sand has sunk to near the stream level, but it re-appears in the cutting at Wootton Station, and also at several points on the eastern side of the gravel ridge.

Near Briddlesford two pits have been dug in sand, the one in the hollow showing at least 10 feet of very fine white sand and sandy loam. Further east this bed forms several small outliers on the hills around Haven Street. South of Binstead, in the upper part of Stroud Wood Brickyard, more than 7 feet of fine white sand overlie red and mottled clay, and the bed is probably of considerable thickness. On the high land at Upton Mill an old pit has apparently been dug for brick-earth. A boring (B.H. 185) at the bottom of this pit showed a considerable thickness of sand, but no fossils. In another outlier, at East Ashey, the sand has been dug, and it can also be well seen in several parts of the large outlier near Brading, especially in the road cutting between Ricketshill and New Farm.

The beds overlying the sand in the East Medina only extend over the western part of the area, the marine beds being confined to a small portion of the high ridge between Wootton and Downend. Unfortunately at the time of writing there are no open sections of this part of the Hamstead Series, though wells and trial borings yielded plenty of evidence of their occurrence. At Staplers, where evidently a considerable thickness of clay lies above the sand-bed, two borings (B.H. 109, 110) were made on the top of the hill, to ascertain if any representative of the marine beds existed there. The height of this hill is nearly 300 feet, but the highest beds reached seem to be equivalent to those seen in Parkhurst Forest (B.H. 10), and at Noke Farm (B.H. 91). The thickness of the capping of gravel makes it difficult to bore at Staplers Hill, but possibly other beds a few feet higher may be represented there.

Crossing to the parallel ridge further east, we find the beds much hidden by gravel, but fortunately during the progress of the Survey a number of wells were being sunk in this neighbourhood. The most southerly of these, at some new cottages at the northern end of Little Lynn Common, showed :—

		Feet
Drift	Gravel and clay - - - - -	7
Upper Hamstead Beds.	Blue and green clays with *Cerithium* - -	12½
	Hard white seam full of *Melania inflata*, &c. -	½
		20

The fossils, though abundant, belong to few species, the following being all that could be found by J. Rhodes :—

Cytheridea Mülleri.	Hydrobia Chasteli.
	Melania inflata.
Cyrena semistriata.	
Sphærium (Cyclas).	Fish bones.
Cerithium elegans.	Crocodile (scute of).
———— plicatum.	

The fossiliferous clays evidently belong entirely to the *Cerithium plicatum* beds, though from the thickness of the strata one would expect the base of the more truly marine *Corbula* beds to be also represented. These latter very probably do occur in the upper part of this well, but so much weathered as to have the fossils entirely destroyed.

Less than a quarter of a mile north-east of Little Lynn Common another well was sunk, at Dorehill. Unfortunately this was finished and bricked before we heard of it. It showed blue shelly clay, resting on red clay, water being obtained from a running loam at 52 feet. The exact thickness of the different strata, and the depth to the base of the *Cerithium* beds could not be learnt. However the fossils found in the spoil heap seem to show that probably the base of the *Corbula* beds is also preserved. The species found were :—

Carpolithes ovulum.	Cerithium elegans.
Seeds.	————— plicatum.
	Hydrobia Chasteli.
Cytheridea Mülleri.	Melania inflata.
	Paludina (impressions).
Corbula pisum.	
————— vectensis.	
Cyrena semistriata.	

Nearly half a mile north-west of Dorehill, at Briddlesford Lodge, another well shows the beds with *Cerithium plicatum* and *Melania inflata*. The details are :—

			FEET.
Drift	- Clayey gravel - - - - -		4½
Upper	⌠ Yellow clay, much weathered - - -		5
Hamstead	⎨ Dark-blue shelly clay, full of *Cerithium plicatum*		
Beds.	⎩ and *Melania inflata* - - - -		1
Lower	⌠ Green loamy clay - - - - -		1
Hamstead	⎨ Green clay - - - - -		8½
Beds.	⎩ Green clay, with faint red mottling - -		3½
			23½

This well stands in the middle of the farm buildings, and commences at a height of 181 feet above the sea. Another well was sunk at the south-east corner of the farm at a height of 190 feet. It showed some curious bands of broken-up or reconstructed clay. The section was as follows :—

		FT.	IN.
	⌠ Mottled light-grey and dark-red clay - -	8	0
	⎮ Yellow and brown mixed clay—perhaps a recon-		
	⎮ structed shaly clay - - - -	2	0
Lower (?)	⎮ Greenish-blue clay - - - -	1	0
Hamstead	⎨ Tenaceous blue clay - - - -	6	0
Beds.	⎮ Sand parting - - - - -	0	1
	⎮ Reconstructed clay - - - -	0	11
	⎮ Mottled green and red clay, slightly carbonaceous	7	0
	⎩ Blue carbonaceous clay, full of *Unio* - -	5	0
		30	0

Though these wells lie only two chains apart, and apparently ought to penetrate the same beds, their sections are quite different. No trace of the layers with *Cerithium plicatum* could be found in the higher well, and the beds that were found are of such exceptional character as to render it uncertain to what horizon they belong. Judging by the dip of the strata shown in these wells, still higher beds ought to be found on the top of the hill immediately west of Dorchill—unless the gravel is exceptionally thick. This seems to be the only place in the East Medina where there is any likelihood of the *Ostrea callifera* beds being found, but there is too much gravel to allow of a trial boring being made.

At Wootton Station the cutting through Quarrels Copse shows :—

	FEET.
Light-blue clay, with much ' race ' and concretionary stone with casts of *Unio* - - - - - - - -	5
Mottled red and blue clay - - - -	about 20
Fine sand, with water.	
	25

Several wells in the neighbourhood also penetrate the layers immediately overlying the sand, but these beds are very sparingly fossiliferous and yield little but bones of turtle.

Many of the peculiar fossils of the Hamstead Series have already been mentioned, and it only remains to give an outline of the general character of the fauna and flora, and of the conditions under which the beds were deposited. The main mass of the Hamstead Series consists of mottled clays, probably deposited in brackish-water lagoons. These, as is usually the case with the mottled clays of the Oligocene groups, yield few fossils, except bones of Turtle and Crocodile, and drifted plants. Interbedded in the mottled clays, however, we find occasional seams of *Melania* or *Unio*, or laminated clays with plants.* .

The blue clays are much more fossiliferous, yielding abundance of shells—principally *Unio, Cyrena, Paludina, Melania, Melanopsis,* and *Nematura,* with the addition of a few more estuarine forms, such as *Cerithium, Modiola,* and *i Iya,* on certain horizons. These, with myriads of fruit of *Folliculites thalictroides* and *Carpolithes ovulum* and seams of Entomostraca are the fossils commonly met with in the Lower Hamstead Beds.

The marine bands yield a much more characteristic fauna, including a number of species quite unknown in the beds below. It must be remembered, however, that there is no real break, but that the next marine seam—that at the base of the Bembridge Marls—is fully three hundred feet lower, and its fauna is so little known that we cannot compare the two. The only marine beds that can be fairly compared are at the top of the Hamstead Series and in the middle of the Headon Series—nearly five hundred feet apart.

Among the more abundant or peculiar of the marine shells may be mentioned *Ostrea cyathula* and *O. adlata,* both confined to this horizon ; *Cytherea Lyellii, Corbula pisum, C. rectensis, Cuma Charlesworthii, Voluta Rathieri, Streblocerus* and some species of *Cerithium,* such as *C. plicatum, C. Sedgwickii, C. inornatum.*

---

* *See also* J. S. Gardner, *Report Brit. Assoc.* for 1887, p. 414.

The plants of the Hamstead Beds are little known, and only the following short provisonal list can be given :—

Andromeda reticulata, *Ett.*	Cyperites Forbesi, *Heer.*
*Arthrotaxis (Sequoia) Couttsiæ, *Heer.*	Nelumbium, sp.
Carpolithes Websteri, *Brong.*	Chara, 2 sp.
——— globulus, *Heer.*	

As far as one can judge by the character of the mollusca the temperature of the sea appears to have been very uniform during the deposition of the Oligocene beds. There is nothing in the character of the Headon or Hamstead fauna to mark the one as having lived in a colder or warmer sea than the other.

---

* During a recent visit to the Isle of Wight (in Aug. 1889) Mr. Gardner and I obtained cones of the so-called *Sequoia*, which showed clearly that here, as Mr. Gardner had already proved for the Hordwell specimens, the abundant coniferous twigs belong to the shrubby *Arthrotaxis* of Tasmania, not to the gigantic *Sequoia*. The foliage of the two is very similar, but the cones are quite different.—C.R.

CHAPTER XIII.

PLEISTOCENE AND RECENT DEPOSITS.

CLASSIFICATION.

The boundaries of these deposits have now for the first time been drawn on the one-inch map of the Isle of Wight. In the course of this examination some problems of great interest in connection with the physical history of the Island have been opened up; among them the question of the relative age of the older gravels of the south of England and of the Glacial Deposits, the age of the river valleys, and the date of the separation of the Island from the main land.

The classification of the superficial deposits presents considerable difficulty, for though the gravels of different areas indicate a similar sequence of events, yet the events in any two areas may not have been contemporaneous. The period, moreover, during which the gravels have been forming, though undoubtedly prolonged, does not seem to have been broken up by any marked changes of physical conditions, so that no classification can be proposed in which the deposits of one group shall not overlap in time those of another. Yet the position and character of the oldest gravel bring before us a picture of physical conditions so entirely different to those of the present day, that some classification by age becomes necessary.

In the first place, an important series of gravels occurs near and often on the watersheds by which the existing valleys of the Island are divided, and forms well-marked plateaus. Though we have no guide as to the relative age of the separate patches of these gravels, except the doubtful test of height above the sea, yet the similarity in their mode of occurrence justifies their being grouped together under the title of Plateau Gravels. These gravels were obviously laid down before the valleys in their present form had been excavated. Yet their distribution and the direction of the slopes on which they rest point to a drainage system bearing some relation to that which now exists.

A second group of gravels is arranged as terraces along the sides and lower parts of the valleys, and though, like the Plateau Gravels, now undergoing removal by the modern streams, yet showing an obvious connection with their valleys.

Lastly, come the alluvial and peaty deposits still in process of formation along the courses of the streams, or such as might have been formed by the existing streams.

Three principal groups may thus be established in the Superficial Deposits, capable of being arranged in chronological order. But other deposits of importance occur which cannot be placed in any one of these groups. Such is the angular flint-gravel of the Downs, which has probably been in process of formation from the time when the Chalk was first exposed to sub-aërial denudation up to the present day, and therefore runs through all three groups. But inasmuch as it provided the materials from which the Plateau Gravels were constructed, we may conveniently take its description first. The following table gives the sequence of the groups in descending order, the numbers indicating the order of their descriptions in the following pages : —

IV. Deposits now in course of formation or of recent date (Alluvium, Peat, Blown Sand, Tufa, Chalk Talus, &c.).

III. Deposits formed after the present valleys came into existence (Valley Gravels and Brick Earth).

II. Deposits formed before the present valleys existed (Plateau Gravels).

I. Deposits partly earlier than, partly contemporaneous with Groups II., III., and IV. (Angular Flint Gravel of the Chalk Downs).

I.—ANGULAR FLINT GRAVEL OF THE CHALK DOWNS.

This is a deposit of very indefinite age. It occurs on the tops of all of those Downs in which the Chalk dips at a small angle, probably because of the expanse of nearly level ground being greater than in the narrower Downs, where the dip is high. The deposit is unstratified, and closely packed with unworn flints or fragments of flints, imbedded in a loose gritty or sometimes a brown clayey matrix. In three instances near Brading, it contained a large proportion also of perfectly rounded flint-pebbles, mixed with angular flints, but probably derived from some Tertiary pebble-bed.

This deposit is no doubt of sub-aërial origin, the flints, together with a portion of the matrix, representing the insoluble residue of a great thickness of Upper Chalk. But there occur also materials in the matrix which could not have been derived from any part of the Chalk, viz., the grains of quartz and other rocks, which give the gritty character to the gravel; and also the completely rounded pebbles alluded to above. The occurrence of such materials makes it certain that other beds besides the Chalk, presumably some of Tertiary Strata, have been laid under contribution.

The thickness of rock that has been removed since this sub-aërial deposit began to form has undoubtedly been very great. The gravel not only oversteps the present limits of the Chalk-with-flints, but occurs on hills in which no beds so high even as the Middle Chalk now occur, as, for example, on St. Catherine's Hill. In such cases, the gravel seems to have been gradually lowered by the slow solution of the chalk beneath it.

If this view of its origin be correct, some portion of the gravel must date back from a time when all the strata, both Tertiary and

Secondary, extended far beyond their present limits, and must be much older than any of the other gravels in the Island. On the other hand, the formation of the gravels seems to be still proceeding (though far too slowly to admit of observation), for it is impossible to draw any hard-and-fast line between it and the gravelly soil, which is being formed on the outcrop of the Chalk-with-flints by weather and agricultural operations.

The most important patch of this gravel is that which caps the western end of St. Boniface Down, and which supplies great quantities of road-metal to Ventnor. But similar patches occur also on Stenbury and Shanklin Downs. The patch on St. Catherine's Hill is small, and interesting only from its position, far away from, and far below the flinty Chalk ; small pockets of gravel occur also here and there in the Chalk Marl at the edge of the cliff.

The extensive Downs between Calbourne, Chillerton, and Carisbrook are very generally overspread by angular gravel, the boundaries of the deposit following those of the flinty Chalk, but always overlapping them. There are many shallow gravel-pits along the southern edge of the Downs from Westover Down to near Shorwell.

The three patches above alluded to as containing many rounded pebbles occur on Mersley and Brading Downs. No sections can be seen there at the present time, but the gravel has formerly been dug to a depth of about 2 feet for road-metal, and the abundance of beach-pebbles is striking. Except in containing these pebbles, which have probably been derived from some Tertiary Bed, the patches do not seem to differ from the others that have been described.

II.—PLATEAU GRAVELS.

Their Age.

These gravels are so called from their habit of capping flat-topped hills. They occur generally as small patches, separated by deep and broad valleys, and deeply cut into by the action of springs, so as to present the sinuous outline generally found only in beds of much older date. The complete alteration which the features of the country have undergone since these gravels were laid down indicates the great antiquity of the deposits.

Though these outliers have clearly been isolated by denudation, yet they do not seem to have belonged to one continuous sheet ; for they occur at different levels. More probably they represent successive stages in the process of development of the existing system of valleys. In some cases even, the Plateau Gravels run continuously down from the highest part of a watershed nearly to the level of the Valley Gravel, thus tending to link together the two groups. In the slopes of such outliers we have evidence of the position of the lines of drainage at an early date.

This point was first noticed by Mr. Codrington,* who remarked that the high-level plains of the New Forest and the country between Poole and Southampton Water, generally covered with gravel or brick-earth, are portions of a table-land with a gradual southern slope. He further observed that the gravel covering the hills from St. George's Down to Norris "coincides with a plain having a uniform slope to the north," thus giving proof that the excavation of the Solent Valley was in progress during the deposition of the Plateau Gravels.

The great antiquity of parts of the Plateau Gravels is forcibly brought to mind when we study the vast amount of denudation that has been effected since their deposition, and the question naturally arises whether these gravels may not be in part contemporaneous with the Glacial Deposits of the north of England. This question cannot be fully answered until the mapping of the gravels on the main land is completed, but it will perhaps not be premature to point out how far the evidence in the Isle of Wight goes in support of such a supposition.

In the first place, though no organic remains have been found in the Plateau Gravels, the mammoth (*Elephas primigenius*) and and Rhinoceros have been found in the Valley Gravels, which are unmistakeably later in date.

Secondly, the amount of denudation which has taken place in the Isle of Wight, since the Plateau Gravels were laid down, is fully as great as that which the Glacial Deposits have undergone in other parts of England; the valleys which cut up the former into outliers are as broad and as deep as those which have been excavated in the Glacial Beds. To quote a single example—the gravel plateau of St. George's Down terminates southwards and westwards in a bold bank at a height of 363 feet above Ordnance Datum, or at a height of no less than 313 feet above the bottom of the valley, this amount therefore representing the depth of valley cut out since the plateau formed part of the general surface.

Thirdly, the gravels are precisely similar in their mode of occurrence, and in the amount of denudation they have undergone, to those which overspread the chalk hills on the northern side of the Thames valley, in Buckinghamshire and Hertfordshire. A part of these gravels is known to be of Glacial Age by the fact that they underlie outliers of Boulder Clay in the neighbourhood of Watford and Finchley. The others to the west are inferred to be of the same age from the similarity in their character and position.†

Lastly, the gravels and the older strata on which they immediately rest, are sometimes contorted or disturbed in a manner strongly suggestive of the action of ice. Such appearances have been seen below the older gravels only.

---

* On the Superficial Deposits of the South of Hampshire and the Isle of Wight Quart. Journ. Geol. Soc., vol. xxvi. pp. 528–551. 1870.

† They are described in the Memoir on the Geology of London, &c., by W. Whitaker. 1889. Chap .19.

St. George's Down to East Cowes and Osborne.

The sinuous outlier of gravel which spreads over the edges of the highly inclined Chalk and Greensands in this Down is one of the most remarkable in the Island, partly on account of its height above the sea and above the neighbouring valleys, and partly on account of the bold feature it presents to the south. The gravel, being thick and coarse, and having been partly cemented into a hard rock by iron oxide, forms an escarpment rivalling that of one of the older sandstones, while its even surface, slanting gently away to the north, resembles a dip-slope. On the north side, the central part of the outlier has been deeply notched by a number of springs, each forming a combe, and producing scenery of remarkable beauty. The gravel stretches away far to the north, both on the east and west sides, along the nearly level tops of ridges composed of all the rocks up to the Chalk-with-flints.

The original limits of the sheet of gravel, of which these outliers are remnants, are difficult to determine owing to the vast amount of denudation which they have undergone. On the eastern side, the boundary of the deposit may have run at the foot of the rising slope of Chalk which forms the east end of Arreton Down. On the western side, we find no corresponding feature nearer than the Down beyond Carisbrook. The gap between these two features is nearly three miles broad, and was probably the route by which the enormous masses of flint-gravel and Greensand chert of the neighbourhood of Cowes passed the Downs.

It should be remembered that the Medina valley, which follows the same general line, is of much later date. It was during the process of its excavation that the old gravels were so extensively eroded, and the features of the old valley were nearly obliterated. North of the Downs it is scarcely traceable, except by the slight eastward or westward inclination of the gravels towards the Medina. The absence of any definite limits here arises partly from denudation, but partly also from the spreading out of the gravels into wide sheets which range along and slope down towards the Solent.

The gravels rest on a plain which slopes north, as mentioned above. The amount of slope may be calculated as follows:—In the western arm of St. George's Down the level falls about 90 feet in a mile, or at the rate of 1 in 60. This arm, however, trends towards the Medina; but if a line is taken parallel to the Medina we find that the fall is less. At St. George's Down the height is about 320 feet; nearly two miles to the north it sinks to 280 feet—a fall of about 1 in 260. The Whippingham outlier continues the slope down to about 120 feet, giving a general fall of 200 feet in 6 miles, or about 1 in 160. It is noticeable that the rate of fall tends somewhat to decrease as the gravels are followed further from their source.

Taking next a parallel line about a mile further east we arrive at similar results. In the eastern arm of St. George's Down (including the Downend outlier) the level falls from 315 feet above the sea to 200 feet in 2¼ miles, or at the rate of 1 in 104. A mile and a half to the north it has fallen to 170 feet, or at the rate of 1 in 264. These measurements give a general fall of 1 in 136 in a distance of 3¾ miles. North of Palmer's Farm the outlier of Plateau Gravel trends to the east and falls rapidly in the same direction, being apparently connected with the valley now occupied by Wootton Creek and not with the valley of the Medina.

The gravel of St. George's Down is composed almost entirely of flints with a few fragments of chert and ironstone. A noticeable feature in it is the occurrence of rolled flints, a few completely rounded, and probably derived from Tertiary pebble beds, but many only partly water-worn. In this respect the Plateau Gravel differs from the Angular Gravel of the Chalk Downs, in which the flints are quite unworn.

The cementing of the gravel into blocks by a ferruginous cement has already been noticed. These blocks occur in abundance all along the southern boundary of the outlier, and are found also in several distant spots, having probably been carried off for rockeries, or building. The rain which is absorbed by the gravel naturally travels down the northerly slope, and is given off in the springs previously alluded to, but there is one spring on the south side, close to the house which is so conspicuous on the brow of the hill, known as the Dropping Well. The water oozes from a layer of cemented gravel, and is never known to fail.

A great number of pits has been opened in the outlier, the gravel being brought down from the southern and western parts by inclined planes, and from the northern parts by road to Shide. Some of the pits show upwards of 30 feet of rough stratified gravel, but the greatest thickness in the outlier is probably considerably more than this. No bones or implements have ever been found in this or any other outlier of the Plateau Gravels.

As the gravels are traced northward from St. George's Down the only noticeable change in them is that they become somewhat more water-worn, but their composition remains the same. Commencing with the outliers nearest the Downs, we find shallow pits near Staplers, which show 5 or 10 feet of gravel resting on an irregular surface of Oligocene clay. Nearer Newport two small outliers seem to fill hollows in the clay.

A mile to the north an outlier stands on Mount Misery at a height of only 170 feet above the sea. Here the clays are in constant downward movement, and continue to slip so steadily towards the Medina that the low position of the gravel may have no connection with its original height.

At Downend a brickyard exhibits the following section :—

	FEET
Reddish brick-earth with scattered chips of flint	- 15
Rough sand.	

Other parts of the pit show this brick-earth resting on the flint-gravel ; it apparently belongs to the same period, but like the gravel, is entirely devoid of fossils and appears to have been decalcified. Various other pits have at different times been opened in this outlier, but the only one at present worked is at Little Lynn Common. At the cross roads further north the gravel is said to be as much as 16 feet thick, though the usual thickness is about 7 feet.

The Whippingham and Osborne outlier occupies about two square miles, but though the gravel sometimes reaches as much as 20 feet in thickness, ridges of clay constantly rise through it, and make the working very uncertain. A good section occurs at Whippingham, and another above Norris Wood. The latter shows over 10 feet of subangular gravel, more rolled and more distinctly bedded than in the pits further south. From this sheet of gravel the water-supply of Osborne is obtained.

The Wootton outlier is similar to the one just described. A large pit about a quarter of a mile west of Wootton Lodge, shows 10 feet of worn flint and chert gravel. Another pit near the northern end of the outlier gives a section of similar gravel with numerous well-worn flint pebbles.

Parkhurst Forest to West Cowes.

West of the Medina, the gravels have the same general northerly fall, combined with a slight inclination towards the Medina. At the same distance from the Downs and from the Medina we find gravels like those near Downend, and at about the same height. The outlier in Parkhurst Forest, at the Signal House, is 260 feet above the sea; the southern end of the North-wood outlier is 213 feet and the northern end at 120 feet, giving a fall of 140 feet in 3 miles, or 1 in 113.

The outliers in Parkhurst Forest are a good deal worked, but call for no special description. The Northwood outlier is much more important. for not only is it extensively worked, but it has also yielded till lately a sufficient supply of water for Cowes. The principal pits are two near Northwood Church, both worked to a depth of 13 feet; Place Brick-yard, which shows 5 or 6 feet of gravel overlying the clay; a pit close to the cliff north-west of Northwood Park and just above the 100-foot contour; and a pit at the east end of Tinker's Lane. These all contain gravel of the ordinary character; but a pit on the north side of Ruffin's Copse, of greater interest, shows :—

	FEET.
Gravel and mottled clay, mixed - - - - -	5
Fine white sand with black specks, about - -	10
Gravel (now hidden), said to be - - - -	2
	17

A trial boring made a few hundred yards further east, for the purpose of testing the water supply, is said to have penetrated the following deposits :—

		FEET.
Gravel	- - - - - - - -	11
Sand	- - - - - - - -	20
To clay	- - - - - - - -	31

The sand crops out in Ruffin's Copse, and there yields a considerable supply of water.

The resemblance of this sand to that found in Goodwood Park, near Chichester, is so great, and the height (130 feet) coincides so exactly, that careful search was made here for marine shells. Nothing, however, could be found, the bed appearing to have been thoroughly decalcified; it has no impervious covering like that which has preserved the deposit with its shells at Goodwood.

Returning to the neighbourhood of the Downs, we find close to Gunville a mass of flint shingle at a height of 140 feet. This does not appear to have any connexion with the Oligocene or Eocene Beds, neither does it seem to belong to the ordinary Plateau Gravels. Its true position must at present be left uncertain for want of sections.

For three miles west of Gunville no gravels occur near the Downs, and denudation has been so great that the outliers near the Solent, thoroughly isolated, cannot be traced to their place of origin.

Thorness and Rew Street.

The only pit now open in the Rew Street outlier is one in its south-east corner. This, however, does not show much of the gravel, but has been opened for sand, like that three-quarters of a mile further east in Ruffin's Copse. This sand has been exposed to a depth of 12 feet, but no fossils could be found. Its height above the sea is slightly over 100 feet.

The outlier east of Great Thorness shows no section. Its height is about 130 feet. The larger outlier west of Great Thorness is worked to a depth of 15 feet, and slopes markedly to the eastward, not to the west, where the larger valley lies.

Hamstead.

The sheet of Plateau Gravel at Hamstead appears to have no connexion with the present system of drainage. At the highest point, close to Hamstead Farm, it reaches 200 feet, but in every direction except the north-west, where it is cut off by the cliff, it quickly sinks to the 100-feet contour, or even lower. This sheet is composed of partly-worn flint gravel, with many quartz pebbles and occasional blocks of greywether sandstone. Greensand chert was not observed in it.

Calbourne.

Some gravels near Calbourne seem to belong to this series, though they are probably somewhat newer than the outliers of Hamstead and Headon Hill. They range in height from 200 feet at Westover to 120 feet near Newbridge.

A small outlier caps the highest part of the hill near Norton-green, apparently unconnected with the present valleys.

Headon Hill.

Another outlier, on Headon Hill, is perhaps the most puzzling of any. It reaches a height of 390 feet, but is separated from the Downs by a deep valley, and is cut off on the west and north by sea-cliffs. The gravel is exceptionally thick, appearing sometimes to measure 30 feet. It is composed of unworn flints and sand with pieces of ironstone, but no chert or foreign rocks could be found in it.

Wootton Bridge to Ryde.

Returning to the East Medina, east of Downend, we find no trace of Plateau Gravel on the Tertiary area anywhere near the Downs. The whole of the country through which the lines of railway pass consists of low ground which has suffered great denudation in more recent times. One gap through the Downs, that through which the eastern Yar passes, is probably of ancient date, but no gravels lie in it and the continuity of the plateaus north and south of the Down is lost. It therefore only remains to describe the belt of Plateau Gravel which ranges parallel with the coast between Wootton and Bembridge.

The outlier east of Wootton Bridge consists of partly rounded flint and chert gravel, rising to a height of 170 feet towards the south, but sinking below the 100-foot contour on the north, and below 70 feet towards Ashlake. The lowness of the gravel towards Ashlake, however, may be mainly due to a landslip which has also affected the position of the Hamstead Beds.

East of the outlier just described, the character of the gravel changes in a marked manner, and the beds have all the appearance of true beach-shingle. The first pits in which this character presents itself occur close together south-west of Binstead Lodge. The Ryde outlier evidently consists of similar materials, though at present no sections of it can be seen.

Ryde and St. Helen's.

The large sheet east of Small Brook deserves special study, for the sections are curious and some of the pits may ultimately yield fossils. The southern and eastern branches of this mass show no sections, but well-worn shingle is seen in the fields. The western branch descends to within about 30 feet of the sea-level and shows fine sands like those of Ruffin's Copse. Close to Preston in a large brick-yard and gravel-pit the subjoined section may be seen :—

	FEET.
Shingle and mottled clay, contorted together ·	2 to 6
Fine sand with seams of loam and scattered flints - - - - -	9

Several other pits between this brick-yard and Oakfield show similar beds, the sand always lying below the gravel. Search was made there for fossils, but none could be found.

The large irregular outlier at St. Helen's consists also of shingle, but offers no sections, except in the cliff above Priory Woods. Unfortunately the exact heights of the outliers east of Ryde cannot be given as no contours are found on this part of the map.

Bembridge.

The last outlier to be described is the sheet of shingle between Bembridge and the Foreland. This mass, well seen in the cliffs, rests on a surface of Bembridge Marl sloping to the north-east, so that the gravel descends almost to the sea-level in that direction. To the south-west it rises rapidly, but instead of disappearing gradually it seems to abut against a steep bank of clay near Howgate Farm. At the same time the boulders become much larger, so that between the Foreland Inn and the old cliff the gravel consists of a mass of coarse flint shingle, 25 feet thick, with current-bedding dipping to the north-east. Towards Tyne Hall and East Cliff Lodge the shingle is finer and has a thickness of about 15 feet. Though this gravel consists mainly of flint pebbles, mixed with them there is a noticeable quantity of Greensand chert and sandstone, ironstone, a small proportion of greywether sandstone, and occasional pebbles of veined grit and quartz.

The shingle just described is so similar, both in position and character to that found at Selsey in Sussex, 12 miles to the east, that search was made here for the associated bed of marine shells which has yielded so large a fauna in Sussex. Unfortunately the Bembridge gravel is so full of water and slips so much over the clay that it is generally impossible to examine its bottom, and no shell bed was met with. As the shells at Selsey only occur in local patches under the shingle, some section exposed by a storm may yet show a relic of this curious marine bed in the Bembridge peninsula. This bed should be searched for whenever the base of the gravel is exposed.

So greatly do the gravels in the north-eastern portion of the Isle of Wight resemble the lower series at Brighton, Goodwood, and Selsey in position, materials, and arrangement, that they not improbably belong to the same period. The curious change the Plateau Gravels undergo when traced westward seems to point to the higher portions being sub-aërial continuations of the lower marine beds. How these angular Plateau Gravels were formed still remains uncertain.

Blake Down, Newchurch, Alverstone, and Sandown.

The features above described in St. George's Down are reproduced, but on a smaller scale and at a lower level, in the gravel patch of Blake Down and the series of patches which runs north-ward to near Blackwater.

Blake Down, forming the watershed between the Medina and the eastern Yar, and the highest ground in what has been called the Bowl of the Island, is capped with a deposit of gravel similar to, though not so thick as, that of St. George's Down. The slope of the plain on which it rests falls in this case towards the east, that is down into the valley of the Yar, and, as before, the springs break out at the lower margin of the gravel, and have cut it back into a sinuous outline.

The highest point of the gravel outlier occurs at its south end, where it is 278 feet above the sea; towards the north the plateau slants down to a level of 230 feet. But the gravel runs down two of the low ridges, which project eastwards, to a point 125 feet above the sea, and only about 20 feet above the Valley Gravel of the Yar. This is the nearest approach we get to an actual connection between the Plateau Gravels of subdivision II., and the Valley Gravels of subdivision III.

Many gravel pits are dotted over Blake Down, showing stratified flint-gravel with a few fragments of chert, and an occasional band of gritty sand. Sometimes a layer of loam 1 to 3 feet thick, lies above the gravel, but nothing that could be mapped as brick earth.

The series of outliers extending northwards from Blake Down are clearly portions of a once continuous sheet. A line drawn along their western margins forms a regular curve, and probably corresponds approximately with the original boundary of this area of gravel. But on the eastern side the sheet has been deeply eroded by the streams draining into the Blackwater. Two small patches of gravel occur on the west side of the Medina, but they lie at a lower level, contain more chert than those last described, and are probably of later date.

Excluding these two patches we find the level of the upper margin of the series of gravel outliers falling northwards from 278 feet at Blake Down to 200 feet near Blackwater, and with such regularity as to convey the impression that the gravel must have been deposited along one continuous valley. Though the present watershed between the Medina and the Yar passes right across this line of gravels, yet it is so low, being only about 25 feet above the alluvial level of the Yar, that physically the valley may be said to run on continuously, along the line indicated. We may suppose that the stream from Niton and Whitwell, which now forms the head water of the Yar, formerly continued a northerly course by Blackwater to the Medina, instead of, as now, making a sharp bend across the normal direction of drainage at Budbridge. Such alterations in the course of a river are not unknown else-where, and have generally been brought about by the eating back of one of the sources of the one river until it taps the waters of the other.

The date of the change must have lain between the deposition of the Plateau Gravels and that of the Valley Gravels. For while the former follow the original valley, the latter have been carried along the new course of the river. It may be noted that when the terraces of Valley Gravel were formed, the bed of the Yar must have been about 20 feet higher than now, that is at about the level of the watershed.

In some parts of the broad tract of Lower Greensand which runs eastwards to Sandown, the remains of an old gravel-covered plain are very striking. They occur at a fairly constant level, but there are scattered patches also at a variable height on the sides of the hills. South-west of Arreton, for example, several patches of gravel, associated with brick-earth, occur in an irregular manner on the flanks of St. George's Down. They are clearly intermediate in age between the Plateau Gravel on the hill-top, and the Valley Gravel of Horringford, and, as might have been anticipated, contain a larger proportion of Lower Greensand material than does the older gravel. The best sections are to be found in three road cuttings west-south-west of Arreton.

Near Newchurch good examples of gravel-covered plateaus may be observed. One extends through the village and along the top of the steep bank overhanging the alluvial flat, showing in its course a tendency to slope down towards the north, that is towards the valley of the Yar. Another, cut by denudation into a sinuous outline, is well exposed at Skinner's Hill, on the road from Newchurch to Borthwood, and is worked in many places for gravel. These patches, more stony than those near Arreton, are associated also with brick-earth in an irregular manner, which makes it impossible to draw a hard and fast boundary for this deposit.

The hill near Sandford is capped with a conspicuous outlier of these gravels at a height of 200 feet above the sea; and similar but very thin patches occur near Apse and Apse Heath. At Alverstone the gravel caps the top of the steep bank which bounds the modern alluvial flat, as at Newchurch.

Two more patches belonging to this same series of outliers occur on the top of the cliff between Shanklin and Sandown. In the more southern of the two, at Little Stairs Point, may be seen at different points on the cliff, sand and loam with flints, 9 feet thick; flint gravel, 12 feet thick; and loam and brick-earth 6 feet, with flint gravel 1 foot thick underneath.

Lastly a few small patches occur on the north side of the Yar between Alverstone and Yarbridge. Their mode of occurrence is precisely similar, except that the ridges on which they lie slope to the south, and more rapidly than those on the south of the Yar slope to the north.

It will be gathered from this disposition of the deposits that the lowest part of the ancient valley in which this sheet of gravel was laid down occupied about the same position as the bottom of the existing valley, and that then, as now, the ground rose rapidly to the north towards the Central Downs. Judging from their mode of occurrence, we may infer that the gravels of Blake Down, Newchurch, Alverstone, and the Sandown Cliffs were approximately contemporaneous.

Brook.

The greater part of the series of gravels and brick-earth which caps the cliff at Brook and Brixton belongs to a later group, and will be described under the head of Valley Gravels, but four small patches may be referred with more probability to the Plateau Gravels.

The Valley Gravels, it will be noticed, follow an old line of valley, which runs nearly parallel with the coast. The encroachments of the sea have removed the south side of this valley, except for a distance of about a mile between Brook and Chilton Chines, where the slight convexity of the coast leaves room for just the lower slopes of some hills which formed the south side of the old valley. The cliff section shows that the valley deposits thin away against these slopes, leaving the Wealden Beds bare, but on mounting the slopes we find another series of gravels of a different character coming on at a higher level. The section is similar to that above described, where the Plateau Gravel of Blake Down runs down nearly to the valley gravel of the Yar, leaving only a strip of bare Lower Greensand between. The difference between the two gravels at Brook consists in the comparative absence of brick-earth and stratification in the higher and older set, and especially in the peculiar contortions which appear both in the older gravel and in the Wealden Clays on which it rests. The clays have been bent and puckered, and the gravel forced into the puckers so as to occur in pockets, while the beds of loam or sand in the gravel are doubled up and bent, or dragged over towards the west. There are four places only where the cliff rises high enough to reach these older gravels, and their thickness barely reaches 8 feet. The contortions are best seen in the patches at the east and west ends respectively. As mentioned before, these contortions are regarded as probable evidence of the action of ice during the deposition of the gravels, perhaps in the form of frozen soil, or of masses imbedded in the gravels.

III.—THE VALLEY GRAVELS AND BRICK-EARTH.

Mode of Occurrence.

We have already mentioned that these deposits differ from the Plateau Gravels in having been distributed along the lower parts of the existing valleys. They were no doubt made up principally of the materials of the older gravels, redistributed after the excavation of the valleys to nearly their present depth.

They occur as terraces, often nearly level, bordering the modern Alluvium, but at a variable height, up to 50 feet, above it, and often separated from it by a steep bank. The streams having lowered their beds below the base of the gravel, the greater part of this bank is formed by rock in place, usually the Lower Greensand. This is particularly the case along the upper part of the eastern Yar, where, as may be seen on the map, a narrow

strip of Greensand nearly always intervenes between the gravel
and the Alluvium. The greater age which this difference in level
indicates, together with the difference in character, justifies the
placing of the gravels and the Alluvium in separate groups. It
will be seen also that great changes in the physical geography of
the Island have taken place since the gravels were deposited.

The Valley Gravels are most fully developed in the valleys of
the two Yars at the eastern and western ends of the Island
respectively. Those of the Medina are comparatively unimportant.

The Valley Gravels of the Eastern Yar.

The longest feeders of this river descend from Whitwell and
Niton, and from Wroxall. From near Whitwell northwards an
almost continuous terrace of gravel borders the Alluvium on one
side or the other. The gravel ranges in thickness up to 10 feet,
and is generally loose and stony, but occasionally consists in the
upper part of loam. Small pits for road metal may be seen
almost everywhere, and a good section occurs at Beacon Alley in
a road-cutting.

The gravel of this part of the valley has doubtless been derived
from the Blake Down plateau, and from the continuation of it,
which is indicated by the small patches north of Whitwell. The
terraces cease at Budbridge, and the streams which descend from
Godshill, where there are no Plateau Gravels, are entirely devoid
of gravel terraces.

The Wroxall feeder, on the other hand, draining a country in
which outliers of Plateau Gravel form a marked feature, is bordered by the most extensive gravel terrace in the Island. The
terraces near Sandford are narrow, but the gravel is well seen in
several pits. A little further north the valley widens out into a
nearly level space a mile broad, and about 1½ miles long, uniformly overspread with gravel, except in the sides of the channels
which the river and its tributaries have cut in it. This gravel has
been extensively dug at Horringford in a siding from the railway,
where the cuttings show well the irregular surface of Lower
Greensand on which it rests.

From Horringford eastwards the terraces occur on the north
side of the river only. The gravel appears repeatedly on the top
of the bank of Lower Greensand, at a height of only about 6 feet
above the Alluvium.

In the lower part of the Yar there are no terraces, but the
tributary which descends from Apse has formed a large gravel flat
near Black Pan. The gravel, dug near Ninham, and near the
high road to Sandown, contains much chert and greensand, but
has no doubt been principally formed from the old Plateau Gravel
of which patches still remain on the neighbouring hill-tops, as
previously described.

North of the Downs patches of stony brick-earth at Bembridge,
near Howgate Farm, and in the valley south-east of Sea View,

may be referred to this series, or may be considered as thick deposits of rainwash. Such local deposits of loam are common over the Tertiary area, but can seldom be mapped, as without sections they are indistinguishable from the older Tertiary clays. In the upper part of the patch at Howgate Farm Mr. Codrington found a palæolithic implement—the only one yet found in the Isle of Wight.

Wootton Creek.

There are now no sections visible in the brick-earth of this locality, and it has been found impossible to map such small patches in the absence of sections. The following account is taken from Forbes' Memoir, but, since it was written, a large bone has been dug out of the brick-earth from a well close to the Baptist Chapel at Wootton Bridge. This bone has not been satisfactorily determined. It has been described as a tusk of elephant, but its discoverer, Mr. Newbury, says it was pointed at each end.

"Along the western side of Wootton Creek, on the slope of the banks, are considerable deposits of rich umber-brown sandy clay, with scattered, small, and but slightly worn fragments of flints. This clay is of considerable thickness in places, varying from 6 and 8 to 20 or 30 feet. It shows only very slight evidence of successive deposition ; it extends to a height of 30 feet or more up the slope of the hill, and appears to be distributed in extensive patches. It ceases altogether before the lower edge of the gravels that cap the hill above is reached, the interval being occupied by Eocene clays. Patches of brick-earth occur also, though apparently more sparingly, on the eastern side of the creek ; it may be seen along the edge of the shore of the Solent at Fish-house, at the eastern angle of the creek. It is highly prized as a brick-earth, and was in requisition for the bricks used in the new fortifications at Sconce."

Medina Valley.

There is apparently little gravel or brick earth in the Medina valley, the only patches of importance lying between Newport and Shide.

At Shide the brick-earth was formerly dug, but all the pits are now closed. On the west side of St. John's Road a large pit, still worked, extends as far south as Elm Grove. The upper end of this pit was opened for sand (Lower Bagshot Sand), but the part now worked lies in brick-earth with carbonaceous seams. No fossils have been found here. At first sight this sheet of brick-earth might be expected to underlie great part of Newport, but drainage works showed Oligocene Beds so near the surface as to suggest that the loam must occupy a lateral valley extending towards Carisbrooke.

A short distance further north gravel has been dug on both sides of the Medina. The patches are interesting, inasmuch as they contain a much larger proportion of Greensand chert than is found in the plateau gravels. It seems clear that in this case the gravel is derived directly from the Greensand, and not from the plateau gravels, though the present stream with its slight fall is incapable of transporting such coarse material.

Near Coppin's Bridge loam comes on again, overlying the gravel.

The Western Yar.

The most remarkable fact in connection with the valley gravels of this tract is the entire disappearance of the river by which they were deposited. For nearly the whole of the southern side of the valley of the Yar, as well as a large part of its drainage basin, has been removed by the encroachment of the sea, so that the old river gravels have come to occupy the position of a terrace of gravel capping the sea cliff, while the small streams, which drain what is left of the basin of the old Yar, now find their way direct to the sea by deep notches or chines cut in this cliff. The evidence on which this gravel terrace is attributed to such a river was first recognised by Mr. Codrington in 1870,* and is singularly impressive.

The breach in the Chalk range at Freshwater is out of all proportion large in comparison with the stream which now occupies it. Moreover, the river gravels conclusively prove the valley to have once formed the channel of a river comparable in size to the Medina, or eastern Yar. The distribution of these gravels further shows that this river, like the others, flowed from south to north, draining lands which, lying to the south of the Chalk range, have since been washed away. We may further assume that some of the sources of the river lay in the direction of St. Catherine's Down, in the area which has formed the principal watershed of the Island from a very early period.

The gravels at Brook occur in the line which the old river might have been expected to take, and at such a height above those of Freshwater Gate, as would be required to allow a gradient for the stream. When we add to this that the gravels and brick-earths bear every appearance in themselves of being old river deposits, there is left no room for doubt that they mark the course of the old Yar.

The occurrence of teeth of *Elephas primigenius* in these gravels at Freshwater has long been known ; remains of the same animal have been recorded also from Brook Chine and Grange Chine by Mr. Codrington (*op. cit.*, p. 539).

The continuous section afforded by the cliff gives unusual opportunities for examining these gravels. In describing the section, we will commence in the upper part of the valley and proceed westwards to Freshwater.

Gravel first makes its appearance on the top of the cliff between Blackgang and Atherfield. It is seen as a band 2 to 4 feet thick underlying a considerable depth of alluvial deposits and blown sand (*see* p. 234), and is composed principally of chert. It may

---

be contemporaneous with the far thicker deposits about to be described.

But the principal deposit consists of brick-earth resting on stratified flint-gravel and sand. It commences at Grange Chine, the easternmost patch being on the east side of the chine, near Brixton Mill. On the west side of the chine, a slip shows brick-earth, 5 feet thick, resting on 3 feet of gravel, and in the field close by is a shallow pit from which bricks were made for the viaduct of the Military Road. These deposits seem to have been laid down by the stream which now runs in Grange Chine, at the point where it joined the Yar, for at the cliff close by they spread themselves westwards, and attain a great thickness. Remains of *Elephas primigenius* have been observed at a point 100 yards east of Grange Chine at 60 or 70 feet above the sea (Codrington, *op. cit.*).

The sections seen in the cliff between Grange Chine and Chilton Chine are as follows:—

400 *yards west of the Stream of Grange Chine.*

	FEET.
Brick-earth	4
Gravel	4—5
Loam, dark and clayey in parts, with bands of flint gravel, containing some ferruginous sandstone	18
	27

250 *yards west of the preceding Section.*

	FEET.
Brick-earth	2
Gravel and loam	7
Blue silt and clay, with fragments of wood	4
Gravel and sand	6
	19

Near Chilton Chine.

	FEET.
Brick-earth, thinning away near the chine	6
Gravel	8
	14

Four hundred yards to the west of Chilton Chine the cliff rises a little in height, and is bare of gravel for a distance of 300 yards. This slight rise, like those referred to in the description of the Plateau Gravels (p. 218), evidently formed the foot of the slopes which enclosed the Yar valley on the south. In observing the thinning away of the river deposits against the slope it will be noticed that the brick-earth passes beyond the limits of the gravel, so as to rest directly on the Wealden Beds, before it also thins out.

There are likewise variations in the thickness of brick-earth due to erosion, for the small stream which now follows the old valley has cut out its smaller valley in the old deposits of the larger one. The sand and gravel beneath are fairly constant in thickness. The following sections were noted :—

On the west Side of Chilton Chine.

		FEET.
Brick-earth - - - - - - -		0—4
Gravel - - - - - - . - -		8
		12

Half-a-mile west of Chilton Chine.

		FEET.
Red and yellow loam - - - - -		2
Do. with flints - - - -		1
Sand - - - - - - -		6
Gravel - - - - - - -		4
		13

We now reach the parts of the cliff which were described on p. 220, as being capped with Plateau Gravel. The Valley Gravel, it will be noticed, runs to the edge of the cliff between the low hills on which the Plateau Gravels rest, so that the relations of the two can be conveniently studied. Remains of *Elephas primigenius* have been recorded from a point half a mile east of Brook Chine, about 96 feet above the sea.[*] Apparently they must have occurred in what has been described as Plateau Gravel, but the point is uncertain.

On the east side of Brook Chine gravelly loam, 6 to 8 feet thick, rests on 4 feet of well-bedded sand and gravel ; but at the chine, and for a few yards west of it, the gravel has been re-arranged and will be described among the more recent deposits (p. 231, Hazel-nut Gravels).

At Hanover Point the Valley Gravels thin away against a slope of Weald Clay rising to the south, as near Chilton. On the east side of the point the following section was noted :—

		FEET.
Brick-earth - - - - -		8
Bright buff sand - - - - -		4
Grey sand, with some gravel - - - -		4
		16

At Shippard's Chine the Hazel-nut Gravels re-appear, but 200 yards to the north-west of the chine we find the following section :—

		FEET.
Gravel made up of ferruginous sandstone (recent) -		1–2
Brick-earth - - - - - -		4–6
Gravel and sand - - - - -		8–10

---

[*] Codrington, *Quart. Journ. Geol. Soc.*, vol. xxvi. p. 539.

200 *yards north-west of the preceding Section.*

	FEET.
Brick-earth - - - - - - -	3–4
Laminated sand and loam - - - - -	12

Lastly, in a small chine, 350 yards north-west of Shippard's Chine, we are presented with the section illustrated by the accompanying woodcut.

FIG. 79.

Section in Valley Gravels at the east end of Compton Bay.

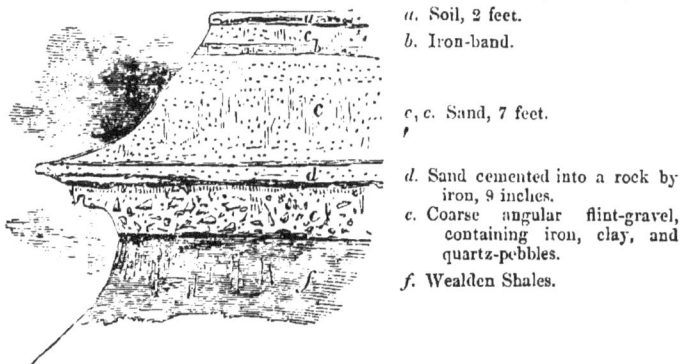

a. Soil, 2 feet.

b. Iron-band.

c, c. Sand, 7 feet.

d. Sand cemented into a rock by iron, 9 inches.

e. Coarse angular flint-gravel, containing iron, clay, and quartz-pebbles.

f. Wealden Shales.

This is the last section in the Valley Deposits, for 50 yards further on they thin away against the rising slopes of Afton Down, and do not touch the coast again till we reach Freshwater Gate.

At Freshwater Gate the cliff cuts across the old valley at right angles, giving a clear section of all the river deposits, except the modern Alluvium which lies at and below the sea-level. The section has long been noted for the finding of two teeth of *Elephas primigenius* in the gravel as described in detail by Mr. Godwin Austen.*

On the west side of the valley (Fig. 80) the lower part of the gravel is composed of large partly worn flints, with chert and ironstone, and is stained and partly cemented by iron-oxide. Above this rock a grey stratified chalky loam overlaps the flint-gravel, and runs up the slopes of chalk above it, much as a rain-wash would do. Nearer the middle of the valley this chalky loam is overlain with brown loam and brick-earth, but, still lower down, thins out, leaving the brown loam resting on the flint-gravel. The section now exposed at the Bath House shows—

	FEET.
Brown loam - - - - - - -	3
Flint gravel, with a few bands of sand or grit -	about 20

* See Geological Survey Memoir on the Tertiary Fluvio-Marine Formation of the Isle of Wight, p. 2. (1852.)

FIG. 80.

Freshwater Bay from the East. From a Sketch by Prof. E. Forbes.

On the east side of the valley, of which a view is given in Fig. 15, p. 74, a thin spread of flint-gravel and chalky loam occupies the top of the cliff for a considerable distance, and forms a small outlier, now rapidly crumbling away, on the sea-stack known as the Stag Rock. These deposits rapidly thicken into the valley, where behind the new esplanade the subjoined section may be seen :—

	FEET.
Soil	1–2
Flint gravel	1–2
Lenticular mass of stratified chalky loam, with fragments of flints	0–6
Flint gravel	4+

The lower beds of flint gravel, on the two sides of the valley, have probably been derived from older gravels that once lay on lands to the south, since washed away. The flint fragments in the upper part have a fresher and less water-worn appearance, and have probably been washed out of the chalk of the Fresh-water Downs. No fragments of chalk, it will be noticed, occur in the lower or far-derived flint gravel, the wear and tear of transport having been too great for their survival. In the upper beds on the east side of the valley Mr. Godwin Austen observed considerable numbers of *Pupa muscorum* and *Succinea oblonga*, the latter now extinct in the Isle of Wight.

"The Elephant remains found at Freshwater consist of two molar teeth, of which the first was met with on the west side of

the valley, in a excavation on the site of the lower hotel, and
where the specimen is now preserved; the other was procured
from the beds on the east side."[*]

North of the gap through the Downs the Gravels have not
yielded fossils, though they form sheets of considerable extent.
From the scarcity of sections it is also difficult to say whether these
deposits belong to one period or mark successive stages in the
denudation of the valley.

In the sheet of gravel which extends to Freshwater Bay a pit
has been opened at Easton at a height of about 50 feet above
the Alluvium, but the gravel slopes continue down to the Marsh.
On the opposite side of the Yar the gravel occupies a plateau
from 30 to 50 feet above the sea, and a pit shows 25 feet of coarse
gravel resting on Bagshot Sands. In Afton Park a large pit was
opened to supply ballast during the construction of the railway.
It showed about 6 feet of gravel, resting in one place on shelly
clay—probably Barton Clay—but the gravel itself yielded no
fossils. The sheets further north show no sections, and are
only interesting as fringing the present estuary.

IV.—BEDS NOW FORMING, OR OF RECENT DATE.

In this group we include Alluvium, Peat, Blown Sand, Chalk
Talus, Tufa, &c. Chronological arrangement being impossible
among such beds, the Alluvial Deposits will be taken in the
geographical order of the streams with which they are associated.

ALLUVIUM AND PEAT.

a. *The Western Yar, and the Coast from Freshwater to Yarmouth.*

The small stream which now follows the old valley of the Yar
takes its rise at Freshwater Gate in a spring known as the Rise
of Yar, situated on the eastern edge of the Alluvium at a
distance of 200 yards from high-water mark. Though fresh, this
spring ebbs and flows coincidently with the tide. In dry weather
it ceases to flow soon after the tide begins to fall.

The Alluvium, consisting of peat, silt, and marsh clay, extends
continuously southwards to the foreshore, where, however, it is
almost always covered with sand and shingle. In digging a
foundation for the sea-wall, this peaty deposit was excavated to a
depth of 10 feet without the bottom being reached, and was
found to be abundantly charged with fresh water. The ponding
back of this water by the rising tide is probably the cause of the
spring alluded to above.

The tide flows up the Yar as far as Freshwater, where it is
stopped by a dam. Formerly the whole of the marsh must have
been part of the estuary, for shells of the common cockle occur
abundantly just below the peat opposite Afton House.

---

[*] The Tertiary Fluvio-Marine Formation, &c., p. 5.

A deposit of tufa and tufaceous marl lying on the top of the cliff at Widdick Chine has attracted a good deal of attention. This tufa is a deposit from the springs given out by the Headon Limestone immediately above. There is nothing to point to its being of any great antiquity, for the stoppage of the springs is merely due to the recession of the cliff, by which they have been tapped at another point. The section is now almost entirely overgrown. These deposits were first noticed and described by the late Mr. Joshua Trimmer (*Quart. Journ. Geol. Soc.*, vol. x. p. 53 (1854)), and were subsequently referred to in greater detail by Professor Forbes ("Memoir on the Tertiary Fluvio-marine Formation," p. 8), and in notes by Mr. Bristow appended to his Memoir. When the first edition of this Memoir was published this deposit could be seen to occupy the upper part of the cliff in Totland Bay for a distance of nearly 350 yards, at about 60 feet above the sea. On the top (Fig. 81) lay an unequal thickness of brown loam, containing

FIG. 81.

Tufaceous deposit of Totland Bay.

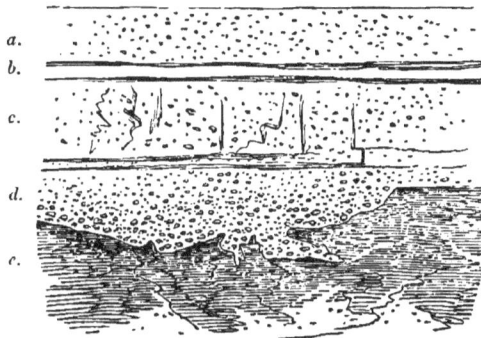

a. Ferruginous brown sandy loam.
b. Brown clay and perished shells.
c. Fine tufa.
d. Coarser tufa.
e. Potamomya sands of the Upper Headon Beds.

a few scattered angular flints, beneath which was a layer of brown clay and decayed shells, resting on four or five feet of calcareous tufa (with a few black lines derived from decomposed vegetable matter), sometimes equalling fluvio-marine limnæan limestone in hardness. This tufa was finest in the upper part, and became gradually coarser towards the bottom, where it was full of round calcareous concretions of various sizes, and of what seemed to be the twigs and stems of plants, which having fallen into water highly charged with carbonate of lime became incrusted with it. The concentric concretions were largest at the base of the deposit, and decreased in size in an upward direction, the whole deposit resting on an uneven surface of the Potamomya sands, which underlie the limnæan limestone of Totland Bay. Occasionally a layer of small angular flints intervened between the tufa and the sands.

Helix nemoralis, H. rotundata, Cyclostoma elegans, with occasional *Bulimus lubricus* and *Pupa muscorum* are the most abundant land-shells, and occur throughout; in the loam are *Succinea* and *Limnæa,* and in the lower part a small *Planorbis* and fragments of *Unio.* In addition to the above, the following shells were noticed by Prof. E. Forbes, viz., *Helix arbustorum* (or *nemoralis*), *H. pulchella, H. ericetorum, H. cellaria, H. hispida, H. hortensis, Achatina acicula, Clausilia, Pisidium, Limnæa palustris, Succinea oblonga, Cyclas,* &c.

The only other deposit of similar character is a small patch of shelly tufa immediately below the limestone a quarter of a mile further east. This tufa is seen in the road cutting east of York's Farm, but occupies so small an area that it cannot be placed on the map.

b. *The Coast from Freshwater to Blackgang.*

It has been previously explained that the streams which now empty themselves into the sea between Freshwater and Blackgang have once been tributaries of the old river Yar. In consequence of the encroachment of the sea by which the river was intercepted, some curious anomalies have been brought about in the position of the alluvial deposits.

It will be noticed that a long strip of Alluvium which commences near Chilton Chine, only 50 yards from the edge of the cliff, winds away westwards parallel to the coast, catching a little land drainage in its course. At Brook it passes out to the edge of the cliff, and the water from it, cutting through the Alluvium and deep into the Wealden Beds, escapes by the chine so formed to the sea. But a few yards west of Brook Chine another strip of Alluvium appears on the top of the cliff, and, winding round Hanover Point, passes out to the cliff again at Shippard's Chine. This latter isolated strip is, without much doubt, the continuation of the other which runs westward from near Chilton Chine. The separation of the two strips has resulted from a comparatively recent encroachment of the sea in Brook Bay.

The alluvial tract follows the centre of the Valley Deposits of the old Yar, coinciding in position with what must have been the course of that river. That any part of the Alluvium dates back to the time when this river ran through the Freshwater Valley is hardly probable. But it was probably deposited by a diminished representative of the old Yar, gathering the drainage of Brook, Chilton, and still earlier of Brixton and Shorwell, and falling into the sea somewhere a little further south and west than Shippard's Chine.

The section of this Alluvium at Shippard's Chine has long been noted for the occurrence in it of timber and the shells of nuts. These were first noticed by Mr. Webster, who described them as follows:—

" It was near to this place, that I had been informed, fossil fruits had been found in great abundance, and which were regularly called in the island, Noah's nuts. . . . Near the top of this cliff lie numerous trunks of trees, which, however, were not lodged in the undisturbed strata, but buried eight or ten feet deep under sand and gravel. Many of them were a foot or two in diameter, and ten or twelve feet in length. Their substance was very soft, but their forms and the ligneous fibre were quite distinct: round them were considerable quantities of small nuts, that appeared similar to those of the hazel. None of the wood nor fruits were at all mineralised. . . . No hazel whatever now grows upon the island. . . . Pieces are sometimes found so fresh as to bear being worked into furniture."*

FIG. 82.

Sketch of Gravels with Hazel Nuts in Shippard's Chine.

	Inches.		
a. Ferruginous loam	- 6	f.	Angular flint gravel, hardening into conglomerate.
b. Black clay	- 6	g.	Coarse sand, with fragments of fine sandstone, nuts, twigs, branches, &c.
c. Pale ferruginous clay	- 6		
d. Black carbonaceous clay	- 6	h.	Red mottled clay of the Wealden.

The sketch forming Fig. 82 was made in the southern side of Shippard's Chine in June 1856. The upper two feet consisted of black peaty clay and ferruginous pale clay, overlying ferruginous loam, which rested on angular flint gravel, sometimes hardening into conglomerate, beneath which was a coarse sand enclosing fragments of fine sandstone. This sand, based upon

---

* Sir H. Englefield's Isle of Wight, p. 152.

red mottled Wealden clay, contained numerous shells of nuts, and the remains of beetles mixed with matted fragments of the twigs and branches of trees. The latter, which were sometimes coated with phosphate of iron, retained their original shapes and general appearance, and were saturated with water, which on evaporation left a light shrivelled substance behind. The largest fragments did not exceed two or three inches in diameter.

In more recent years a causeway has been made on the north side of the chine, and in the approach to it the following beds have been cut through :—

	Ft.	In.
Brick earth, a reddish loam - - - - -	6	0
Grey silt, with much soft and blackened wood and bark, and black, brittle nut-shells	0	6
Hard cemented gravel	2	6
Dark earth, with much wood, as above - - -	0	6
Gravel - - - - . . -	1	0
Vegetable layer, not continuous . . . -	0	2
Gravel - . . . . . -	2	0
Wealden Clay	—	
	12	8

On the opposite side of the cutting a still more recent alluvial peat and rootlet bed, about 18 inches thick, lies above the brick-earth of this section, probably the black peaty clay seen in 1856.

On the west side of Brook Chine also there occurs a peaty layer in gravels of the same age as those at Shippard's Chine, and probably once continuous with them, as previously mentioned. A large tree trunk is to be seen sticking out of the bed in an inaccessible position near the top of the cliff.

It has already been explained that the gravels in which these vegetable remains occur are later than the Valley Gravels of Group III., which cap the neighbouring cliffs. The newer series was no doubt made up from the washing of the older, and it is difficult to draw a hard line dividing the gravels of the two ages. The later or "hazel-nut gravels" clearly form part of the alluvial deposit which commences near Chilton Chine (p. 230).

The stream, which has cut out the great ravine known as Grange Chine, is fed by the two powerful springs of Bottlehole Well and Shorwell. The alluvial flat of the former consists of peat where the stream runs over the Lower Greensand, that of Shorwell of silt, sand, and fine gravel. The chine begins where the two streams join at Brixton, and has been of course cut through the Alluvial Deposits deep into the variegated beds of the Wealden series.

The water, which enters the sea by way of Shepherd's Chine (Cowleaze Chine on the former edition of the one-inch map), is principally derived from springs issuing at the foot of the escarpment which we described on p. 44 as running past Pyle and Kingston. The springs being highly charged with iron, the alluvial flat at Atherfield contains much ochre ; the broad flat west of Corve is peaty. The stream meanders through

Little Atherfield bordered by a narrow alluvial flat, which however in the area underlain by clay (the Atherfield Clay and Wealden Beds) widens out, and becomes indefinitely bounded.

The chine commences at Combtonfield as a small notch, but slants down towards the sea so as to gain a depth of about 90 feet at the sea-cliff. The chine being cut along the middle of the alluvial flat, gives a section along both its banks of the alluvial deposits, which have thus come to occupy the curious position of being 90 feet above the stream which formed them.

The mouth of the chine up to the year 1810, when the old edition of the Ordnance Map was published, was situated 350 yards further north than its present position. Before Fitton visited the spot a change had taken place which he thus describes. The streamlet " was very tortuous near the shore, and formerly came close to the edge of the cliff near its present outlet, but made its way to the beach at Cowleaze; till the soft and narrow barrier at top having been cut through, the water soon deepened the chasm, and formed a new chine, leaving its previous bed, with Cowleaze Chine itself, deserted and dry."*

The change is reported to have been hastened at the last by a shepherd having dug through the narrow barrier of shale, whence the name of Shepherd's Chine for the new mouth. The old ravine of the stream remains much as it was, except that the sides are overgrown. It runs near, and roughly parallel to the sea-cliff, and is separated from it by a long and narrow but flat-topped ridge, capped with two small outliers of Alluvium; a remarkable position in which to find remains of such a deposit. The stream has greatly deepened the new chine since it gained an exit by the shorter route,—a result which followed naturally from the temporary steepening of the gradient, and the consequent temporary increase in the rate of erosion. The case is precisely analogous to those of Brook Chine and Shippard's Chine described on p. 230.

The following sections in the Alluvium were noted:—

On the south side of Shepherd's Chine.

	FEET.
Loam	2-5
Gravel and sand	2-6

On the north side of Shepherd's Chine, near Chine.

	FT. IN.
Sandy loam	2 0
Flint gravel	2 6
Grey loam and grit, with many small fragments of stems and nut-shells	1 6
Flint gravel, with many fragments of Wealden Shales, and with fragments of wood	4 0
	10 0

* On the Strata below the Chalk. *Trans. Geol. Soc.*, Ser. 2, vol. iv. p. 197. 1836 (read 1827).

In an outlier between the cliff and the old course of the stream.

	Ft.	In.
Brown loam -	2	0
Light blue silt	2	0
Grey silt, with stones -	0	8
Flint gravel -	1	6
Gravel, chiefly of fragments of Wealden Beds	3	6
	9	8

The mode of occurrence of this deposit leads to the inference that it is of the same age as the Alluvium at Shippard's Chine, where also nut-shells are imbedded.

Whale Chine forms the outlet for a small stream taking its rise in the western slopes of St. Catherine's Down. The sides of this extremely precipitous ravine are capped, like those of Shepherd's Chine, with an alluvial deposit, consisting of loamy beds above, and gravelly beds below, the majority of the stones in the latter being chert and ferruginous sandstone. The subjoined section may be seen at the top of the cliff, on the north side of the chine :—

	Ft.	In.
Loam -	9	0
Black peaty seam	0	3–4
Grey silt, with bands of chert gravel below -	4	0
Chert gravel -	4	0
	17	4

On the north-east side of the Military Road, the chert gravel comes to the surface, and has been dug for road-metal. On the south side of the chine it is overspread by Blown Sand, which will be described subsequently, but the gravel can be traced beneath this covering in the face of the cliff for about three-quarters of a mile, rising south-eastwards from about 145 feet above the sea at Whale Chine to about 200 feet at Walpen Chine. The following sections were noted in it :—

At Ladder Chine (see also p. 237).

	Feet.
Blown sand, variable -	6–15
Yellow loam -	2
Chert gravel -	3

South side of Walpen Chine.

	Feet.
Blown sand, grey -	15–20
Do. brown -	5–10
Coarse angular chert gravel, resting on slightly bent beds of Lower Greensand -	3

100 yards south of Walpen Chine.

	FEET.
Blown sand, with fragments of shale and a few small stones	15
Blown sand, brown	3
Grey silt	1
Peaty layer	$\frac{1}{4}$
Ochry layer and silt	$\frac{1}{2}$
Grey silt	2
Chert gravel	2

The last section visible in the undercliff formed by the thick clay which lies next below the Sandrock Series (p. 30), exposes the following strata :—

	FEET.
Blown sand	6–8
Yellow loam	4–6
Chert gravel	0–2

South of this undercliff, the Blown Sand rests directly on the rock.

This large spread of gravel is clearly not the product of the small stream of Whale Chine, or of the still smaller one of Walpen Chine, but may perhaps have been deposited by the upper waters of the old Yar, of which the present streamlets were tributaries.

c. *The Medina.*

The Alluvium of the River Medina commences at Chale Green, and forms a long strip of marsh land, gradually widening to about 200 yards in the part known as the Wilderness and near Gatcombe, but narrowing down as it passes the projecting spur of Upper Cretaceous Rocks of Gossard Hill, and those of the central range of the Island. The alluvial deposits are generally marsh-clay and silt, with a black peaty soil on top.

On the other hand the Alluvium of the tributary which joins the Medina at Blackwater is principally peat, as perhaps the name indicates ; its boundaries on the low watershed near Merston are extremely indefinite, as described on p. 218. Below Newport the Alluvium consists of estuarine clay and silt.

d. *The Eastern Yar.*

The Alluvium of the two longest feeders of this river, namely, those which descend from Whitwell and Wroxall, consists superficially of a narrow strip of marsh-clay spread over the bottom of a shallow trough cut through the Valley Gravels into the Lower Greensand. The alluvial flat is bounded for some miles by a low bank of Greensand with a thin covering of gravel. But the streams which rise on the north side of Godshill, and join the river above Horringford, drain some extensive peaty flats and are bordered by peaty land, until they join the Yar. The development of peat has resulted from the form of the ground and the issue of the springs which mark the outcrop of a clayey bed in the Lower Greensand, as described on p. 45.

Below Newchurch the alluvial flat is bounded by steep banks of ferruginous sand (Lower Greensand), and is extremely irregular in its boundaries, the river in its wanderings having undermined first one bank then the other. The soil is of the usual dark character, but there is no great thickness of peat.

At Sandown the river must have been formerly joined by an important tributary, for the alluvial flat, known as Sandown Level, which branches off to the south, is at least as broad as that of the main river. This tributary Alluvium runs only half a mile before it is cut off abruptly by the sea, so that nearly the whole of the basin of the river which formed it has disappeared. The streams of Shanklin and Luccomb Chines were probably some of the head waters of the river, and a little patch of gravel on the south side of Shanklin Chine may have formed part of its valley deposits. The tract of land on which Yaverland and Bembridge are situated is isolated from the rest of the Island by this alluvial flat and that of the Yar, and would be literally an island at high tide in certain winds, but for the artificial bank along the seaward margin of Sandown Level. It corresponds curiously to the "Isle of Freshwater" at the opposite extremity of the Isle of Wight.

Brading Harbour was continually inundated at high water until the end of February 1880, when the sea was finally shut out by the present permanent embankment, which encloses an area of 600 acres. Sir Hugh Middleton, in the time of James I., employed a number of Dutchmen to recover it from the sea by embankments. 7,000l. were expended in the work; but, partly by the badness of the soil, which proved a barren sand, partly by the choking of the drains for the fresh water, by the weeds and mud brought by the sea, but chiefly by a furious tide which made a breach in the bank, they were obliged to desist, and put a stop to their expensive project (See Pennant's Isle of Wight, vol. ii. p. 149).

Near Lane End, Bembridge, a hollow in the older gravel contains a newer peat and gravel. It was impossible to separate the two gravels on the map and no determinable fossils were observed in the peat, but these deposits seem to be merely the Alluvium of the small stream which now flows through Lane End.

The alluvial deposits of the smaller streams that flow into the Solent consist of marsh-clays with trunks of trees, but in the absence of clear sections there is little to be said about them. It may be pointed out, however, that the Alluvium of all the streams descends far below their present beds. Though we have no means of telling the full depth, yet judging by analogy, we should expect that the old channels of the larger streams have been cut fully 40 feet deeper than their present ones, as is the case in most parts of England. This indicates that their excavation dates back to a period when the land stood at a considerably higher level.

BLOWN SAND.

The largest area of Blown Sand in the Isle of Wight is to be found on the top of the vertical cliff between Atherfield and Chale, at a height of 150 to 250 feet above the sea. The sand is blown up from the face of the cliff, not from the beach below, and consists merely of disintegrated Lower Greensand. Several sections in it have been noted above in describing the gravel below it (p. 234); the greatest thickness of it seen was about 20 feet, but it probably exceeds this in parts of the line of dunes which it forms along the edge of the cliff. It extends also for some hundreds of yards inland in the form of a thin covering of dusty sand. The most westerly patch of this sand lies on the outcrop of a bed of iron-sand, and contains vast quantities of spherical grains of iron-oxide derived from it.

On either side of Ladder Chine the sand is piled up in small hummocks or dunes, and, if we descend into the chine, the source of the sand becomes sufficiently obvious. The chine appears to have commenced its existence as a small notch cut by the surface-drainage from the adjoining fields. The wind, especially that from the south-west, entering the notch has gradually widened it out into a beautifully symmetrical amphitheatre, leaving the harder beds and concretions standing out in tiers of benches, but whirling every loose particle of sand up over the top of the cliff. The chine thus provides an interesting illustration of wind-erosion, comparable on a small scale to the scenery of parts of the desert region of Western America.*

Very small spits, consisting partly of blown sand, extend half way across the alluvial flats of the western Yar and of the Newtown estuary. At the mouth of the eastern Yar a more extensive tract of Blown Sand rises here and there into small dunes, used for the Golf Links, and serves to protect Bembridge Harbour on the north-east side. The sand travels in all cases from west to east.

CHALK TALUS.

At the foot of the slopes of the chalk hills a gravelly detritus of chalk has accumulated to a considerable thickness. It is well seen in Compton Bay, where the steepest part of the cliff in which the Upper Greensand crops out is formed by a stratified chalk talus, or rain-wash, from the slopes of Afton Down. The deposit here reaches a thickness of 20 feet, and is compact enough to stand in a vertical cliff. The second exposure is seen in the road-cutting between Brixton and Calbourne, where the talus has spread itself over the Upper Greensand, and become hardened. The third occurs on St. Catherine's Hill, on the summit of Gore Cliff. In this locality the deposit consists of hard calcareous mud, attaining a thickness of about 9 feet, and becoming harder and

---

* As was remarked to the writer by Mr. G. K. Gilbert, of the United States Geological Survey, during an excursion to this locality.

darker towards the lower part. It contains numerous existing land-shells, among which are *Helix aspersa, H. nemoralis, H. ericetorum, H. virgata, H. rotundata, Bulimus lubricus,* &c.*
It rests on the northern slopes of a small outlier of the Chalk Marl, but extends a few yards beyond the boundary of the Chalk, so as to touch the Upper Greensand. It is made up almost entirely of small fragments of Chalk and Chalk mud, but contains a little Upper Greensand, and a very few fragments of chert. It is clearly a rain-wash from the slopes of a hill of Chalk, which must have once existed to the south, but of which the small outlier is the only surviving fragment. The remainder of the hill has slipped down to various positions in the Undercliff, one of the most striking features of which is the great slices of Chalk and Upper Greensand, still retaining their relative positions.

The inland limits of the deposit are altogether indefinite, but presumably tend to follow the boundary of the Chalk, though slightly overlapping it as in the cliff. Similar deposits would probably be seen along the greater part of the base line of the Chalk, were there any sections to show them. Agriculturally they are important, for they produce a chalk-soil over the outcrop of the Upper Greensand. In the same way the guttering down of the Gault, described on p. 58, has spread a clay-soil over the outcrop of the Carstone, and part of the Sandrock Series.

---

* *Helix aperta* also appeared in the list in the 1st edition of the Memoir. But as the authority is not forthcoming, and the occurrence of this continental shell is improbable, it is now omitted.

CHAPTER XIV.

DISTURBANCES AND FAULTS.

Of the movements of the strata which produced the almost unique geological features of the Isle of Wight, the most marked was that which brought the Chalk up in a nearly vertical position in the central range. The fold of the strata thereby effected is found, however, on close examination to consist of two separate anticlinal axes, the one dying out as the other increases; while other lines of lesser disturbance run nearly parallel, each having its influence on the structure of the Island.

Before describing in detail the various folds observable in the Isle of Wight we will briefly notice the great series of nearly parallel anticlinal and synclinal axes of the south and south-east of England, of which they form part. These axes, taken in order from north to south, are as follow :—

1st. The great syncline of the London Basin, which extends from Marlborough in the west, and is lost under the German Ocean to the east.

2nd. The great anticline of the Weald of Kent, which commences in the west as two separate anticlines, the one near Devizes, the other near Petersfield, passes under the English Channel, and terminates about 14 miles east of Boulogne.

3rdly. The syncline of Chichester, which passes north of Portsdown to the sea near Worthing, and eastwards along the coast by Brighton.

4thly. The anticline of Portsdown and High Down, which runs under the sea at Worthing.

5th. The syncline of the Isle of Wight, which runs from near Dorchester in the west through the Tertiary area of the Island and out to sea near Brading.

6th. The double anticline of the Isle of Wight, which commences off the coast of Devon, strikes the shore near Weymouth, runs along the Dorset coast near St. Albans Head, through the Cretaceous area of the Isle of Wight, and out to sea near Sandown.

These axes are not strictly parallel. The London axis, for example, runs a little north of east ; the Weald axis curves round considerably south of east in its eastern part ; the Chichester and Portsdown axes are nearly parallel to that of the Weald, but are inclined a little more to the south; while the synclinal axis, and the two nearly coincident anticlinal axes of the Isle of Wight, run nearly east and west. The want of parallelism in these great folds is not sufficient, however, to invalidate the assumption that they form part of a single series, and were formed contemporaneously.

They have, moreover, this property in common, namely, that the north side of every anticline is much steeper than the south side. Thus the strata rise gently towards the north for a varying distance, and then, reaching the crest of the fold, plunge suddenly down, slowly to rise again. This sudden downward plunge is seen in the Hog's Back, in Portsdown, and in the central Downs of the Isle of Wight, which form the northern sides of the respective anticlinal folds, enumerated above.

We may next notice that these folds do not run for an indefinite distance either east or west, but die away, each syncline being truly an elongated basin, and each anticline an elongated dome. The two ends of a fold are visible in one instance only, viz., in the anticline of the Weald, but the western terminations of all the others, excepting the Isle of Wight (Brixton) anticline, can be seen, and in this case we find the eastern termination of the fold near the centre of the Island. The Sandown anticline, which commences where the Brixton anticline dies away, probably itself disappears a short distance east of Sandown; for, as previously pointed out, the strike of the Chalk in the southern Downs is such as to cause this range to meet the central range at an oblique angle. Similarly we have evidence of the eastern termination of the Isle of Wight syncline off Selsea Bill.

In respect of their relative positions to one another these folds show this peculiarity, that while they run east and west (approximately), as if formed by a force acting from the south, they are arranged en échelon along a line running a little north of east. This can be most easily rendered intelligible by drawing a line through the whole system of folds touching the area of maximum movement in each fold. Such a line starting from near Weymouth, runs between Cowes and Newport, near Portsdown and Chichester, a little north of Battle, and thence out into the German Ocean, where presumably the deepest part of the London syncline is situated. The line thus traced has a direction of east 10°–15° north, and, what is deserving of remark, is not very far from being parallel to the great Chalk escarpment across England.

The Palæozoic Rocks on which the Secondary strata rest in the north-west of France, and which doubtless pass under the south-east of England are known to be intensely contorted, and thrust over one another, the strike of the folds being about west-north-west, turning to east and west where they emerge at the surface in Devon and Somerset. The Carboniferous Rocks of Valenciennes also tend to assume this strike towards the west. But though there is this approximate agreement in direction between the folding of the Secondary and Tertiary Rocks, and that of the Palæozoic Rocks, it must not be concluded that any connection exists between them. The Palæozoic Rocks had already been folded when the Secondary Period commenced, while the folds with which we are concerned were produced in a late Tertiary age. It is, however, possible that the direction of the later folds was influenced by that of the earlier set, for the old rocks may have yielded more readily along the former lines of flexure, than along new lines crossing these obliquely.

We have already noticed the sudden downward plunge of the beds on the north side of all the anticlines. This form of fold seems to be the first stage in the formation of a thrust-plane or slide-fault. For though in the Isle of Wight the movement has not usually gone further than to produce verticality of the beds, yet on following the fold across to Dorsetshire, that is nearer the area of greatest movement, we meet an instance of an actual thrust-plane in the Chalk. This dislocation was first noticed by Mr. Webster in 1811, and described and figured by him in Englefield's Isle of Wight (pp. 164-168, Pl. 26 and 27). The cliff of Handfast Point is formed in the southern part of vertical beds, and in the northern of nearly horizontal beds of chalk. The horizontal strata, as they approach the vertical series, turn upwards in a great curve, forming nearly the quarter of a circle. A fracture has taken place, exactly following one of the curved bedding-planes, and the curved and gently inclined beds have been pushed bodily over the edges of the vertical beds, so as now to rest upon them with an appearance of an extreme unconformity. The bedding of the vertical strata seems at a distance to be regular, with the lines of flints in their usual condition. But on a closer view, the chalk is found to be entirely reconstructed. The flints are not only broken to fragments, but the fragments are more or less separated from one another, while the entire mass of chalk is traversed by veins of calc-spar, and by planes of slickenside filled in with secondary flint. The chalk, moreover, has been hardened to the consistency of limestone.

No trace of a similar thrust-plane is found at either end of the Isle of Wight, but at Ashey the close proximity of fossiliferous strata, probably representing the middle part of the Bracklesham Series, to the basement bed of the London Clay, shows that a strike fault of a peculiar character must there be present. The bedding on each side of the presumed line of fault is perfectly vertical, and to account for the absence of about 400 feet of clays and sands the simplest explanation seems to be that adopted in the new edition of Sheet 47 of the Horizontal Sections now in preparation —that at Ashey a thrust-fault occurs, and that its form and effect on the beds correspond closely with what we know is found on the mainland. Even at a considerable distance from the belt of highly inclined rocks, in the Tertiary Beds of the Isle of Wight, small thrust-planes are occasionally found in the harder strata (see Fig. 83).

FIG. 85.

Section between How Ledge and Colwell Chine. (Scale 20 feet = 1 inch.)

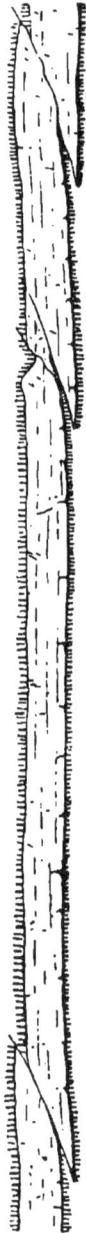

The date of the disturbances of which we have a part in the Isle of Wight is known to have been subsequent to the deposition of the Hamstead Beds (Middle Oligocene) by the fact that these strata share in the tilting up of those along the central range. There is no evidence of the movement having commenced in an earlier period. Had such been the case, there would have been a tendency in the Tertiary formations to thin away against the anticlinal folds. On the other hand, the movements have been proved to have been earlier than the Pliocene. For on the North Downs, near Lenham,* we find Lower Pliocene deposits resting directly on the Chalk, the absence of all the older Tertiary strata being clearly due to the denudation that resulted from the upheaval of the Weald anticline. The date of the disturbances may therefore be assigned approximately to the Miocene Period.

As will be presently seen, the fixing of this date is of special interest, for the production of the folds directly determined the courses taken by most of the South Country rivers.

As is often the case where beds have undergone much folding, there are comparatively few faults in the Isle of Wight. The few which have been observed produce only a trifling effect on the position of the outcrops, and have had no share whatever in producing the physical features of the Island. They have been noted in the course of the detailed descriptions of the sections, but we may enumerate them here for the purpose of comparing their directions. The amount of throw is uncertain in every case, but always insignificant, except at Ashey (p. 114).

Ashey.	Chillerton Down,
Compton Bay ?	W. 30° S. and S.
(p. 8).	25° W.
Carisbrook, W. 15°	Culver Cliff, W. 30° S.
S.	Little Stairs Point,
St. Catherines,	E. 20° S.
W. 11° S.	

Commencing with the northern half of the Island, we see at once from the map that the most important feature in that district is

---

* Clement Reid, On the Pliocene Deposits of North West Europe. *Nature*, vol. 34, p. 341. 1886.

the broad flattened syncline occupied by the Hamstead Beds. On the north side of this trough the strata rise at a gentle angle—probably never more than 5°. On the south they rise abruptly at a high angle, so that near the Chalk they are nearly always vertical, sometimes even slightly inverted. Other minor folds occur, but these are all of comparatively slight importance.

It has already been pointed out that the anticlinal and synclinal folds in the south of England form ovals elongated in an east and west direction. This syncline is no exception to the rule, for if the base of the Hamstead Beds be followed by means of wells and borings, it is found to lie below the sea-level for about 14 or 15 miles, but then to rise rapidly towards the east, so that the Bembridge Limestone lies at or close to the beach on the coast. On the west the syncline must die out rapidly beneath the Solent, for neither Hamstead nor Bembridge Beds have yet been detected on the mainland, though the bottom of the Hamstead Beds descends well beneath the sea-level at Hamstead and Bouldnor.

Owing to the absence of any hard rocks in the greater part of the Oligocene series the exact position of the synclinal axis cannot be easily traced, but the various trial-borings made during the progress of the Survey enables us to fix it within narrow limits. Its centre follows a curved line passing through Bouldnor Cliff, Shalfleet, the southern part of Parkhurst Forest, Dorehill, Ashey, Ricketshill, and Brading Harbour.

On the northern side of this syncline traces of several minor undulations may be detected, but it is difficult to reduce them to any definite system. Headon Beds are brought up near Norris and Osborne in a rather peculiar manner, but this seems to be mainly due to the increase in the rapidity of the dip along a line parallel with the coast between Osborne and Ryde. The occurrence of the Bracklesham Beds on the opposite coast shows that such an increase must take place, while at Norris the coast projects somewhat beyond the general line, so that the strata are there brought within the influence of this increased dip.

In Thorness Bay the Bembridge Limestone sinks beneath the sea-level, so that Hamstead Beds are seen in the cliffs. In Newtown Bay, on the other hand, the Osborne Beds rise. These two folds show a tendency to follow east and west lines, but nothing more can be said about them.

Following next the southern margin of the Tertiary basin, we find that in Whitecliff Bay, while all the lower beds are vertical, the Bembridge Limestone, after dipping at a very high angle in the upper part of the cliff, suddenly flattens into a horizontal reef on the foreshore, the Bembridge Marls being only slightly affected. This structure continues as far as Brading, where not only is the Bembridge Marl affected, but the lower part of the Hamstead series is also tilted slightly. Near Nunwell and Ashey the pressure of the terrestial movement that plicated the strata seems to have reached its maximum, for all the beds as high as the Lower Hamstead series are tilted at high angles and so

compressed that the breadth of their outcrop is less than their
true thickness.

Between Ashey and Newport the strata below the Bembridge
Limestone still continue nearly always vertical, though not quite
so much compressed. In several places there seems to be a
tendency to develop small secondary plications parallel to the
main fold but not traceable more than a few hundred yards.

West of the Medina the Bembridge Beds, though dipping at
high angles, are seldom or never vertical, but the borings at
Gunville show that the Hamstead Beds are still much tilted.

These variations in the extent to which the Tertiary formations
have been influenced by the monoclinal curve seem at first to point
to variations in the height or sharpness of the curve. But they
may be otherwise simply explained. If a series of curved strata
were cut through at different levels the age of the beds most
strongly tilted would be found to vary considerably. If the
country at Ashey were lowered one or two hundred feet by
denudation the Bembridge Limestone at the surface would only
dip at low angles, and more of the Secondary beds would appear
to be affected by the disturbance. Any such lowering of the
surface leads to an apparent shifting of the verticality into a lower
geological horizon and an apparent shifting of the line of greatest
disturbance towards the south. Thus the apparent dying out of
the syncline eastward may really be the result of an upward tilt
in that direction, causing the curve to be cut through at a lower
level, but not affecting the thickness of the beds affected or the
real height of the curve. (Fig. 84.)

FIG. 84.

*Diagram Section to show variation in the dip of the Strata as
the Surface is lowered.*

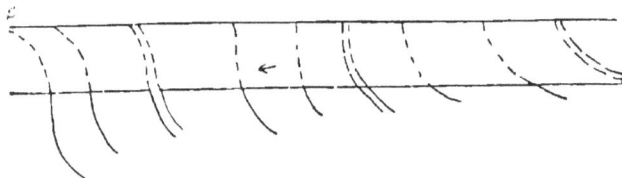

Where the strike of the rocks turns sharply southward at
Calbourne, the angle of dip rapidly lessens and the width of
the several outcrops correspondingly increases. At Shalcombe,
however, where the former strike has been resumed, the lower
beds are again vertical, while the Osborne and Bembridge Beds
occupy long dip slopes. The sudden curve of the strata to the
south and re-appearance of the high angles along a new line
is connected with the dying out of an anticline, which cannot
be traced west of Calbourne, except perhaps in the lower angles
of the dips in the southern part of the long slope of Bembridge
Limestone. A slight indication of the flattening of the beds may
also be found in Freshwater, and even as far as Totland Bay.

On the south side of the central range of Downs, the strata rapidly curve over and assume a horizontal position at a distance varying from one to two miles south of the region in which they are vertical. They never present, however, so sharp a fold as that seen in the Bembridge Limestone in Whitecliff Bay.

The central range consists of two separate axes, which may be conveniently named the Sandown and the Brixton anticlines.

The first appearance of the Sandown anticline in the Tertiary area west of Calbourne has already been referred to, and its subsequent course eastwards by Arreton to Sandown Level was noted in describing the Lower Greensand (pp. 42–44). The axis runs in a nearly straight line due east and west as far as Newchurch, but then bends round to about E. 18° S., its direction being definitely given by the line of Downs from Ashey to Culver. The strike of the strata forming the Southern Downs is a little north of east, and the two ranges therefore, if prolonged in these directions, would meet at no great distance from the coast near Sandown. The central point between the east and west ends of the axis lies perhaps not far from the centre of the exposure of the Wealden beds, where the strata are horizontal.

The dip on the north side of the arch formed by the anticline ranges from 60° to verticality, on the south side from 2° to 3°, in accordance with the general rule previously alluded to that the north side is the steeper in all these anticlines.

A little north of Shanklin, a gentle anticline, accompanied by a fault, probably is the continuation of that which has been noted near Gossard Hill, though it cannot be traced through the Lower Greensand area. This hill itself stands on the northern side of the anticline, the beds showing a northerly dip of about 10°. On the other hand, the large chalk pit on Chillerton Down lies south of the axis, which must therefore run very near the south side of Gossard Hill. The distance between the axes of this and the Sandown anticline amounts to two miles both here and in Sandown Bay.

The fault at Little Stairs Point occurs on the north side of this anticline and runs about E. 20° S. for the small distance it is seen in the cliff. It trends therefore nearly parallel to the axis of the anticline, and perhaps replaces it. In order, however, to effect a displacement of the beds corresponding to that of a fold it should have a downthrow to the north, but in reality it throws the strata in the opposite direction.

South of this small anticline the beds gently roll over and assume the south-easterly dip which prevails through the Southern Downs. The direction of dip is not constant, but ranges locally from south-east to south and even south-west, as in the reef of Yellow Ledge near Luccomb. But the general direction may be ascertained by taking the levels of the base of the Chalk at various points. In St. Catherine's Down this base is about 620 feet above the sea at the north-west end and 500 feet at the south-east, the dip being south-east. In Appuldurcombe Down the base lies at 600 feet, but falls in a south-south-easterly direction to

300 feet above the sea in the top of the cliff above Steephill. Similarly it is at a height of nearly 600 feet on the north side of St. Martin's Down but falls to the south-south-east till it is 300 feet above Ventnor. It is clear that the cliff from Blackgang to Bonchurch does not give the line of strike, in as much as the base of the Chalk falls from 500 feet at the former to 300 at the latter. The true strike may be traced by drawing a line through the points in each contour, at which it is intersected by the base of the Chalk. Taking the 600-foot contour first we find such a line touches the northernmost point of the Chalk on St. Catherine's, Appuldurcombe, and St. Martin's Downs. A similar line drawn through the 500-foot contour is almost exactly parallel to this, and at a distance of a little more than 1,000 yards from it, from which it may be calculated that the average dip amounts to 1 in 31 or a little less than 2°. Lastly a line drawn in the same way with reference to the 400-foot contour runs approximately parallel to the other two, but with a less decided bend, and therefore more nearly parallel to the coast between Niton and Bonchurch.

It will be noticed that the strike from Appuldurcombe westwards is south-west, curving round to the south-south-west in St. Catherine's Hill, while from Appuldurcombe eastwards it is only a little south of west. The difference is clearly due to the influence of the Brixton and Sandown anticlines; in St. Catherine's we have the remains of an escarpment of Chalk which must once have closed in this end of the Brixton area, while St. Martin's Down forms part of a long escarpment which formerly bounded the Sandown anticline on the south, eventually joining itself on to the continuation of the Central Downs, as already suggested.

It has been remarked that there is evidence of the dip becoming rather steeper in the Undercliff than it is in the Downs immediately to the north. If to the three lines of contour above enumerated we add a fourth, viz., the 300-foot contour, which the base of the Chalk touches at Ventnor and Bonchurch, we shall find that there is less distance between this and the 400-foot contour, than there is between the 400- and 500-foot contours, or that, in other words, the gradient of the Chalk increases towards the coast.

The Brixton anticline first makes its influence perceptible in the strike of St. Catherine's Down, as already mentioned. Further west it becomes more marked, and the position of its axis is indicated by the southward sweep of the Atherfield Clay and Wealden Beds, but the axis itself lies just outside the coast line. It seems to run about west-north-west, but curves round to due west at Freshwater, and to W. 14° S. at the Needles. Here it passes out to sea, but re-appears on the coast of Dorset, everywhere throwing the beds into a nearly vertical position along its north side, and eventually bringing Oolitic rocks up to the surface near Weymouth.

It will be seen that the Brixton and Sandown anticlines form two members of the great system of folds which have been described as replacing one another along a line ranging a little north of east. It is therefore in accordance with this rule that the Sandown anticline lies a little to the north, as well as to the east, of that of Brixton, and that the one dies out where the other begins. The actual region in which this replacement of the one fold by the other takes place lies between Calbourne and Gatcombe. For west of the longitude of Gatcombe the Sandown anticline gradually flattens out, so as to allow the Chalk to extend itself southwards, but at the same time the extreme southerly point of the spread of Chalk, viz., Chillerton Down, assumes a dip which is obviously due to the Brixton anticline. The increase in this dip along the southern border of the Downs proceeds *pari passu* with the flattening of the Sandown anticline along the northern border, until at Calbourne the latter is scarcely recognisable, while the former has become fully pronounced. Between these two lines, viz., about Newbarn, Rowborough, and Idlecombe the beds are nearly horizontal, but assume a westerly dip further west, which increases to 17° near Calbourne

CHAPTER XV.

PHYSICAL GEOGRAPHY.

Having already noted the leading physical features of the Isle of Wight, we will now proceed to show that they have been produced by denudation acting along lines of drainage which were determined by the formation of the anticlines and synclines described in the previous Chapter. Though the modern rivers still follow the courses so determined, the actual surface-features produced by the movements of the strata have long since disappeared; and, as in the case of the Weald, the anticlinal areas of the Isle of Wight show that the regions of greatest upheaval in past times are often those of least elevation at the present. In studying the physical features of the Island, one of the most prominent facts that strikes us is the comparative insignificance of the central chalk ridge or back-bone as a watershed. Both in past times and at the present day, the principal rivers of the Island cut right across it, ignoring, as it were, the easier passage which seems to exist for them along it either to the east or west. The explanation has been already found in the case of the Weald, and it will be sufficient here to point out the similarity between the two districts.

The existing watersheds of the Isle of Wight are complicated by the fact that there are so many small streams having a separate existence. Ignoring the minor watersheds between these, we will trace that which separates the water draining south into the Channel, from the water which runs north into the Solent. This line runs from the cliff near Sandown over Shanklin and Boniface Downs to the cliff above Ventnor, and thence over Rew Down. Westwards it keeps close to the cliff edge as far as St. Catherine's Down, along which it runs, turning down south of Chale Green to Kingston, and thence along the southern brow of the Downs to the Needles.

It has been shown that the streams which run into the sea in Brixton Bay were within a geologically recent period tributaries of the western Yar, and that similarly the streams of Shanklin Chine and Luccombe Chine were tributaries of the eastern Yar. The separation of these streams from their original drainage-basins has been due to the encroachment of the sea, and if we trace the watershed as it existed before the separation took place, we find that it must have run south of the whole Island, excepting only a small portion of Week, Rew, and Boniface Downs. That is to say that the whole drainage of the Island, excepting the short and steep heads of valleys in the south side of these downs, must have made its way northwards, the water from the area now occupied by the Lower Greensand all escaping in this direction across the high central ridge of chalk.

The physical geography of the Weald has been too fully described* to need more than an allusion here. The rivers of that area rise in what is now the area of least elevation, and make their way to the north and south through gaps cut in the bold escarpments of the North and South Downs. The watershed, however, though now low, follows the axis of the anticline, that is the line of greatest upheaval in past times, and in this fact is provided the key to the history of the rivers not only of the Weald and Isle of Wight, but of all the part of England affected by the synclinal and anticlinal folds described above.

For we find that without exception the main lines of drainage follow the synclinal axes, while the tributaries flow at right angles to, and off the anticlinal axes. The first land to emerge from beneath the sea was that formed by the crests of the anticlinal folds, and each of these thereupon became a watershed, and has so remained. The last land to emerge was the deepest line of each synclinal fold, and along this was collected the drainage from the anticline to the north and south of it. The lines of drainage and watershed, thus initiated, have been maintained, though the form of surface due to the original movements has been lost. It thus happens that the watersheds have little relation to the hill-ranges of the present day.

The two leading examples of rivers following the synclinal troughs are the Thames and the Frome. Part of the Thames, with its tributary the Kennet, form a line of drainage running the whole length of the London syncline. On the north side it collects the rivers which run down the back of the Chalk escarpment of Berkshire, Buckinghamshire, and Hertfordshire; on the south side it gathers the streams which descend from the anticlinal axis of the Weald and its continuation on the north side of Salisbury Plain.

Similarly, in the Hampshire Basin, we find the Frome following the synclinal axis, and forming an exact counterpart of the Kennet. On its north bank it receives rivers which flow down the south side of the anticline named above, and on its south side it must have received the drainage of the Isle of Wight anticline until the Hampshire Basin was invaded by the sea.

In the alterations brought about by the encroachment of the sea, lies the principal difference between the rivers of the London and the Hampshire Basins. The Frome now enters the sea near Poole, but it is clear that, before the sea made the great breach in the Chalk escarpment which separates Dorset and the Isle of Wight, this river must have followed the syncline eastwards. For this breach, though probably commenced as a river valley, can hardly have been the course followed by the Frome, for in such a case the river must have turned from following a syncline to cut directly across an anticline. On the other hand, we have in the Solent, and the arm of the sea at Spithead, an old valley and

---

* W. Topley. Geology of the Weald (Memoirs of the Geological Survey), chapter 16.

estuary exactly in the position, which we should, by analogy with the Thames, have ascribed to the ancient river Frome.

Among the tributaries of this ancient Frome we may mention on the north side the Stour, the Avon, the Anton, and the Itchin; on the south side, the small stream which traverses the Chalk escarpment at Corfe, the three rivers of the Isle of Wight, and in all probability a tributary between the Needles and the coast of Dorset, in the great gap now occupied by the sea. The northern boundary of the basin of this old river can be traced without difficulty, but of the southern boundary a very small portion only is left. It runs south of Dorchester, across the Isle of Purbeck, and reappears in the extreme south point of the Isle of Wight. The valleys in the south side of Rew, Week, and Boniface Downs are therefore almost the only survivors of another river system next on the south to that of the Frome.

This small portion of watershed does not follow the crest of either the Brixton or Sandown anticline, but lies among the Downs where a southerly and south-easterly dip has fairly set in. An explanation of this fact would probably be forthcoming, could we tell what was the form of the land which once lay to the south.

CHAPTER XVI.

ECONOMIC GEOLOGY.

The Isle of Wight has no mining industries and few quarries or pits, except those for freestone, chalk, sand, and brick-earth. Hydraulic cement is made at the West Medina Cement Works from chalk and Oligocene clay, and at Brading Cement Works from the Bembridge Limestone and Marl. The Wealden Shales are used for brickmaking at Sandown, as well as deposits of brick-earth, associated with gravel, near Borthwood. At Shanklin a bed of clay in the Lower Greensand is dug by the side of the railway for the same purpose (p. 46). The Gault is worked at Bierley, Rookley, and by the side of the railway between Wroxall and Shanklin (p. 64). An extensive deposit of brick-earth near Brixton has received little attention from the remoteness of that district; the bricks for the viaduct of the Military Road over Grange Chine were manufactured from this deposit (p. 224). Brick pits are opened in various parts of the superficial and Oligocene Beds, but curiously enough the bed that would probably make the best brick-earth—the free-cutting decalcified loam so often met with in trial borings low down in the Hamstead Series—has not been used. Tiles and coarse pottery can be manufactured out of the Reading Beds. The white pipe-clay in the Bagshot Beds is no longer worked, the bed being thin and so nearly vertical that it can only be reached by mining.

The Bembridge Limestone was formerly much used as a building stone, but the principal quarries are now worked out, and brick is generally preferred. The limestone stands the weather very well, though the large cavities left by the fossils are often objectionable and much of the stone is too soft for use. The sea-walls round the northern portion of the Island are generally built of Bembridge Limestone. A better building stone is obtained from the four-foot freestone of the Upper Greensand, described on pp. 64–72. This bed has been worked from time to time through a larger part of its outcrop in the central and southern parts of the Island, but the principal quarries, now in use, lie around Shanklin, Bonchurch, and Ventnor. Road metal is obtained from the Angular Flint Gravel on St. Boniface Down (p. 210), from the Plateau Gravel on St. George's Down (p. 212), and from the Valley Gravel at Horringford (p. 221). There are many smaller pits scattered about, which have been referred to in the description of these gravels in chapter xiii.

For a short time the coal-seam in the Bagshot Beds in Alum Bay was worked, but it is of very little value. Alum was formerly manufactured in the Island from the clays of Alum Bay, and as early as 1579 at works in Parkhurst Forest. The Crown used formerly to monopolise the whole of the alum, and proper people were appointed to gather and preserve it for Government. This practice commenced with Queen Elizabeth, who sent a mandate to Richard Worsley, then Captain of the Isle of Wight, in order to ascertain the truth of what she had heard, and a warrant was issued, dated 7th day of March 1561, to search for " certan Oure of Alume."

Iron pyrites was collected on the shore about Shanklin, and carried by boat to London, during the last century.* The clay ironstone, which is found in considerable quantities lying loose upon the shore at the foot of the cliffs between Yarmouth and Hamstead ledge, was collected on the beach and sent to Swansea, to be smelted into iron.

PHOSPHATIC NODULES.

Reference has frequently been made in the preceding pages to the occurrence of phosphatic nodules at various horizons, but more especially in the Cretaceous Rocks. In consideration of the great economic importance of such nodules, it is proposed to devote a few lines to describing their mode of occurrence and composition.

The Wealden Beds.

Phosphate of lime occurs in these beds, but in small quantities only, in the numerous fragments of lignite, which are found at almost all horizons in the variegated marls. The wood is similar in appearance to that which occurs in the Lower Greensand, and which is stated by Messrs. Paine and Way† to be rich in phosphoric acid. They remark that the fossil forest at Brook Point is probably impregnated with phosphoric acid. It should be noted, however, that most of these lignites are encrusted with, or traversed by threads of iron pyrites. They are moreover too thinly scattered through the clay to be profitably mined.

The Lower Greensand.

A specimen of the fossil wood which occurs sporadically in so many of the beds of the Lower Greensand was analysed by Dr. T. L. Phipson with the following result :—‡

---

* Warner. History of the Isle of Wight, pp. 261 and 263. 1795.
† On the Phosphoric Strata of the Chalk Formation. Journ. Roy. Ag. Soc. England, vol. ix. p. 82. (1848.)
‡ Chemical News, vol. vi. p. 194. (1862.)

Water	-	-	-	- 11·00
Organic matter	-	-	-	6·62
Sand &c.	-	-	-	- 4·40
Lime	-	-	-	- 38·52
Magnesia	-	-	-	- 1·00
Phosphoric acid	-	-	-	32·43
Fluorine	-	-	-	- 3·90
Chlorine	-	-	-	traces.
Sulphuric acid	-	-	-	traces.
Oxide of iron, of uranium, pyrites, and				
loss	-	-	-	- 2·13

                                     100·00

Specific Gravity, 2·71.

The specimen showed crystals of wavellite and iron pyrites here and there.

The fossil remains of animals also in these beds have been found by Messrs. Paine and Way (*op. cit.*, p. 84) to be very rich in phosphoric acid. Among these may be particularized the blocks of fossils in the Scaphite, the Lower Crioceras, and the Second Gryphæa beds ; besides the casts of Ammonites and Scaphites which lie upon the beach. Some nodular masses of shells of a dark iron colour in the cliffs near Shanklin are stated by Mr. Nesbit to contain phosphoric acid to the extent of at least 15 per cent. The whole of the substances examined contained likewise organic matter and fluorine, at times in large quantities.[*]

In the upper part of the Lower Greensand, at Redcliff near Sandown, there occurs the band referred to as the " Coprolite Bed " (p. 37), and as was pointed out, phosphatic nodules occur at about the same horizon in Compton Bay. The nodules in the " Coprolite Bed " are probably richer in phosphate than any others in the Island. They are of a brown or yellow colour and about ½ to ¾ inch in diameter. The band in which they occur, however, is only 4 inches thick. It rises from beneath the beach about 160 yards from the centre of the gully formed by the Gault. Those in the Compton Bay section are small and few and far between.

The Upper Greensand.

The only attempts hitherto made in the Isle of Wight to extract phosphatic nodules, were commenced in the Chloritic Marl, on St. Catherine's Down. The nodules are of a pale brown colour, friable, and of rather a low specific gravity. They are scattered through about three feet of sand, and are nearly all casts of shells, principally *Ammonites varians.* The workings, which seem to have been soon abandoned, were commenced about the year 1851 on the brow of Gore Cliff at the north end of the outlier of chalk. The following analyses by Mr. J. C. Nesbit are quoted from the Notes on the Geology and Chemical Constitution of the various Strata in the Isle of Wight by Captain Ibbetson, p. 36.

---

[*] *Quart. Journ. Geol. Soc.*, vol. iv. p. 262. (1848.)

*Phosphoric Acid, etc., in Nodules and Casts of Shells, in the
Chloritic Marl, St. Catherine's Downs.*

	Insoluble Matter per cent.	Phosphoric Acid* per cent.	Phosphate of Lime per cent.	Amount = 100 tons of Bones.
				Tons.
Cast of Turrilite - - -	5·00	24·26	49·79	90
Cast of Ammonite - - -	6·00	21·28	43·68	103
Small spongite nodule - - -	17·00	20·20	41·60	108
Small spongite nodule - - -	9·60	19·13	39·26	114
Cast of Ammonite - - -	10·00	23·06	47·32	95
Cast of Ammonite - - -	9·60	23·44	48·10	93
Cast of Turrilite - - -	21·00	17·23	35·36	127
Green calcareous sand immediately encasing ditto - -	21·00	5·36	11·05	409
Small nodule - - - -	4·40	20·07	41·60	108
Green sand or hassock in which the fossils occur - - -	26·50	1·23	2·53	178
Large nodule - - -	Not determined.	13·81	28·34	158
Large nodule, portion near exterior -	17·00	7·98	16·38	274
Portion from interior - - -	18·00	7·85	16·12	274
Large nodule, interior - - -	6·00	10·86	22·29	201
Ditto, near exterior - - -	12·90	9·56	19·63	229
Ditto - - - -	7·00	7·72	15·86	283
Calcareous green sandy coating of nodule	50·00	9·44	19·37	230
Large nodule - - - -	7·00	11·65	22·92	200
Ditto, interior - - -	13·00	14·82	30·42	148
Large nodule - - - -	22·00	16·60	34·06	132
Ditto, near exterior - - -	6·70	9·18	18·85	232

* Good Cambridge coprolites contain about 26 per cent. of phosphoric acid.

Chalk.

A very thin, but well-marked band of nodular chalk, known as the Chalk Rock, runs through the whole of the central range of the Island, as described on pp. 75–89. The nodules are slightly phosphatic as shown by the following analysis, made by M. Duvillier for M. Barrois.*

Nodules from the Chalk-rock of Shalcombe Down.

Insoluble matter, clay	- -	2·43
Soluble silica	- -	0·72
Oxide of iron	- -	0·89
Phosphate of lime	- -	4·4
Carbonate of lime	- -	91·25
		99·77

SOLUBLE SILICA.

The Upper Greensand of the Undercliff was examined by Messrs. Way and Paine for the purpose of comparing it with a bed of the same age in the neighbourhood of Farnham, in which silica in the soluble form existed in large proportions. They found however that the Upper Greensand of the Isle of Wight was comparatively poor in this form of silica, as shown by the following table. It should be stated that the silica, which of course formed one of the largest constituents of the sandstones, occurred as quartz, &c. in the insoluble form.

* Craie de l'Ile de Wight, 1875, p. 19.

Upper Greensand of the Undercliff.

	Soluble Silica.	Carbonate of Lime.
Cherty flint	3·11	1·34
Blue limestone (rag)	1·71	66·00
Rubbly rock	2·82	5·80
The Freestone Bed	3·20	14·04
" False freestone "	5·94	12·54
Sand with occasional chert	9·64	8·75
Light-brown or cream-coloured " malm "	8·56	12·50
Dark malm (passage-bed into the Gault)	4·84	8·30
Best Farnham Malm-rock, up to	72·00	0·00

ANALYSES OF CHALK.

Chalk from the East Quarry, Ashey Down.*

1. Analysis made at Tennant's Works, Manchester, 1874.
2.   Do.   do. by Dr. Voelcker, 1875.
3 and 4. Do.   do. by Mr. Pattison of Newcastle-on-Tyne.
The specimen No. 3 was taken moist.

	(1.)	(2.)	(3.)	(4.)
Carbonate of lime	98·53	98·05	93·64	98·01
Sulphate of ,,	—	—	—	·06
Moisture	·31	·39	4·73	·2
Silica (" siliceous matter " in 3)	·17	1·28	·5	·4
Alumina and oxide of iron	·44	·21	·16	·6
Magnesia (carbonate)	—	·07	1·01	·94

Analysis of Glauconite from the Lower Greensand of Compton Bay (see p. 22), by MR. J. HORT PLAYER, F.G.S., F.C.S. April 12th, 1889.

Loss by ignition	9·2
Silica	56·6
Alumina	9·6
Ferric oxide	14·6
Ferrous oxide	2·7
Lime	·9
Magnesia	1·1
Potash	4·9
	99·6

" The substance used in the above analysis when examined under the microscope is seen to consist of glauconite with a very small admixture of quartz. The grains of glauconite vary in size from about ·25 mm. to ·5 mm. They are of fairly uniform dimensions in the different directions, but many of them shew more or less rounded protuberances. They are opaque by transmitted light except at the edges. Small particles produced by crushing the grains are grass-green by ordinary light and give minute aggregate polarisation. There is no trace of any definite structure in the substance of the grains."—J. J. H. TEALL.

* Communicated by the late Mr. J. Young to Mr. Bristow.

APPENDIX I.

THE MEAN ANNUAL AND MONTHLY RAINFALL OF THE NORTH AND SOUTH SIDES OF THE ISLE OF WIGHT.

	Jan.	Feb.	March.	April.	May.	June.	July.	August	Sept.	Oct.	Nov.	Dec.	Total.
Osborne, from 1858 to 1887, inclusive. Height above sea level, 172 feet. Diam. of gauge, 8 inches. Height above ground, 8 inches. Observer, Mr. J. R. Mann.	2·99	2·22	1·75	1·61	1·85	1·88	2·04	2·22	3·35	3·30	3·01	2·90	29·21*
St. Lawrence, from 1866 to 1885, inclusive. Height above sea level, 75 feet. Diam. of gauge, 5 inches. Height above ground, 12 inches. Observer, Rev. C. Malden.	3·21	2·50	1·91	1·89	1·57	1·72	2·11	2·30	3·22	3·66	3·35	3·14	30·58†

* During the period of 22 years 1866–87 at Osborne the least monthly rainfall was in May 1876 with 0·15 inches, and the greatest in September 1866 with 8·66 inches. The wettest year was 1872 with 39·38 inches, and the driest 1870 with 21·96 inches.

† During the period of 20 years, 1866–85, at St. Lawrence the least monthly rainfall was in May 1871 with 0·6 inches, and the greatest in November 1877 with 9·13 inches. The wettest year was 1872 with 39·95 inches, and the driest 1870 with 21·99 inches.

APPENDIX II.

TABLES OF FOSSILS.

TABLES I.—III. CRETACEOUS.

In these lists the Survey Collections have been supplemented by those of the various authors, whose names are indicated. The specimens collected for the Survey, previous to 1887, were identified by Messrs. H. W. Bristow, F.R.S., and R. Etheridge, F.R.S. A further collection was made for the Survey in the years 1887–88 by John Rhodes, and the specimens have been identified by Messrs. G. Sharman and E. T. Newton, F.G.S., who have also corrected the whole of the lists for the synonymy. The names formerly in use, but now discarded, are printed in Italics, with a reference to those by which they have been replaced.

The lists of plants from the Wealden Beds and Lower Greensand have been revised by Mr. W. Carruthers, F.R.S., and that of the Ostracoda from the Wealden Beds by Professor T. Rupert Jones, F.G.S.

The authorities, by whom the fossils have been recorded, are indicated by letters as below :—

Ba. Barrois. Recherches sur le Terrain Crétacé Superieur, *Lille*, 1876, and Craie de l'Ile de Wight, *Paris*, 1875.
Be. Bell. Monograph of the Fossil Malacostracous Crustacea. *Pal. Soc.* for 1862.
C. Carruthers. *Trans. Linn. Soc.*, vol. xxvi. p. 690, 1870. *Journ. Bot.*, vol. v. *Geol. Mag.*, vol. iii. p. 542, 1866; Dixon's Geology of Sussex. 2nd ed.
D. Davidson. *Pal. Soc.* for 1855.
Fi. Fitton. *Trans. Geol. Soc.*, ser. 2, vol. iv., p. 103, 1836, and *Quart. Journ. Geol. Soc.*, vol. iii. p. 289, 1847.
Fo. Forbes. *Quart. Journ. Geol. Soc.*, vol i. pp. 190, 237, 345, 1845.
G. Gardner. *Geol. Mag.* for 1875 and *Rept. Brit. Assoc.* for 1876.
H. Hulke. *Quart. Journ. Geol. Soc.*, vols. xxvi.—xxx.; xxxii.; xxxiv.–xxxvi.; xxxviii. *Phil. Trans. Roy. Soc.*, vol. xxxi., 1881 *Proc. Roy. Soc.*, vol. xxxiii.
I. Ibbetson. On the Geology and Chemical Constitution of the various Strata in the Isle of Wight, *London*, 1849.
Jo. Jones. *Quart. Journ. Geol. Soc.*, vol. xli. p. 333, 1885. *Geol. Mag.* for 1878, pp. 110, 277, and for 1888, p. 534.
L. Lydekker. *Quart. Journ. Geol. Soc.*, vol. xliv. p. 54, 1888. Catalogue of the Fossil Reptilia in the British Museum (*in the press*).
Ly. Lycett. *Pal. Soc.*, 1872–79 (Trigonia).
Ma. Mantell. Geological Excursions round the Isle of Wight, 3rd ed., 1854.
Mo. Morris. Catalogue of British Fossils, 2nd ed., 1854.
N. Norman. Geological Guide to the Isle of Wight, 8vo. *Ventnor*, 1887.
O. Owen. On the Fossil Reptilia, *Pal. Soc.*
Pa. Parkinson. *Quart. Journ. Geol. Soc.*, vol. xxxvii. p. 370, 1881.
Pr. Price. Monograph of the Gault, p. 27.
S. Geological Survey Collections previous to 1887.
Sur. „ „ „ during 1887–8.
Se. Seeley. *Quart. Journ. Geol. Soc.*, vol. xxxi. p. 461, 1875; vol. xxxix. p. 55, 1883; vol. xliii. 1887.
Sh. Sharpe. On the Mollusca of the Chalk. *Pal. Soc.* for 1853.
W. Wright. On the British Fossil Echinodermata, *Pal. Soc.* for 1864–78.

TABLE I. WEALDEN.

The letters refer to the authorities by whom the fossils have been recorded. See p. 257.

The localities are indicated by numbers as below:—

1. Isle of Wight, exact locality not specified.	9. Grange Chine and Brixton.
2. Compton Bay.	10. Between Brixton and Atherfield.
3. East side of Compton Bay.	11. Barnes.
4. Shippard's or Compton Grange Chine.	12. Between Barnes and Cowleaze Chine.
5. West of Brook Point.	13. Cowleaze and Shepherd's Chine.
6. Brook Point.	14. Atherfield.
7. Brook Bay.	15. Sandown Bay.
8. Sedmore.	

The specimens marked thus Sh. are from the Wealden Shales. Those from the variegated Wealden beds are indicated by V.

Plantæ.

Abietites. See Pinites.
V.   Bennettites saxbyanus, *Carr.* 6 C.
Sh.   Carpolithes sertum = the impressions of parts of Equisetites. Burchardti, *Dunker.*
     Chara? 10 S.
V.   ? Clathraria Lyellii,* *Mant.* ? 7 Ma.
V.   Cycadeostrobus† crassus, *Carr.* 6 C. 15 Ma., S.
V.   ,,   elegans, *Carr.* 6 C.
V.   ,,   ovatus, *Carr.* 6 C.
V.   ,,   truncatus, *Carr.* 6 C.
V. ?   ,,   tumidus, *Carr.* 6 C. ?
V.   ,,   Walkeri, *Carr.* 6 C.
Sh.   Endogenites erosa, *Mant.* 12 C.
Sh.   Equisetites Burchardti, *Dunker.* 12 C.
?   Fittonia squamata, *Carr.* 1 S., C.
Sh., V. Lonchopteris Mantelli, *Brong.* 6 Ma. 4, 13, 14 Sur.
V.   Pinites Carruthersi, *Gardn.* 6 G.
V.   ,,   Dunkeri, *Mant.* 7 S., C.
V.   ,,   valdensis, *Gardn.* 6 G.
Sh.   Seeds. 13, 14 Sur.
V.   Thuytes (fruit of) 9 Sur.
V.   Zamia crassa ?, *Lindl. & Hutt.* See Cycadeostrobus crassus, *Carr.*

Crustacea.

Ostracoda.

Sh.   Candona Mantelli, *Jones nov. sp.* 12 Sur.
V.   Cypridea Austeni,‡ *Jones* ? 7 Ma.?
Sh., V.   ,,   Dunkeri, *Jones.* 4 Sur. 5, 9, 14, 15 Jo.

---

* The occurrence of this plant in the Wealden series is doubtful. The specimen so named by Mantell was found by him in the shingle of Brook Bay.

† Mr. Carruthers remarks in Dixon's Geology of Sussex, 2nd ed. p. 280, that " in the Wealden at Brook Point, numerous cones of Cycadeæ occur. They are converted into jet, and are largely charged with iron pyrites. . . . That they are Cycadean fruits there can be no doubt ; but to what living genus they have relations, or to what fossil stems or foliage they may belong, it is impossible to say, I accordingly proposed for them the generic name Cycadeostrobus, by which I intended to convey nothing more than that they were Cycadean cones (Seeman's Journ. Bot., v. p. 8)."

‡ This species is figured in Mantell's Isle of Wight Ed. 3, 1854, p. 223, under the name of C. valdensis, *Fitt.* from Brook Bay. But the figure was copied from Fitton Pl. 21, fig. 1, which shows C. Austeni, and may have been wrongly used for the specimens from Brook Bay. See *Geol. Mag.* for 1878, p. 277.

Sh., V. Cypridea spinigera, *Sow. sp.* 4, 13, 14, Sur.; 2, 14, 15 Jo.; 7 Ma.
Sh.   „   tuberculata, *Sow. sp.* 4 Sur.; 13 Jo.; 14 Fi.   Punfield, Fi.
Sh., V.   „   valdensis, *Fitton sp.* 2, 4, 13, 14, Sur.; 15 Fi.; 2, 4, 5, 13, 14, 15 Jo.; 6, 11 Fi.   Punfield, Fi.
Sh.   Cyprione Bristovii, *Jones.* 4 Sur.
Sh.   Cypris cornigera, *Jones nov. sp.* 14 Sur.
          „   *faba, Sow.*   See Cypridea valdensis.
Sh.   Darwinula leguminella, *Forbes sp.* 13 Su.; 14, 15 Jo.
Sh., V. Metacypris Fittoni, *Mant. sp.* 2, 4, 13, 14 Sur. 5, 9, 14, 15 Jo. Punfield, Sur.

Cythere.   See Metacypris.

Lamellibranchiata.

Sh.   Cardita ? 14 Sur.
Sh.   Cyrena major, *Sow.* 2, 3, 14 Fi.
Sh.   „   media, *Sow.* 11, 13, 14, 15 Fi.; 13, 14 Sur.
Sh.   „   „   (large variety), 14 Sur.
Sh.   „   membranacea, *Sow.* (? = C. media) 11, 13, 14, 15 Fi.; 14 Sur.
Sh.   „   subquadrata, *Sow.* 13 Sur.
Sh.   „   *sp.* 2, 4, 13 Sur.
Sh.   „   *sp.* 14 Sur.
Sh.   Exogyra Bousingaultii, *D'Orb.* 14 Sur.
Sh. ? Modiola, 15 Fi.
Sh.   Ostrea distorta, *Sow.* 14 Sur.
Sh.   ,,   *sp.* 3, 11, 15 Fi.
Sh.   Potamomya ? 14 Sur.
Sh.   Unio antiquus, *Sow.* 13 Sur.
Sh.   „   Gualteri ?, *Sow.* 14 Sur.
V.   „   valdensis, *Mant.* 6 Ma.; 9 Sur.; 8 N.

Gasteropoda.

Cerithium.   See Vicarya.
Sh.   Paludina elongata, *Sow.* 3, 11, 13, 14, 15 Fi.; 4, 13 Sur.
Sh.   „   fluviorum, *Sow.* 15 Fi.; 4, 13 Sur.
Sh.   „   *sp.* 2, 4, 13 Sur.
Potamides.   See Vicarya.
Sh., V. Vicarya (Melania) strombiformis, *Schloth.* (= Potamides carbonaria, *Auct.*) 14 Sur., 7 Ma.; 2, 14 N.
Sh.   „   *sp.* 15 Fi.

Cephalopoda.

Sh.   Ammonites (a derived specimen) 15 Fi.

Pisces.

Sh., V. Hybodus basanus, *Eg.* 14 Ma.; Sur.; 7 S.
Sh.·   „   subcarinatus, *Ag.* 4, 13, 14 Sur.
V.   Lepidotus Fittoni, *Ag.* 8 Ma.
V.   „   Mantelli, *Ag.* 6, 7 S.
Sh.   Various fish-remains, 2, 14 Sur.

Reptilia.

By E. T. NEWTON, F.G.S.

Deinosaurian remains have been obtained in Sandown and Brixton Bays from various horizons in the lower variegated strata of the Wealden, and from a bed in the upper Wealden Shales, which forms the floor of Cowleaze Chine and rises to the top of the cliff west of Barnes High. The bones were at first referred in most cases to the Iguanodon, but have since been made the subject of special study, chiefly by Hulke, Huxley, Lydekker, Owen, and Seeley.   The latest revision of this work is given in the "Catalogue of the Fossil Reptilia in the British Museum" by Mr. R. Lydekker, which has been made the basis of the following list.

In this Catalogue will be found full information as to the various changes in name undergone by many of the bones. The names which have been used, but are now discarded, form the second of the two following lists :—

V. Aristosuchus pusillus, *Owen.* 7.
? Calamospondylus, Foxi, *Lydekker.* I.
V. Cetiosaurus brevis, *Owen.* 7.
? Cœlurus Daviesi, *Seeley.* 1.
V. Goniopholis. 7.
V. Heterosuchus valdensis, *Seeley.* 7.
V. Hylæochampsa vectiana, *Owen.* 7.
V. Hylæosaurus Oweni, *Mantell.* 7, 9.
Sh. Hypsilophodon Foxi, *Huxley.* 12.
? Icthyosaurus?
V. Iguanodon bernissartensis, *Boulanger.* 7.
?   ,,   Dawsoni? *Lydekker.* 1.
?   ,,   Mantelli, *Meyer.* 1.
V.   ,,   *sp.* 9.
? Megalosaurus Dunkeri, *Koken.* 1.
V. Oolithes obtusatus, *Carr.* (Reptile eggs), 9.
V. Ornithocheirus nobilis, *Owen.* 7.
V. Ornithopsis Hulkei, *Seeley.* 7, 9, 15.
V. Pelorosaurus Conybeari, *Mantell.* 15.
V. Pholidosaurus. 15.
Plesiochelys Brodiei, *Lydekker.* 1.
Plesiosaurus.
Sh. Polacanthus Foxi, *Hulke.* 12.
V. Saurian bones, various. 2, 3, 9 Sur.
? Sphenospondylus gracilis, *Lydekker.* 1.
V. Suchosaurus cultridens, *Owen.* 15.
V. Titanosaurus. 7.
V. Tretosternum Bakewelli, *Mantell.* 7.
V. Turtle, bones of, 8 Sur. ? carapace of, Norman, 15.
V. Vectisaurus valdensis, *Hulke.* 9.

Synonyms.

Bothriospondylus magnus, *Owen,* to Ornithopsis Hulkei, *Seeley.*
Ceteosaurus Bucklandi, *Meyer.* to Megalosaurus Dunkeri, *Koken.*
Ceteosaurus or Pelorosaurus tooth, to Ornithopsis Hulkei, *Seeley.*
Cheirotherium footprints to Iguanodon.
Chondrosteosaurus gigas, *Owen,* in part to Ornithopsis Hulkei, *Seeley.*
      ,,   magnus, *Owen,* in part to   ,,   ,,
Crocodilus to Goniopholis.
Eucamerotus, *Hulke,* to Ornithopsis, *Seeley.*
Iguanodon Seeleyi, *Hulke,* to Iguanodon bernissartensis, *Boulanger.*
Megalosaurus Bucklandi, *Meyer.* to Megalosaurus Dunkeri, *Koken.*
Ornithopsis eucamerotus, *Hulke,* to Ornithopsis Hulkei, *Seeley.*
Pelorosaurus ? tooth, *Owen* to Ornithopsis Hulkei, *Seeley.*
Poikilopleuron Bucklandi, *Meyer,* to Megalosaurus Dunkeri, *Koken.*
      ,,   pusillus, *Owen,* to Aristosuchus pusillus. *Owen.*
Streptospondylus major, *Owen,* to Iguanodon bernissartensis, *Boulanger.*
Thecospondylus Daviesi, *Seeley,* to Cœlurus Daviesi, *Seeley.*
Trionyx Bakewelli, *Mantell,* to Tretosternum Bakewelli, *Mantell.*

? Aves.

V. Ornithodesmus cluniculus, *Seeley,* Brook.
(It has been thought that this may be an Ornithosaurian. See *Quart. Journ. Geol. Soc.,* vol. xliii., p. 206, 1887.)

TABLE II. LOWER GREENSAND.

The letters refer to the authorities by whom the fossils have been recorded. See p. 257.

The locality, when not otherwise stated, is Atherfield.

The horizons are indicated by numbers as below:—

1. Perna Bed.	4. Sand-rock Beds.
2. Atherfield Clay.	5. Carstone.
3. Ferruginous Sands.	

Plantæ.

Bennettites gibsonianus, *Carr.* Luccombe Chine, C.

„ maximus, *Carr.* 3 S. (Shanklin).

Coniferous wood, 1, 3 Fi.; 3 Sur. (Shanklin); 5 Sur. (Bonchurch and Dunnose).

Fucoid, 1, 2 Sur.; 3 Fi.

Lonchopteris Mantelli, *Brong.* 1 Sur.; 3, 4 Fi.

Pecopteris reticulata, *Mant.* See Lonchopteris Mantelli, *Brong.*

Pinites Leckenbyi, *Carr.* 3 C. (Shanklin).

Incertæ sedis.

Actinophyllum *sp. ?* Beckles.

Conis contortuplicata, *Lons.* Atherfield, Lonsdale (*Quart. Journ. Geol. Soc.*, vol. v. p. 63. 1848).

Actinozoa.

Cyathophora ? elegans. See Holocystis elegans.

Holocystis elegans, *Lons.* 1 Sur.; 1 Sur. (Sandown); 3 S. (Shanklin).

Isastræa haldonensis, *Dunc.* S.

„ neocomiensis, *Tomes.* Tomes (Geol. Mag. for 1885).

„ reussiana, *E. & H.* Tomes.

Parastræa ? 1 Sur.

Pleurosmilia neocomiensis, *E. de From.* Tomes.

Turbinoseris de Fromenteli, *Dunc.* 1 Sur. (= *Leptophyllia anglica*, Tomes).

Echinodermata.

Cardiaster Benstedi, *Forbes*, 3 W. (Atherfield and Shanklin); Sur. (Compton Bay).

Catopygus vectensis, *Wright*, 3 W. (Shanklin).

Clypeopygus Fittoni, *Wright*, 3 W. (Shanklin).

Echinospatagus Renevieri, *Wright*, 1, 2, 3 Fi.; 3 W. (Shanklin).

Enallaster Fittoni, *Forbes.* 1 W.; 2 Sur.; 3 W. (Shanklin); 5 Sur. (derived); S. (Sandown).

Fragment, 5 Sur. (Bonchurch).

Hemipneustes. See Enallaster.

Holaster complanatus, Fitton. See Echinospatagus Renevieri, *Wright.*

Peltastes Wrightii, *Desor.* (= *Salenia punctata*, Forbes), S. W. (Sandown and Atherfield).

Pseudodiadema Fittoni, *Wright* (at first incorrectly identified with Diadema autissiodorense, *Cotteau*), 3 W.

„ Ibbetsoni, *Forbes*, 3 W.

„ Malbosi, *Ag. and Desor.* (= *Diadema Mackesoni*, Forbes), 3 W.

„ *sp.* S.

Annelida.

Serpula antiquata, *Sow.* 1, 2, 3 Fi.

„ filiformis, *Sow.* S.

„ gordialis, *Schloth.* S. (Sandown and Atherfield).

„ plexus, *Sow.* 1, 3 Fi.; S.

„ quinque-angulatus, *Röm.* S.

„ *sp.* 2 Sur.

Vermicularia polygonalis, *Sow.* 1, 2 Sur.; 3 Fi.

„ *sp.* 1 S.; 1 Sur. (Sandown).

Crustacea.

Astacodes falcifer, *Phil.* S.
,,   vectensis, *Bell.* 3 Fi.
Hoploparia longimana, *Sow.* 3? S. ; 5 Sur. (Dunnose and Sandown?).
Meyeria magna, *M'Coy.* See M. vectensis, *Bell.*
,,   vectensis, *Bell.* 2 Sur.
,,   Willettii, *H. Woodw.* (*Geol. Mag.* for 1878, p. 556).
Mithracites vectensis, *Gould.* 3 Sur.
Xanthosia, *sp.* 3 S. (Shanklin).

Polyzoa.

Chisma furcillatum, *Lons.* 3 Fi.
Choristopetalum impar, *Lons.* Mo.
Diastopora, *sp. nov.* (*Lons. mss.*), 1 Fi.
Entalophora irregularis, *D'Orb.* 1 Sur.
Siphodictyon gracile, *Lons.* 3 Fi. ; 3 S. (Shanklin).

Brachiopoda.

Lingula truncata, *Sow.* 1, 3 Fi. ; 3 S. (Sandown); D. (Shanklin).
Rhynchonella cantabrigensis, *Dav.* D.
,,   depressa, *Sow.* 1 Sur. S. (Sandown) ; D. (Shanklin).
,,   gibbsiana, *Sow.* 1, 3 Fi.; 1 Sur. (Sandown).
,,   latissima, *Sow.* S.
,,   nuciformis, *Sow.* D. (Shanklin).
,,   parvirostris, *Sow.* D. (Shanklin).
,,   sulcata, *Park.* 3 Fi.
,,   ,,   *var.* parvirostris, *Sow.* 3 Fo. (Shanklin).
,,   *sp.* 1 Sur. ; 3 Sur. (Compton Bay).
Terebratella Davidsoni, *Meyer.*
,,   Fittoni, *Meyer.* S.
,,   oblonga, *Sow.* 1 Fi. ; Fo.
Terebratula depressa, *Lam.* D. (Shanklin).
,,   microtrema, *Walker.* D. (Shanklin).
,,   prælonga, *Sow.* D. (Shanklin).
,,   sella, *Sow.* 1, 3 Fi. ; 1, 2, 3 Sur. ; 1 Sur. (Sandown).
Waldheimia (Terebratula) celtica, *Morr.* 3 S. (Shanklin and Sandown).
,,   Morrisii, *Meyer.* 3 S. and D. (Shanklin).
,,   tamarindus, *Sow.* 3 S. (Shanklin).
,,   Wanklyni, *Walker.* D. (Shanklin).

Lamellibranchiata.

Monomyaria.

Anomia convexa, *Sow.* 3 Fi. ; Fo. (Shanklin).
,,   lævigata, *Sow.* 3 Fi. ; S.
,,   radiata, *Sow.* 3 Fi. ; Fo.
Avicula depressa, *Forbes.* 3 Fi. ; Fo.
,,   ephemera, *Forbes.* 3 Fi. ; Fo.
,,   lanceolata, *Forbes.* 2, 3 Fi. ; Fo.
,,   pectinata, *Sow.* 3 Fi. ; S.
,,   *sp.* 5 Sur. (Bonchurch).
Exogyra conica, *Sow.* 3 Fi. ; Fo.
,,   harpa, *Goldf.* 1, 3 Fi. ; Fo.
,,   laciniata, *Mills.* S.
,,   plicata, *Lam.* S.
,,   sinuata, *Sow.* 1, 2, 3 Fi. ; 1 Sur. ; Fo. ; 1 Sur. (Sandown).
,,   subplicata, *Röm.* 1 Sur. (Sandown).
,,   tombeckiana, *D'Orb.* 1 Sur.
,,   *sp.* 2 Sur. ; 5 Sur. (Blackgang and Sandown).
Gervillia alæformis, *Sow.* 1, 3 Fi. ; 1 Sur. ; 1 Sur. and Fo. (Sandown).

Gervillia anceps, *Desh.* 3 Fi.; S.; Fo.; W. [G. aviculoides].
,,   *aviculoides.* See G. anceps.
,,   *forbesiana,* D'Orb. See G. solenoides.
,,   linguloides, *Forbes.* 2 Sur.; 3 Fi. and Fo.
,,   solenoides, *Defr.* 1, 2, 3 Fi.; S. Fo. (Shanklin).
Gryphœa. See Exogyra.
Hinnites Leymerii, *Desh.* 1, 3 Fi.; S.; Fo.
Inoceramus concentricus, *Park.* Fo.
,,   gryphæoides, *Sow.* (?=T. concentricus, *Park.*), 3 Fi.
,,   neocomiensis, *D'Orb.* 3 Fi.; Mo.
Lima cottaldina, *D'Orb.* 1, 2, 3 Fi.
,,   dupiniana, *D'Orb.* 2 Sur.
,,   elongata?, *Sow.* Fo.
,,   semisulcata, *Sow.* 1, 3 Fi.
,,   undata, *Desh.* 1 Fi.
,,   *sp.* 3 Fi.; 1 Sur. (Sandown); 5 Sur. (Blackgang).
Ostrea *carinata, Sow.* See O. frons.
,,   frons, *Park.* 1, 3 Fi.; 1 Sur.; Fo.
,,   Leymerii, *Desh.* 1, 3 Fi.; Fo.
,,   macroptera, *Sow.* See O. frons.
,,   *prionota, Forbes.* See O. frons.
,,   retusa, *Sow.* 3 Fi.
,,   *sp.* 2 Sur.
Pecten cinctus, *Sow.* S.
,,   circularis, *Forbes*, may be the P. cottaldinus of D'Orbigny.
,,   cottaldinus, *D'Orb.* 1 Sur.; 3 Fo.
,,   interstriatus, *Leym.* 1, 3 Fi.; 1 Sur.; 1 Sur. and Fo. (Sandown).
,,   *obliquus.* See P. interstriatus.
,,   orbicularis, *Sow.* 3 Fi.; 1 Sur.; 1 Sur. and Fo. (Sandown); Mo.
        (Shanklin); 5 Sur. (Bonchurch, Dunnose, and Sandown).
,,   quinquecostatus, *Sow.* 1, 3 Fi.; 1 Sur.; 3 Fo. (Shanklin); 1 Sur.
        (Sandown); 5 Sur. (Sandown).
,,   robinaldinus, *D'Orb.* Mo. (Shanklin).
,,   *sp.* 2 Sur.
Perna *alœformis.* See Gervillia.
,,   Mulleti, *Desh.* 1 Fi. and Sur.; 1 S. (Compton Bay); 1 Sur.
        (Sandown).
,,   ricordiana, *D'Orb.* 1 Fi.; Mo. (Sandown and Shanklin).
,,   royana, *D'Orb.* S.
Pinna Galliennei, *D'Orb.* 3 Fi.
,,   *restituta, Forbes.* See P. tetragona.
,,   robinaldina, *D'Orb.* 2, 3 Fi.; S.
,,   tetragona, *Sow.* S.; Fo.
,,   *sp.* 1, 2 Sur.
Plicatula carteroniana, *D'Orb.* Mo. (Shanklin); 5 Sur. (Sandown).
,,   placunæa, *Lam.* 1 Fi.; Fo.

Dimyaria.

Anatina Agassizii, *Pict. and Rona.* S.
,,   Carteroni, *D'Orb.* S.
Arca Carteroni, *D'Orb.* 1 Fi.; Fo.; S. (Sandown ?).
,,   cornueliana, *D'Orb.* 3 Fi.; Fo.
,,   dupiniana, *D'Orb.* S.
,,   Raulini, *Leym.* 1, 2, 3 Fi.; 1, 2 Sur.; 1 Sur. (Sandown); Fo.
,,   robinaldina, *D'Orb.* S.
,,   securis, *Leym.* 1, 3 Fi.; Fo.
Astarte Beaumontii, *Leym.* S.; Fo. (Sandown).
,,   multistriata, *Sow.* 1 Fi.
,,   numismalis, *D'Orb.* 1, 3 Fi.
,,   obovata, *Sow.* 1 Fi.; 1 Sur. and Mo. (Sandown).
,,   *sp.* 5 Sur. (Sandown).
,,   striato-costata, *D'Orb.* Mo.
,,   substriata, *Leym.* 1 Fi.; Fo.
Cardita fenestrata, *Forbes*, 1, 2, Fi. 1 Sur.

Cardita neocomiensis, *D'Orb.* S.; Mo.
  ,,   quadrata, *D'Orb.* S.; Mo.
Cardium Austeni, *Forbes*, (Hemicardium, 1, 2, 3 Fi.); 1 Sur.; 3 S.
     (Shanklin); Fo.
  ,,   cornuelianum, *D'Orb.* 3 Fi.; Fo.
  ,,   Ibbetsoni, *Forbes*, 3 Fi.; 2 Sur.; 3 Fo.
  ,,   imbricatorium, *Desh.* Fo.
  ,,   peregrinosum, *D'Orb.* 3 Fi.; Fo.
  ,,   raulinianum, *D'Orb.* 3 Fi.
  ,,   *sp.* 5 Sur. (Bonchurch and Sandown).
  ,,   sphæroideum, *Forbes*, 1 Fi.; 1 Sur. and Fo. (Sandown).
  ,,   subhillanum, *Leym.* 3 Fi. and Fo.; S. (Shanklin).
  ,,   Voltzii, *Leym.* 3 Fi.
Corbis (Sphæra) corrugata, *Sow.* 1, 3 Fi.; Fo.; 1 Sur. (Sandown).
  ,,   ? fibrosa, *Forbes*, Fo.
Corbula incerta, *D'Orb.* 1, 2 Fi.
  ,,   striata, *Sow.* 1 Sur. (Sandown).
  ,,   striatula, *Sow.* 1, 2, 3 Fi.; 2 Sur.; Fo.
Cucullæa exaltata, *Nilss.* 1, 3 Fi.; 1 Sur.; Fo. (Sandown).
  ,,   gabrielis, *Leym.* S.
Cypricardia ? undulata, *Forbes*, 2 Fi.; Fo.
Cyprina angulata, *Flem.* 1 Sur.; 3 Fi.; Fo.; 1 Sur. (Sandown).
  ,,   ,, *var.* rostrata, *Sow.* 3 Fi.; Fo.
  ,,   elongata ?, *D'Orb.* S. (Sandown).
Cytherea caperata, *Sow.* 3 Fi.
  ,,   parva, *Sow.* 1, 3 Fi.; S.; Fo.
Gastrochæna dilatata, *Desh.* Fo.
  ,,   *sp.* 3 Fi.
Goniomya mailleana, *D'Orb.* S.
Isocardia ? ornata, *Forbes.* 3 Fi.; Fo.
  ,,   Sur. (Sandown).
Leda scapha, *D'Orb.* 2, 3 Fi.; S.; Fo.; 5 Sur. (Sandown) ?
  ,, spathulata, *Forbes* 3 Fi.; Fo.
  ,, *sp.* 1 Sur.
Lithodomus oblongus, *D'Orb.* 3 Fi. and Fo.
Lucina arduennensis, *D'Orb.* S.
  ,,   dupiniana, *Forbes*, 3 Fi.
  ,,   globiformis, *Leym*, 1, 3 ? Fi.; Fo.
  ,,   solidula, *Forbes*, 1, 3 Fi.; Fo. ?
Mactra Carteroni, *D'Orb.* 2, 3 Fi.
Modiola æqualis, *Sow.* 1, 3 Fi.; 1 Sur. (Sandown); Fo. (Mytilus).
  ,,   bella, *Sow.* 1 Fi.; Fo. (Mytilus).
  ,,   (Mytilus) cornuelianus ?, *D'Orb.*, 1 Sur. (Sandown.)
  ,,   reversa, *Sow.* ? 1 Sur. (Sandown).
Myacites. See Panopæa.
Mytilus lanceolatus, *Sow.* 1, 2, 3 Fi.; S. (Sandown).
  ,,   ,, *var.* edentulus, *Sby.* 3 Fi.; Fo.
  ,,   simplex, *Desh.* 1 Fi.; S.; Fo.
Nucula antiquata, *Sow.* 3 Fi.
  ,,   impressa, *Sow.* Fo.
  ,,   obtusa, *Sow.* Fo.
  ,,   *scapha.* See Leda.
  ,,   simplex, *Desh.* 3 Fi.
  ,,   *spathulata.* See Leda.
  ,,   *sp* ? 5 Sur. (Blackgang).
Panopæa arcuata, *D'Orb.* S.
  ,,   elongata, *Röm.* 1, 3 Fi.; S.
  ,,   irregularis, *D'Orb.* 2, 3 Fi.
  ,,   mandibula, *Sow.* 1, 3, 5 Fi.; 1, 2 Sur.; 3 Fo.
  ,,   ,, *var.* obliqua, 3 Fo.
  ,,   neocomiensis, *D'Orb.* 1, 3 Fi. and Fo.; 1 Sur. ?
  ,,   plicata, *Sow.* 1, 2, 3 Fi.; Fo.; 1, 2 Sur.; 1 Sur. (Sandown).
  ,,   *sp.* ? 5 Fi.

Pholadomya Agassizii, *D'Orb.* 1 Fi.
,,   Martini, *Forbes* 1, 3 Fi.; 1 Sur. ?; 3 Fo.
,,   *sp.* 2 Sur.
Solecurtus Warburtoni, *Forbes* 3 Fi. and Fo.; 3 S. (Shanklin).
Tellina angulata, *Desh.* 3 Fi.; S.
,,   *Carteroni, D'Orb.* See T. angulata.
,,   inæqualis, *Sow.* 1, 3 Fi.; Fo.; S. (Sandown).
,,   vectiana, *Forbes* 3 Fi.; Fo.; S.
,,   5 Sur. (Sandown).
Teredolithes. See Gastrochæna.
Thetis gigantea (young), *Sow.* S.
,,   Sowerbii, *Röm.* 1, 3 Fi. (as T. major and minor), Fo. S. (Shanklin),
     1 Sur. (Sandown).
Trigonia *aliformis, Forbes.* See T. vectiana.
,,   carinata, *Ag.* 1 Fi. and Lyc.; Fo. (Sandown); S.
,,   caudata, *Ag.* 1, 3 Fi.; 3 Lyc.; Fo. (Shanklin); S.
,,   *dedalea, Forbes.* See T. nodosa.
,,   Etheridgi, *Lyc.* Fi., I., and Fo. (as T. caudata); 1 Lyc.; S.
,,   nodosa, *Sow.* 1, 3 Fi. (as T. rudis); 1, and Fo. (as T. dedalea);
     1 Lyc. (Sandown), 3 Lyc.; 1 Sur (Sandown) ?
,,   ornata, *D'Orb.* 1 Fi. (as var. spinosa); 1 Lyc.
,,   *spinosa, Forbes.* See T. ornata.
,,   *rudis, Fitton.* See T. nodosa.
,,   vectiana, *Lyc.* Fi., I., Fo. (as T. alæformis); 1 Lyc.; 3 ? Lyc.;
     1 Sur. (Sandown).
,,   *sp.* S. (Shanklin).
Venus brongniartiana, *Leym.* Mo.
,,   *caperata, Sow.* See Cytherea.
,,   cornueliana, *D'Orb.* 1 Fi.
,,   *fenestrata, Forbes.* See Cardita fenestrata.
,,   orbigniana, *Forbes.* 1, 2 Sur.; 3 Fi. and Fo.; 1 Sur. (Sandown) ?
,,   ovalis, *Sow.* var. elongata, 1 Fi. and Sur.; S. (Sandown).
,,   *parva.* See Cytherea.
,,   ricordeana, *D'Orb.* 1 Fi.
,,   sp. ? 5 Fi.
,,   striato-costata, *Forbes.* 1 Fi.; S.
,,   vectensis, *Forbes.* 2, 3 Fi.; S.; 3 Fo.

Gasteropoda.

Actæon affinis, *Sow.* S.
,,   albensis, *D'Orb.* 3 Fi.; Fo.
,,   marginatus, *D'Orb.* 3 Fi.; Fo.
,,   *sp.* 5 Sur. (Sandown).
Aporrhais, 1 Sur. (Sandown).
,,   *calcarata, Auctorum.* See A. (Dimorphosoma) ancylocheila and
     A. (D.) kinclispira.
,,   dupiniana, *D'Orb.* 3 ? (Sandown) G.
,,   Fittoni, *Forbes.* (=: Pteroceras, *Forbes, Quart. Journ. Geol. Soc.*,
     vol. i. p. 351. Figured by Mantell in "Geological Excursions"
     as P. retusa.) 3 Fi.; S.
,,   glabra, *Forbes.* 3 Fi.; S.
,,   Parkinsoni, *Mant.* 3 G. (Shanklin).
,,   robinaldina, *D'Orb.* 1, 3 Fi.; Sur.; Fo. (Shanklin).
,,   (Dimorphosoma) ancylocheila, *Gardn.* 3 G.
,,   ,,   kinclispira, *Gardn.* 3 G.
,,   · ,,   vectiana, *Gardn.* 3 G. (Shanklin).
,,   ,,   *sp.* 3 G. (Shanklin).
,,   (Ornithopus) moreausiana, *D'Orb.* (=Pteroceras retusa of Fitton.
     See Gardner, *Geol. Mag.* for 1875 and Forbes, *Quart. Journ.*
     Geol. Soc., vol. i. p. 350). 3 Fi.
Cerithium aculeatum, *Ms. Forbes.* S.
,,   attenuatum, *Forbes.* 3 Fi.; Fo.
,,   clementinum, *D'Orb.* 3 Fi.; Fo.

Cerithium lallierianum, *D'Orb.* 3 Fi. and Fo.
,,   neocomiense, *D'Orb.* 3 Fi. and Fo. ; S.
,,   Phillipsii, *Leym.* 3 Fi. ; S. ; Fo.
,,   turriculatum, *Forbes.* 3 Fi. and Fo. ; S.
Dentalium cylindricum, *Sow.* 3 Fi. ; Fo.
Emarginula neocomiensis, *D'Orb.* 1 Fi. ; Fo.
Eulima melanoides, *Desh.* 3 Fo. (Shanklin).
Littorina conica, *Sow.* Mo. (Shanklin).
,,   *rotundata, Sow.* See Natica.
Natica cornueliana, *D'Orb.* 3 Fi. ; Fo.
,,   gaultina, *D'Orb.* 1, 3 Fi. ; 1 Sur. ? ; Fo.
,,   *lævigata, D'Orb.* See N. rotundata.
,,   rotundata, *Sow.* 1, 3 Fi. ; Fo. ; S.
Patella *sp.* 3 S. (Shanklin).
Pleurotomaria gigantea, *Sow.* S.
,,   *sp.* ? 5 Sur. (Blackgang).
Pterocera Fittoni, Forbes. See Aporrhais.
,,   *retusa* of Forbes and Fitton. See Aporrhais moreausiana.
Rostellaria. See Aporrhais.
Scalaria dupiniana, *D'Orb.* S.
Solarium minimum, *Forbes.* Fo.
,,   *sp.* ? 5 Fo.
Tornatella. See Actæon.
Trochus, sp. 5 Sur. (Bonchurch).
Turbo munitus, *Forbes.* 1 Sur.
Turritella dupiniana, *D'Orb.* 3 Fi. ; Fo.
Vicarya strombiformis, *Schloth.* (= *Potamides carbonaria, Auct.*) 1 Sur (derived).

Cephalopoda.

Ammonites Beudantii ?, *Brong.* 5 Sur. (Blackgang) ; a fragment.
,,   Carteroni, *D'Orb.* S.
,,   consobrinus, *D'Orb.* 3 Fi.
,,   cornuelianus, *D'Orb.* 3 Fi. ; Fo.
,,   Deshaysii, *Leym.* 1, 2, 3 Fi. ; 2 Sur. ; Fo. ; S. (Sandown).
,,   furcatus, *Sow.* 1 Fi. ; Fo.
,,   Hambrovii, *Forbes.* 2 Sur. ; 3 Fi.
,,   ? inflatus, *D'Orb.**
,,   ? interruptus, *Brong.*†
,,   leopoldinus, *D'Orb.* 1, 2 Fi.
,,   Martini, *D'Orb.* 3 Fi. and Fo. ; S.
,,   nutfieldensis, *Sow.* S. (loc. ?).
,,   (rolled fragments), 1 Sur. (Sandown).
,,   (a fragment) 5 Sur. (Blackgang).
Ancyloceras gigas, *Sow.* 3 Fi. (Scaphites) and W. ; S.
,,   Hillsii, *Sow.* 3 Fi. and W.
,,   matheronianus, *D'Orb.* 1 Mo.
Belemnites *sp.* Fo. (as ? B. lanceolatus).
Crioceras (Ancyloceras, *D'Orb.*) Bowerbankii, *Sow.* 3 Fi. ; S.
Hamites, S. (Sandown).
Nautilus plicatus, *Sow.* S. (loc. ?).
,,   pseudoclegans ? S. (loc. ?).
,,   radiatus, *Sow.* 1, 3 Fi. ; 1 Sur
,,   requinianus, *D'Orb.* 1 W.
,,   Saxbii, *Morris,*‡ 3 Fi.
Scaphites. See Ancyloceras.
,,   *grandis.* See Ancyloceras gigas.

---

* This Ammonite is recorded by Fitton from the Atherfield Clay and Perna Bed, and by Forbes from Atherfield. Its occurrence in the Lower Cretaceous Rocks, however, has not been verified.

† One specimen of this Ammonite was presented to the Museum of Practical Geology by Dr. Fitton, as being from the Lower Greensand of the Isle of Wight.

‡ Ann. Mag. Nat. Hist. 1848.

Pisces.

Edaphodon Sedgwickii, *Ag.* S.
Hybodus basanus, *Eg.* ? 1 Egerton (*Quart. Journ. Geol. Soc.*, vol. i.
  p. 197).
Hybodus, sp. S.
Lamna, 1 W.; 5 Sur. (Dunnose).
Lepidotus, 1 Sur.
Odontaspis, 1 W.
Protosphyræna (Saurocephalus), 1 W.
Various, 1 Sur. (Sandown).

Reptilia.

Chelonia, S. (Shanklin).
Iguanodon Mantelli, *Owen*, 3?, O. and L.
Plesiosaurus *sp.*, O. and Whidborne, *Quart. Journ. Geol. Soc.*, vol. xxxvii.
  p. 480 (Shanklin).

TABLE III.—Upper Cretaceous.

The letters refer to the authorities by whom the fossils have been recorded. See p. 257.

Those fossils marked S. were collected from the "Lower Chalk" as mapped in 1852. That subdivision included the Lower and Middle Chalk of the present survey.

The Middle Chalk of Barrois includes the Chalk Rock and about 20 feet of chalk above it.

The Upper Greensand of Barrois includes about 35 feet of beds now included in the Gault. That of Fitton includes the Chloritic Marl.

The numbers indicate the localities enumerated below :—

Localities.

1. Isle of Wight, exact locality not specified.	13. Brixton Down.	27. Standen.
2. Needles.	14. Calbourne.	28. East Standen.
3. Alum Bay.	15. Rowborough.	29. Arreton.
4. High Down.	16. Apes Down.	30. Messley.
5. Main Bench.	17. Alvington.	31. Knighton.
6. Freshwater.	18. Cheverton.	32. Yarbridge.
7. Afton Down.	19. Shorwell.	33. Bembridge.
8. Compton Bay.	20. Chillerton.	34. Culver.
9. Shalcombe.	21. Gatcombe.	35. The Undercliff.
10. Brook Road cutting.	22. New Barn.	36. Blackgang and Niton.
11. Motteston Down.	23. Bowcombe.	37. Western Lines.
12. Brixton and Calbourne Road.	24. Carisbrook.	38. Bonchurch and East End.
	25. Mount Joy.	39. Frequent.
	26. Shide.	

	Gault.	Malm.	Chert Beds.	Upper Greensand, Sub-division not specified.	Chloritic Marl.	Lower Chalk.	Middle Chalk.	Upper Chalk.
Plantæ.								
Algæ - - -	..	..	..	..	..	..	24 Sur.	
Chondrites fastigiatus, *Sternb.*	38 Fi. & M.	..	..	35 M.				
Clathraria Lyellii, Mant.	..	35 I.	38 Pa.					
Coniferous wood - -	38 Sur.	..	35 N.					
Fucoides Targioni., see Chondrites fastigiatus.								
Spongida.								
Axinella stylus, *Hinde.* -	..	..	35 Hinde.					
Craticularia (Brachiolites) Fittoni, *Mant.*	..	..	..		1 S.	1 S., 1 L.	1 Ba.	
Chenendopora - -	..	..	..	..		..	12 Sur.	
Doryderma - -	..	..	35 Hinde.					
Heterostinia (Chenendopora) obliqua, *Ben.*	..	..	..		1 S.			
Cliona cretacea, *Portl.* -	..	..	..	..	..	..	..	1 S.
Dendrospongia fenestralis, F. Rœm., see Craticularia Fittoni.								

* Steephill and East End.

	Gault.	Malm.	Chert Beds.	Upper Greensand. Sub-division not specified.	Chloritic Marl.	Lower Chalk.	Middle Chalk.	Upper Chalk.
Disthelcs conferta, *F. Roem.*	..	..	..	..	..	1 Ba.		
Hallirhoa agariciformis, *Ben.*	..	..	..	1 S.	10 Sur., 1 I.			
Hippalimus fungoides, Lam.x. see Hallirhoa agariciformis.								
Jerea (Siphonia) Web-steri, *Sow.*	..	35 Ba. & N.	..	37 Fi.				
Plocoscyphia fenestrata, *T. Smith.*	..	..	..	..	8, 36 Sur.	34 Sur.		
labrosa, *T. Smith.*	..	..	8 Sur.	..	1 I., 36 Sur.	32, 36 Sur.?, 1 Ba.		
Plocoscyphia meandrina, see P. labrosa.								
Plocoscyphia reticulata ?, *Hinde.*	..	..	..	..	8, 35 Sur.			
* ,, - - -	..	..	..	..	..	12, 29, 36 Sur.	9 Sur.	7, 24, 29, 32 Sur.
Scyphia Fittoni, see Craticularia.								
Siphonia pyriformis, auctorum, see P. tulipa.								
Siphonia tulipa, *Zittel.* -	..	35 Ba.	1 Pa. & I.					
Spongia meandroides, Ibbetson, see Plocoscyphia labrosa.								
Stauronema Carteri, *Sol-las.*	..	..	..	..	8, 35, 36 Sur.			
,, - - -	..	..	12 Sur.					
Ventriculites moniliferus, *F. Roem.*	..	..	..	..	..	..	..	6, 31, 34 Ba.
Ventriculites -	..	..	..	..	..	..	..	7, 12 Sur.
Hydrozoa.								
Porosphæra (Coscinopora) globularis, *Phil.*	..	..	..	..	..	..	..	7, 12, 24, 29, 32 Sur.
Actinozoa.								
Micrabacia coronula, *Goldf.*	..	..	..	..	35 Pa.	36 Sur.		
Monocarya, see Parasmilia.								
Parasmilia centralis, *Mant.*	..	..	..	..	..	..	..	7 Sur., 5 Ba.
Smilotrochus - -	..	..	..	..	..	35 Sur.		
Trochocyathus - -	..	..	..	..	35 N.			
Echinodermata.								
Bourgueticrinus (Apiocri-nus) ellipticus, *Miller.*	..	..	..	..	..	..	..	5, 6 Ba.
Cardinster fossarius, *Ben.*	..	35 N.	..	1 S.				

* Chalk Rock.

	Gault.	Malm.	Chert Beds.	Upper Greensand. Sub-division not specified.	Chloritic Marl.	Lower Chalk.	Middle Chalk.	Upper Chalk.
Cardiaster latissimus, *Ag.*	..	..	..	1 S.				
* „ pillulus, *Lamk.*	..	..	..	..	..	..	..	34 Sur., 7 Ba.
„ pygmæus, *Forbes.*	..	..	..	..	..	1 S.		
„ - - -	..	35 N.						
Catopygus columbarius, *Lamk.* (*carinatus, Goldf.*).	..	..	..	1 I.				
Cidaris clavigera, *Kœnig.*	..	..	..	..	..	..	..	7, 32 Sur., 5, 6 Ba.
„ dissimilis ?, *Forbes*	..	..	..	..	..	35 Sur.		
„ hirudo, *Sorig.* -	..	..	..	..	..	..	13 25, 34 Ba.	6, 25 Ba.
„ pleracantha, *Ag:*	..	..	..	..	..	..	..	29 Ba.
„ pseudohirudo, *Cotteau.*	..	..	..	..	..	..	..	2, 11 Ba.
„ sceptrifera, *Mant.*	..	..	..	..	..	..	..	5, 6 Ba.
„ serrata, *Desor.* -	..	..	..	..	..	..	..	7, 24 Ba.
„ subvesiculosa, *D'Orb.*	..	..	..	..	..	..	..	13, 31, 34 Ba.
„ vesiculosa, *Goldf.*	..	..	..	..	35 Pa.			
„ (spines) - -	..	..	..	..	..	..	..	29 Sur.
†Cyphosoma - -	..	..	..	..	..	..	..	12, 24, 29, 32 Sur.
Discoidea cylindrica, *Lamk.*	..	..	..	..	..	1 Ba.		
„ minima, *Ag.* -	..	..	..	..	1 Ba.	..	22, 34 Ba.	
‡ „ subuculus, *Klein.*	..	..	..	..	35 Pa., 8, 35 Sur.	..	25 Sur.	
„ - - -	..	..	..	..	..	36 Sur.		
Echinoconus castanea *Brong.*	..	..	..	..	1 Ba.			
„ conicus, *Breyn.*	..	..	..	..	..	..	..	23 Ba.
Echinocorys vulgaris *Breyn* (=Ananchytes ovatus).	..	..	..	..	..	..	..	39 S., Ba.
Echinospatagus (Hemiaster) Murchisoniæ *Mant.*	..	..	..	1 S.				
Goniaster Coombii, *Forbes.*	..	..	..	..	..	1 S.		
„ (ossicle of) -	..	..	..	..	..	..	..	12 Sur.
Hemiaster Morrisii, *Forbes.*	..	..	..	..	..	35 W.		
„ - -	..	..	8 Sur.?	..	1 S.			
Holaster cor-avium, *Ag.* (? sp.)	..	..	..	..	..	..	24 Ba. ?	6, 13 Ba.
„ fossarius, see Cardiaster.								

* In a band of green nodules in the Upper Chalk. See p. 78.   † Chalk Rock.
‡ Melbourn Rock.

	Gault.	Malm.	Chert Beds.	Upper Greensand. Sub-division not specified.	Chloritic Marl.	Lower Chalk.	Middle Chalk.	Upper Chalk.
Holaster lævis, *De Luc.*	..	..	..	.1 S.	1 I.	1 Ba.		
„ nodulosus, *Goldf.*, see lævis.								
„ pillulus *Lamk.*, see Cardiaster.								
* „ planus, *Mant.*	..	..	..	...	....	35 N.	-	
„ subglobosus *Lesk.*	..	..	..	1 S.	1 Ba.	36 Sur., 1 Ba.		
„ treconsis. *Leym.*	..	..	. ..	..	..	32 Sur.? 1 Ba.		
„ sp.	..	..	1 Pa.					
Infulaster major, *Desor.*	..	..	..	..	..	..	..	13 Ba. ?
Micraster breviporus, *Ag.*	..	..	....	...	. ..	..	..	9, 13 Ba.
„ Brongniarti, *Héb.?*	..	..	..	. ..	..	..	..	2, 25 Ba.
„ cor-anguinum, *Klein.*	..	..	..	..	..	..		32 Sur., 23, 28, 29 Ba.
„ cor-bovis, *Forbes.*	..	..	..	..	..	..	..	7, 32 ? Sur.
„ cor-testudin-arium, *Goldf.*	..	..	. ..	..	..	.	.	32 Sur., 30 Ba.
Pentacrinus Agassizi, *Hag.*	..	..	..	..	..	..	..	29, 32 Sur.
„	. ..	..	..	...-	. ..	..	..	13 Ba.
Pseudodiadema variolare, *Brong.*	..	..	..	....	. ..	1 Ba.		
„ (fragments)	..	..	. ..	....	....	..	..	12 Sur.
Ossicles	..	..	..	..	..	..	..	24 Sur.
					..			
				.	. .			
Annelida.					.			
Serpula antiquata, *Sow.*	..	..	1 Pa., 35 N.					
„ fluctuata, *Sow.*	..	..	..	..	..	..	..	6 Ba.
„ granulata, *Sow.*	..	..	..	....	....	..	..	29 Sur.
„ ilium, *Goldf.*	..	..	..	..	..	..	..	7, 29 Sur.
„ obtusa, *Sow.* ?	..	..	..	..	36 Sur.			
„ plana, *Woodw.*	..	..	..	..	..	..	..	29 Sur.
† „ plexus, *Sow.*	..	..	..	..	36 Sur.	.	..	5 Ba., 29, 34 Sur.
„ sp.	..	..	..	..	. .	..	..	7 11, 17 Ba., 24 Sur.
Vermicularia concava, *Sow.*	8 Sur.	8 Sur., 35 Ba.	10, 12 Sur., 35 Ba.	..	8 Sur., 35 Pa.			
„ Philipsii, *Röm.*	..	..	..	1 S.	..			
„ polygonalis, *Sow.*	..	..	..	38 Fi.	. .			
„ umbonata, *Mant.*	..	..	..	37 Fi.	..	1 Ba.		

* Chalk Rock. † In a green nodular band in the Upper Chalk, see p. 78.

	Gault.	Malm.	Chert Beds.	Upper Greensand. Sub-division not specified.	Chloritic Marl.	Lower Chalk.	Middle Chalk.	Upper Chalk.
Vermicularia sp. - -	34 Sur.	..	8 Sur.					
Crustacea.								
Cythere - - -	..	..	10 Sur.					
Cytheridea perforata, *Roem.*	Jo.							
Hoploparia Saxbyi, *M'Coy.*	..	38 Pa.	..	38 Bc.				
Meyeria Willettii, *H. Woodw.*	..	..	..	..	..	35 ?*		
Necrocarcinus Wood-wardii, *Bell.*	..	..	..	..	..	35 Bc.		
Palæocorystes Normani, *Bell.*	..	..	..	..	..	36 Bc.		
Cirripedia.								
Pollicipes - - -	..	..	..	..	..	..	..	29 Sur.
Polyzoa.								
Bidiastopora - -	..	..	..	..	..	..	..	24, ? 29 Sur.
Choristopetalum impar, *Lonsd.*	..	..	..	36 S.?				
Defrancia Michelini, *Hag.* (near to).	..	..	..	..	..	..	..	29, 32 Sur.
Entalophora - -	..	..	..	..	..	..	..	29, 32 Sur.
Eschara - - -	..	..	..	..	36 Sur.	..	..	29, 32 Sur.
Frustellaria confusa, *D'Orb.*	..	..	..	..	..	..	..	29, 32 Sur.
Reptescharella radiata, *D'Orb.*	..	..	..	..	..	..	..	28 Ba.
Semiescharа - -	..	..	..	..	..	..	..	29 Sur.
Brachiopoda.								
Crania - - -	..	35 Ba.						
Kingena lima, *Defr.* -	..	..	..	..	8, 36 Sur.	36 Sur., 1 Ba.		
Lingula subovalis, *Dav.* -	..	..	..	35 Ba.				
Magas pumila, *Sow.* -	..	..	..	..	..	35 Sur.	..	39 Ba.
Rhynchonella *compressa,* see R. dimidata.								
,, Cuvieri, *Sow.*	..	..	..	..	..	..	39 Sur., Ba.	39 Sur., Ba.
,, dimidiata, *Sow.* (=compressa, *Dav.*)	..	35 Ba.	38 Ba.	1 S.				

* Recorded by Dr. H. Woodward from the Chalk near Ventnor, *Geol. Mag.* for 1878, p. 556.
† Chalk Rock.   ‡ Melbourn Rock.

	Gault.	Malm.	Chert Beds.	Upper Greensand. Sub-division not specified.	Chloritic Marl.	Lower Chalk.	Middle Chalk.	Upper Chalk.
Rhynchonella grasiana, D'Orb.	..	..	..	..	8, 12 Sur., 35 Pa.*	1 Ba.		
„ latissima, Sow.	..	1 Pa., 35 N.	10 Sur.	..	8, 36 Sur., ? 1 I.			
„ limbata, Schloth.	..	..	..	..	35 N.	..	..	7 Ba.
†‡ „ mantelliana, Sow.	..	..	..	..	12 Sur.	32, 36 Sur.	25 Sur.	24, 29, 32 Sur.
„ Martini, Mantell.	..	..	..	37 Fi.	35 Sur.	32, 35, 36 Sur.		
„ parvirostris, Sow.	..	..	..	1 S.				
†§ „ plicatilis, Sow.	..	..	..	..	..	..	12, 32 Sur.	7, 32, 34 Sur., 14 Ba.
„ plicatilis var. octoplicata.	..	..	..	..	..	..	..	7 Ba.
„ Schlœnbachi, Dav.	..	..	..	..	36 D.			
„ subduplicata, D'Orb. (? sp.).	..	..	..	..	..	..	..	26 Ba.
„ subplicata, D'Orb. (? sp.).	..	..	..	..	..	..	..	24, 29 Ba.
‡ „ sp. -	..	..	..	..	..	..		7, 29 Sur.
Terebratella pectita, Sow.	..	..	..	1 S.	36 Sur., 35 Pa.			
‡ Terebratula biplicata, Sow.	..	8 Sur., 35 Ba.		..	35, 36 Sur., 5 Pa.	32 Sur.	..	32 Sur.
‡ „ carnea, Sow.	..	..	..	..	..	1 N.	..	7, 9, 32 Sur., 2 Ba.
„ convexa, Sow., see Rhynchonella latissima.								
„ disparilis, D'Orb. (? sp.).	..	..	..	..	..	1 Ba.		
„ lacrymosa, D'Orb. (? sp.).	..	..	..	..	..	1 Ba.		
„ ovata, Sow.	..	35 Ba.	..	..	35 Sur.			
„ pectita, Sow., see Terebratella.								
„ phaseolina, Lamk.	..	..	..	..	36 D.			
„ pisum, Sow., see Rhynchonella Martini.								
†‡ „ semiglobosa, Sow.	..	..	..	..	5 Pa.	10 Sur., 1 Ba.	39 Sur., Ba.	39 Sur., Ba.
„ sp. -	34 Sur.	8 Sur.	..	37 Fi.				
Terebratulina gracilis, Schloth.	..	..	..	..		36 Sur.	12, 24 Sur., 39 Ba.	39 Sur., 13 Ba.
„ striata, Wahl.	..	..	..	..	35 Pa., 36 Sur.	12, 35, 36 Sur., 1 Ba.	13, 26, 34 Ba.	7, 29, 32 Sur., 5, 6 Ba.
„ sp. -	..	..	38 Ba.					

* Derived and indigenous in the Chloritic Marl.   † Melbourn Rock.   ‡ Chalk Rock.
§ In a green nodular band in the Upper Chalk, see p. 78.

	Gault.	Malm.	Chert Beds.	Upper Greensand. Sub-division not specified.	Chloritic Marl.	Lower Chalk.	Middle Chalk.	Upper Chalk.
Lamellibranchiata (*Monomyaria*).								
Anomia - - -	..	..	10 Sur.					
Avicula gryphæoides, *Sow.*	..	..	..	..	8, 35 Sur.			
„ - - -	36 Sur.							
Exogyra canaliculata, *Sow.*	8 Sur.	1 Pa., 35 Ba.	35, 38 Ba.	1 S., 37 Fi.	..	..	..	2, 30 Ba.
„ columba, *Lam.*	..	8 Sur.	10 Sur.	..	36 Sur.			
„ conica, *Sow.* -	..	8 Sur., 35 Ba.	35 N., Pa.	..	35 N.	35 Sur.		
Exogyra haliotoidea, *Lam.*	34 Sur.							
„ laciniata, *Nilss.*	..	..	..	1 S.				
„ undata, see E. canaliculata.								
„ sp. - -	38 Sur.	..	..	..	..	12, 32 Sur.		
Gryphæa vesiculosa, *Sow.*	..	35 Ba.	1 Pa.,35, 38 Ba.	35 Sur., 37, 38 Fi.				
„ sp. - -	8 S., 38 Fi.							
Inoceramus Brongniarti, *Park.*	..	..	..	..	..	..	13, 32 Ba.	7 Sur.
„ concentricus, *Park.*	36 Sur.							
„ Crispii, *Goldf.* (*? sp.*).	..	..	..	..	..	..	..	7, 9, 16 Ba.
„ Cuvieri, *D'Orb.*	..	..	..	..	..	..	6, 13 Ba.	
„ involutus, *Sow.*	..	••	..	..	..	..	..	5 Ba.
„ labiatus, see I. mytiloides.								
„ latus, *Mant.*	..	..	10 Sur.	..	..	35, 36 Sur ?	10, 32, 35, 36 Sur.	
* „ mytiloides, *Mant.*	..	..	..	..	..	..	39 Sur. Ba.	
* „ striatus, *Mant.*	..	..	..	..	..	29 Sur.? 1 Ba.	24, 29, 32 ? Sur.	
„ sulcatus, *Park.*	36 S., 8 N., 1 Pr.							
„ sp. -	34, 38 Sur.	35 Ba., N.	..	1 S.	..	12 Sur.		
Janira. See Pecten.								
Lima archiaciana, *Cor.* - and *Bri.*	..	35 Ba.						
„ aspera, *Sow.* -	..	..	1 Pa.					
„ consobrina, *D'Orb.*	..	..	..	..	35 Sur.			
„ elongata, *Sow.* -	..	..	..	..	1 I.	35 Sur.?		
* „ globosa, *Sow.* -	8 Sur.?	..	..	..	..	..	29 Sur.	32 Sur.

* Melbourn Rock.

	Gault.	Malm.	Chert Beds.	Upper Greensand. Sub-division not specified.	Chloritic Marl.	Lower Chalk.	Middle Chalk.	Upper Chalk.
Lima Hoperi, *Sow.* -	..	..	..	1 S., 37 Fi.	.•	..	..	7 Sur.
„ ornata, *D'Orb.* -	..	..	..	..	.•	35 Sur.?		
„ parallela, *Sow.* -	8, 36 Sur.	..	..	..	..	..		
„ semisulcata, *Sow.*-	..	..	..	..	..	..	..	32 Sur.
* „ spinosa, *Sow.* -	..	.•	..	..	..	..	9, 13, 21 Ba.	24, 29, 32 Sur. 6, 13 Ba.
* „ striata, *Goldf.* -	..	..	..	..	..	..	12 Sur.?	7 Sur.?
„ *sp.* - - -	34 Sur.	8 Sur.	1 Pa.	..	..	32, 36 Sur.	15 Ba.	7, 12 Sur.
Ostræa *canaliculata*, *D'Orb.*, see Exogyra.								
„ *carinata*, *Sow.*, see O. frons.								
„ *conica*, *Lamk.*, see Exogyra.								
„ flabelliformis, *Nilss.* (*? sp.*).	..	..	..	.•	..	..	34 Ba.	
„ frons, *Park* (=*carinata*, *Sby.*).	..	8 Sur., 1 Pa.	..	..	1 S. Ba., 35 N., 36 Sur.	1 Ba., 35 N.		
„ hippopodium, *Nilss.* (*? sp.*).	..	..	..	..	..	1 Ba.	13, 19 Ba.	17 Ba.
„ normanianna ?, *D'Orb.*	..	..	..	..	..	36 Sur.		
„ *pectinata*, see O. frons.								
„ vesicularis, *Lamk.*	..	..	..	..	35 Sur.	35, 36 Sur.	..	32 Sur., 3, 7, 29 Ba.
„ *vesiculosa*, *Sow.*, see Gryphæa.								
„ virgata, *Goldf.* -	..	.•	..	1 S.				
„ *sp.* - -	8 Sur.	..	..	..	..	12 Sur.	12 Sur.	
†Pecten asper, *Lamk.* -	..	35 Ba.	8 Sur.	..	1 I., Ba., 35 N., Pa.	1 Ba., 35 N.		
„ Beaveri, *Sow.* -	..	..	..	..	35 Pa., 36 Sur.			
„ cretosus, *Defr.* -	..	..	..	..	..	..	..	2, 4, 7 Ba.
„ depressus, *Münst.*	..	..	..	..	..	1 Ba.		
„ elongatus, *Lamk.*	..	35 N.						
„ *Galliennii*, see P. interstriatus.								
„ hispidus, *Goldf.* (*? sp.*).	..	35 Ba.						
„ interstriatus, *Leym.*	..	..	8 Sur., 1 Pa.	..	35 Pa.			
„ *laminosus*, *Mant.*, see P. orbicularis.								
„ nitidus, *Mant.* •	..	..	..	..	..	35 Sur.		
„ orbicularis, *Sow.*	8, 34, 36, 38 Su.	1 Pa., 35 Ba.	10 Sur., 35 Ba.	..	8, 35 Sur.	10, 35, 36 Sur.		
„ (Neithea) ornithopsis, *Keeping.*	..	..	..	..	8 Sur.			

* Chalk Rock.   † Derived in the Chloritic Marl according to M⸱ . Parkinson.

S 2

	Gault.	Malm.	Chert Beds.	Upper Greensand. Sub-division not specified.	Chloritic Marl.	Lower Chalk.	Middle Chalk.	Upper Chalk.
Pecten quadricostatus, *Sow.*	1 S.	35 Ba.	..	1 S., 37 Fi.	..			
„ quinquecostatus, *Sow.*	8 Fi., 34 Sur.	8 Sur., 35 Ba.	10 Sur., 1 I., Pa.	..	12, 35, 36 Sur.	8, 35, 36 Sur.	24 Sur.	30 Ba.
„ sp. allied to P. rauliniunus, *D'Orb.*	8 Sur.							
„ *sp.* - - -	..	..	1 Pa.					
Pinna - - -	8, 34 Sur.	35 Ba.						
Plagiostoma Hoperi, see Lima.								
Plicatula inflata, *Sow.* ·	..	..	..	1 S., 37 Fi.	35, 36 Sur.	32, 35, 36 Sur.		
„ pectinoides, *Sow.*	..	1 Pa., 35 Ba.	..	1 S.	8 Sur., 35 Pa.	10 Sur. ?		
„ sigillina, *S. P. Woodw.*	..	35 Ba.	..	..	..	..	..	28, 29 Ba.
„ sp. - -	34 Sur.							
Spondylus æqualis, *Héb.*	..	..	..	..	..	..	..	6 Ba.
„ dutempleanus, *D'Orb.*	..	..	..	..	..	..	..	5, 7 Ba.
„ *spinosus, see Lima.*								
„ striatus, *Sow.*	..	..	..	1 S.				
„ *sp.* - -	..	..	..	..	..	..	19 Ba.	23, 29 Ba.
Lamellibranchiata (Dimyaria).								
Arca Carteroni, *D'Orb.* -	..	..	10 Sur.					
„ mailleana, *D'Orb.*	..	..	8 Sur.	..	35, 36 Sur.			
„ royana, *D'Orb.* ·	..	..	..	..	35 Sur.			
„ - - -	8 S.	..	..	..	..	32 Sur.		
Caprotina, *sp.* - -	..	..	..	..	..	..	..	6 Ba.
Cardita tenuicosta, *Fitt.*	..	35 Ba.						
„ *sp.* - -	8 S., 34, 38 Sur.	..	35, 38 Ba.					
Cardium gentianum, *Sow.*	..	1 I., Pa., 35 N.	..	1 S.				
„ *tuberculatum, see C. gentianum.*								
„ *sp.* - -	38 Sur.	..	1 Pa., 37 Fi.					
Cucullæa carinata, *Sow.* -	8, 34 Sur.	1 I., Pa. 35 Ba.	..	1 S.	..	1 S.		
„ fibrosa, *Sow.* (? sp.)	..	1 Pa.						
„ glabra, *Park.* -	..	1 Pa., 35 Ba., N.	1 Pa.	37 Fi.	8, 12 Sur.			
„ *sp.* - -	38 Sur.							
Cyprina angulata, *Flem.*	..	..	..	1 S.?				
„ *sp.* -	..	..	..	1 S.	36 Sur.			
Cytherea, *sp.* - -	..	..	..	1 S.	36 Sur.?			

	Gault.	Malm.	Chert Beds.	Upper Greensand. Sub-division not specified.	Chloritic Marl.	Lower Chalk.	Middle Chalk.	Upper Chalk.
Isocardia - - -	..	..	·...	·...	6 Sur.			
Lucina tenera, *Sow.* -	8 Sur.				··			
„ *sp.* -	34, 38 Sur.		·		·			
Modiola ligeriensis, *D'Orb.*	..	..	..	1 S.				
„ -	8 Sur.	1 Pa.			··			
Myacites mandibula, Sow., see Panopœa.			·					
Mytilus lanceolatus, *Sow.*	..	36 N.	·		·			
Nucula bivirgata, *Sby.* -	34 Sur.							
Panopœa mandibula, *Sow.*	1 S., 38 Fi.	1 Pa., 35 N., 1 I., 8 Sur.	10 Sur.	1 S.				
„ plicata, *Sow.* -	8 Sur.	1 Pa., 35 N.	..	35 S.	12 Sur.			
„ *sp.* - -	..	35 N.	10 Sur.					
Pholadomya decussata, *Phil.*	..	..	..	..	..	1 N.		
Pholas, *sp.* - -	..	..	..	..	1 I.			
Solen dupinianus, *D'Orb.*	34 Sur.				··			
Thetis *major,* see T. Sowerbyi.								
„ Sowerbyi, *Röm.*	..	1 I.	·..	1 S.				
Trigonia aliformis, *Park.*	..	1 Pa., 35 N.	1 Lyc.	1 S.				
„ archiaciana, *D'Orb.*	..	..	..	35 Lyc.				
„ carinata, *Ag.* -	..	1 I.	..	1 S., 40 Lyc.				
„ *harpa,* see I. carinata.								
„ spinosa, *Park.*	..	1 I.	..	Lyc.				
„ vicaryana, *Lycett.*	..	1 Pa.	..	S.				
Venus, *sp.* -	..	..	..	37 Fi.	35 Pa.			
	..							
Gasteropoda.								
Actæon affinis, *Sow.* -	..	..	..	1 S.	..	10 Sur.		
„	..	..	·.	..	35, 36 Sur.			
Aporrhais Parkinsoni, *Sow.*	..	·.	..	1 S.				
„ *new sp.* -	..	..	..	..	..	20 Bn.		
„ *sp.* - -	34 Sur.	∴.	..	1 S.	10, 36 Sur.	29 ?, 36 Sur.		
Avellana cassis, *D'Orb.* -	..	..	..	..	1 I., Bn.			
„ (Cinulia) -	..	1 Pa.	..	..	12 Sur.?, 35 Pa.	36 Sur.		
Columbellina -	..	..	..	..	36 Sur.			
Dentalium -	8 Sur.							
Emarginula *sp.* -	..	..	..	..	35 Pa.			

	Gault.	Malm.	Chert Beds.	Upper Greensand. Sub-division not specified.	Chloritic Marl.	Lower Chalk.	Middle Chalk.	Upper Chalk.
Fusus? - - -	..	..	..	1 S.				
Gibbula lævistriata, *Seel.*	..	..	..	..	35 Pa.			
Littorina carinata, *Sow.*	..	..	..	1 S.				
Natica gaultina, *D'Orb.*	8 N.	..	..	..	1 I.			
,, - - -		1 Pa.	..	..	..	1 N.		
Pleurotomaria moreausiauns, *D'Orb.*	..	..	..	..	35, 36 Sur.			
,, perspectiva, *Mant.*	.	..	..	..	..	1 S.		
,, Rhodani, *D'Orb.*	..	..	..	..	8 ?, 35 Sur.			
* ,, - - -	..	..	..	..	1 Ba., 35 Pa.	36 Sur.		
Rostellaria, see Aporrhais.								
Solarium conoideum, *Sow.*	..	..	..	37 Fi.				
,, ornatum, *Sow.*	1 Pr.	35 Ba.	..	36,38 S., 37 Fi.		8, 35 Sur.		
Trochus - - -	..	..	..	37 Fi.				
Turbo problematicus, *P. & R.*	..	..	..	1 S.				
,, - - -	..	..	..	1 S.	12 Sur.			
Turritella - - -	8 Sur.?							
Cephalopoda.								
Ammonites auritus, *Sow.*	..	1 Pa.	..	1 S.				
,, Benettiæ, *Sow.* see A. interruptus.								
,, Beudantii?, *Brong.*	38 Sur.							
,, bouchardianus, *D'Orb.*	34 Sur.							
,, catinus, *Mant.*	..	..	..	..	..	35 N.		
,, cenomanensis, *D'Arch.*	..	..	..	..	..	1 Ba.		
,, cinctus, *Mant.*	..	..	..	38 Fi.?				
,, Coupei, *Brong.*	..	..	..	..	39 Sur.	12?, 30, Sur.		
,, curvatus, *Mant.*	..	..	..	..	35 Pa., Sh.			
,, dentatus, see A. interruptus.								
,, deverianus, *D'Orb.*	..	..	..	..	..	10 Sur. ?		
,, falcatus, *Mant.*	..	..	..	..	1 S., Ba.	1 Ba., 35 N.		
,, feraudianus, *D'Orb.*	..	..	..	..	..	35 Sh.		
,, Gentoni, *Brong.*	..	..	..	..	..	1 Ba.		
,, inflatus, *Sow.*, see A. rostratus.								

* Derived in the Chloritic Marl according to Mr. Parkinson.
† Derived and indigenous in the Chloritic Marl according to Mr. Parkinson.

	Gault.	Malm.	Chert Beds.	Upper Greensand. Sub-division not specified.	Chloritic Marl.	Lower Chalk.	Middle Chalk.	Upper Chalk.
Ammonites interruptus, *Brong.*	8, 34 ? Sur., 38 Fi.							
„ laticlavius, *Sharpe.*	..	..	..	..	1 Ba.	1 Sh.		
„ leptonema, *Sharpe.*	..	..	..	..	..	35 Sh.		
„ mammillaris, *Schloth.*	..	..	..	..	1 I.			
„ Mantelli, *Sow.*	..	..	..	37 Fi.?	8, 35, 36 Sur.	10, 12, 32 Sur., 35 N.		
„ *monile,* see A. mammillaris.								
„ navicularis, *Mant.*	..	..	..	..	35 Sur., Ba.	10,32,35, 36 Sur.		
„ octosulcatus, *Sharpe.*	..	..	..	..	..	35 Sh.		
＊ „ peramplus, *Mant.*	..	..	..	..	35 N.	..		9 Ba., 1 Sh.
„ planulatus, *Sow.*	..	..	..	..	38 S.			
„ renauxianus, *D'Orb.*	..	35 Ba.	..	..	..	35 Sh.		
„ Renevieri, *Sharpe.*	..	..	..	..	..	36 Sur., 38 Sh.		
„ rostratus, *Sow.*	1 Pr., 8 Sur.	1 I., Pa., 35 N., Ba.	1 Pa., 35 N.	..	35 N.			
„ rhotomagensis, *Brong.*	..	..	..	..	12 Sur., 1 Ba.	39 Sur., Ba.		
„ Saxbii, *Sharpe.*	..	..	..	..	..	35 Sh.		
„ selliguinus, *Brong.*	..	..	..	38 Fi.				
„ splendens, *Sow.*	..	35 N.	..	..	1 I.			
† „ varians, *Sow.* -	..	35 N., 38 Fi.	..	..	39	39 Sur., Ba.		
„ vectensis, *Sharpe.*	..	..	..	..	1 Ba., 35 Sh.			
⌀ „ Velledæ, *Michelin.*	..	..	..	..	..	35 Sh.		
„ Woolgari, *Mant.*	..	..	..	..	..	..	?	
„ *sp.* - -	..	..	10 Sur.					
„ „ between A. auritus and A. rostratus.	..	1 Pa.						
Baculites anceps, *Lam.* -	..	..	..	..	..	1 N.		
„ baculoides, *D'Orb.*	..	..	..	..	..	1 S., Ba.		
„ Faujasii, *Sow.* -	..	..	..	..	..	35 N.		
„ *sp.* - -	..	..	..	..	38 S., 36 Sur.	29, 32 Sur.		
Belemnitella mucronata, *Schloth.*	..	..	..	..	..	..	..	39 Ba.

＊ Derived in the Chloritic Marl?
† Derived and indigenous in the Chloritic Marl according to Mr. Parkinson.

	Gault.	Malm.	Chert Beds.	Upper Greensand. Sub-division not specified.	Chloritic Marl.	Lower Chalk.	Middle Chalk.	Upper Chalk.
*Belemnitella quadrata, *Schloth.*	..	..	..	..	..	..	..	26 Ba.
Belemnites minimus, *List.*	..	..	..	..	35 N.			
„ ultimus, *D'Orb.*	1 Pr.	..	..	1 S.	1 Ba.,? 36 Sh.			
„ - - -	..	35 Ba.	..	37 Fi.				
Hamites armatus, *Sow.* -	..	1 I., 35 Ba.	..	1 S.	1 S.	1 S., 12 Sur.		
„ attenuatus, *Sow.*	..	..	..	..	35 Pa.	35 N.		
„ elegans, *Park.* -	..	..	..	..	8 Sur. ?			
„ - - -	8 Sur.	..	..	..	12, 36 Sur.			
Nautilus *compressus*, see N. Fittoni, *Sharpe.*								
„ deslongchampsianus, *D'Orb.*	..	..	..	..	1 S.	1 Ba.		
„ elegans, *Sow.* -	..	..	1 Pa.	..	35 Pa.	1 Ba.,N.		
„ expansus, *Sow.*	..	..	..	..	1 S., 35 Pa.	1 N.?		
„ Fittoni, *Sharpe*	..	..	..	37 Fi., 1 Sh.				
„ lævigatus, *D'Orb.*	..	..	..	..	35, 36 Sur.	1 Ba.		
„ largilliertianus, *D'Orb.*	..	..	..	..	1 Sh.			
„ pseudoelegans, *D'Orb.*	..	35 N.	..	..	35 N.	36 Sur.		
„ radiatus, *Sow.* -	..	35 N.	..	1 I.				
„ undulatus, *Sow.*	..	1 I.						
„ sp. - -	..	..	1 Pa., 35 N.	..	..	32 Sur.		
Rhyncholites - -	..	..	..	..	1 S.			
Scaphites æqualis, *Sow.*	..	..	..	..	1 Ba., S. 35 Pa.	35, 36 Sur.		
„ *costatus*, see S. æqualis.								
„ *striatus*, see S. æqualis.								
Turrilites Bergeri, *Brong.*	..	..	..	..	8,35,36 Sur.	36 Sur. 35 N.		
„ bifrons, *D'Orb.*	..	..	..	..	..	35 Sh.?		
„ costatus, *Lam.*	..	..	..	..	1 S.	1S.,Ba., 12 Sur.		
„ gravesianus, *D'Orb.*	..	..	..	..	1 S.	-		
„ Morrisii, *Sharpe.*	..	..	..	..	39 Sur.	12, 35 Sur.		
„ puzosianus, *D'Orb.*	..	..	..	..	38 Sh.			
„ scheuchzerianus, *Bosc.*	..	..	..	..	..	35, 36 Sur.		
„ tuberculatus, *Bosc.*	..	..	..	38 Fi.	35, 36 Sur.	35, 36 Sur.		
„ *undulatus*, see T. scheuchzerianus.								
„ Wiestii, *Sharpe*	..	..	..	..	8, 35 Sur, 35 Pa.	35 Sur.		

* Rare.

	Gault.	Malm.	Chert Beds.	Upper Greensand. Sub-division not specified.	Chloritic Marl.	Lower Chalk.	Middle Chalk.	Upper Chalk.
Pisces.								
Elasmobranch vertebra -	..	..	..	..	1 S.			
Gyrodus - · ·	..	35 N.						
Lamna (teeth and vertebræ).	..	..	..	35 N.	36 Sur., 35 N.	35 Sur.		
Otodus - · ·	..	35 N.	..	..	..	32 Sur.		
Ptychodus paucisulcatus, *Dixon*.	..	..	..	..	..	..	..	1 S.
„ polygurus, *Ag.*	..	..	..	..	..	1 N.		
Various teeth, &c. -	8, 4 Sur., 38 Fi.	..	..	..	8 Sur.	1 N.	12, 24, 35 Sur.	
Reptilia.								
Chelonian remains -	..	35 N., 38 Ma.						
Plastremys lata, *Owen.* -	..	1 Pa.						
Polyptychodon interruptus, *Owen.*	..	..	..	..	..	1 S.		
Titanosaurus, *sp.* -	..	..	..	1 L.				
Various bones - ·	..	..	..	..	35 N.			‘

TABLE IV.—Eocene and Oligocene.

One authority for the occurrence of each species is indicated by the letters :—

E = Edwards, *Monograph of the Eocene Mollusca (Palæontographical Society).*
F = Fisher, *Quart. Journ. Geol. Soc.*, vol. xviii. p. 65. 1862.
G = Gardner, *Geol. Mag.* for 1885, p. 241.
J = Judd, *Quart. Journ. Geol. Soc.*, vol. xxxvi. p. 137. 1880.
K = Keeping and Tawney, *Quart. Journ. Geol. Soc.*, vol. xxxvii. p. 85. 1881.
L = Lydekker, *Cat. Foss. Mammalia in Brit. Museum*
P = Prestwich, *Quart. Journ. Geol. Soc.*, vol. ii. p. 223. 1846.
S = Geological Survey (specimens in the collection, or recorded in the 1st edit. of this Memoir).
W = Wood, *Monograph of the Eocene Mollusca (Pal. Soc.).*
Ww.= Woodward, *Quart. Journ. Geol. Soc.*, vol. xxxv. p. 342. 1879.

MS. species are not included.

As the plants are now being examined by Mr. J. Starkie Gardner it has not been thought advisable to republish the old determinations. Mr. Gardner's account of the flora of the Lower Bagshot Beds of Alum Bay will be found at p. 104.

	Reading Beds.	London Clay.	Lower Bagshot Beds.	Bracklesham Beds.	Barton Clay.	Headon Hill Sands.	Headon Beds.	Osborne Beds.	Bembridge Beds.	Hamstead Beds.
Foraminifera.										
Alveolina fusiformis, *Sow.*	..	..	..	S					•	
„ sabulosa, *Mont.*	..	..	..	F						
Nummulites elegans, *Sow.*	..	..	..	..	S					
„ lævigatus, *Brug.*	..	..	..	S						
„ variolarius, *Lam.*	..	..	..	S	?					
Operculina *sp.*	..	..	..	..	S					
Quinqueloculina Hauerina, *D'Orb.*	..	..	..	..	F					
Rotalina obscura, *Sow.*	..	..	..	F						
Actinozoa.										
Dendrophyllia *sp.*	..	..	..	..	..	..	K			
Lith-aræa Brockenhurstii, *Dunc.*	..	..	..	..	..	..	K			
Madrepora anglica, *Dunc.*	..	..	..	..	..	..	K			
Solenastræa cellulosa, *Dunc.*	..	..	..	..	..	..	K			
Turbinolia Bowerbankii, *E. & H.*	..	..	..	S						
„ Forbesii, *Dunc.*	..	..	..	S						
„ Fredericiana ? *E. & H.*	..	..	..	S						

	Reading Beds.	London Clay.	Lower Bagshot Beds.	Brackleshnm Beds.	Barton Clay.	Headon Hill Sands.	Headon Beds.	Osborne Beds.	Bembridge Beds.	Hamstead Beds.
Turbinolia minor, *Lam.*	..	..	..	S	S					
„ sulcata, *Lam.*	..	..	..	S						
Echinodermata.										
Schizaster D'Urbani, *Forbes.*	..	..	..	..	S					
Annelida.										
Ditrupa incrassata, *Sow.*	..	S								
„ plana, *Sow.*	..	S	..	F	K					
Serpula corrugata, *Sow.*	..	..	..	..	..	..	S			
„ extensa, *Brand*	..	..	..	K						
„ tenuis, *Sow.*	..	..	..	..	..	..	?	S	S	S
Vermicularia bognoriensis, *Mant.*	..	P								
Insecta.										
(*Hemiptera.*)										
Tricephora sanguinolenta, *Scop.*	..	..	..	..	..	..	..	..	Ww.	
Wing of ?	..	..	..	..	..	..	..	..	Ww.	
(*Orthoptera.*)										
Acridiidæ	..	..	..	..	..	..	..	..	Ww.	
Gryllotalpa	..	..	..	..	..	..	..	..	Ww.	
(*Neuroptera.*)										
Agrion	..	..	..	..	..	..	..	..	Ww.	
Hemerobius	..	..	..	..	..	..	..	..	Ww.	
Libellula (wings)	..	..	..	..	..	..	..	..	Ww.	
Perla	..	..	..	..	..	..	..	..	Ww.	
Phryganea	..	..	..	..	..	..	..	..	Ww.	
Termes ?	..	..	..	..	..	..	..	..	Ww.	
(*Diptera.*)										
Tipulidæ	..	..	..	..	..	..	..	..	Ww.	
Wings of ?	..	..	..	..	..	..	..	..	Ww.	
(*Lepidoptera.*)										
Lithopsyche antiqua, *Butler*	..	..	..	..	..	..	..	..	B.	
Lithosia	..	..	..	..	..	..	..	..	Ww.	

	Reading Beds.	London Clay.	Lower Bagshot Beds.	Bracklesham Beds.	Barton Clay.	Headon Hill Sands.	Headon Beds.	Osborne Beds.	Bembridge Beds.	Hamstead Beds.
(Hymenoptera.)										
Camponotus	..	..	..	..	..	..	..	..	Ww.	
Formica	..	..	..	..	..	..	..	..	Ww.	
Myrmica	..	..	..	..	..	..	..	..	Ww.	
Wings of?	..	..	..	..	..	..	..	..	Ww.	'
(Coleoptera.)										
Anobium	..	..	..	..	..	..	..	..	Ww.	
Curculio	..	..	..	..	..	..	..	..	Ww.	
Dorcus (Lucanidæ)	..	..	..	..	..	..	..	..	Ww.	
Staphylinus	..	..	..	..	..	..	..	..	Ww.	
Arachnida.										
Eoatypus Woodwardii, McCook.	..	..	..	..	..	..	..	..	•	
Crustacea.										
(For *Ostracoda*, see table p. 298).										
Balanus unguiformis, *Sow.*	..	..	..	..	..	..	S	..	S	S
Brachipodites vectensis, *H. Woodw.*	..	..	..	..	..	..	..	..	Ww.	
Callianassa Batei, *H. Woodw.*	..	..	..	..	..	..	K	..	..	Ww.
Eosphæroma fluviatile, *H. Woodw.*	..	..	..	..	..	..	..	..	Ww.	
" Smithii, *H. Woodw.*	..	..	..	..	..	..	..	..	Ww.	
Mithracites vectensis, *Gould.*	..	S.								
Pollicipes reflexus, *Sow.*	..	..	..	..	..	..	S			
Xanthopsis Leachii, *Desmarest.*	..	W.								
Polyzoa.										
Membranipora Lacroixii, *Busk.*	..	..	..	..	..	..	S			
Undetermined species	..	..	..	..	..	..	..	..	S	
Lamellibranchiata.										
(Monomyaria.)										
Anomia tenuistriata, *Desh.*	..	..	..	..	..	..	K			
Avicula media, *Sow.*	..	..	..	..	K	..	K			
Lima *sp.*	..	..	..	..	K					
Ostrea adlata, *S. Wood*	..	..	..	..	..	..	..	..	..	S

• See *Proc. Acad. Nat. Sc. Philadelphia*, 1888, pp. 200-202, and *Annals and Mag. Nat. Hist.*, ser. 6, vol. II., 1888, pp. 366-369.

	Reading Beds.	London Clay.	Lower Bagshot Beds.	Bracklesham Beds.	Barton Clay.	Headon Hill Sands.	Headon Beds.	Osborne Beds.	Bembridge Beds.	Hamstead Beds.
Ostrea callifera, *Lam.* -	..	..	..	..	..	..	..	..	..	S
„ dorsata? *Desh.* -	..	..	..	..	F					
„ flabellula, *Lam.* -	..	..	..	F	S	..	S			
„ longirostris, *Lam.* -	..	..	..	..	..	..	..	..	..	W
„ vectensis, *Forbes* -	..	..	..	..	..	..	?	..	S	
„ velata, *Wood* -	..	..	..	..	..	..	S			
„ large *sp.* -	..	S								
Pecten bellicostatus, *Wood* -	..	..	..	..	..	..	K			
„ carinatus, *Sow.* -	..	..	..	..	K					
„ corneus, *Sow.* -	..	..	..	F	F					
„ idoneus, *Wood* -	..	..	..	K						
„ reconditus, *Brand.* -	..	..	..	..	S	..	S			
„ 30-radiatus, *Sow.* -	..	..	..	F						
Pinna affinis, *Sow.* -	..	S								
„ margaritacea, *Lam.* -	..	..	..	F						
(*Dimyaria.*)										
Arca appendiculata, *Sow.* -	..	..	..	K	S					
„ aviculina? *Desh.* -	..	..	..	F	F					
„ bianzula, *Lam.* -	..	..	..	..	..	..	K			
„ lævigata, *Caill.* -	..	..	..	..	..	..	J			
„ Websteri, *Forbes* -	..	..	..	..	..	..	..	..	S	
Astarte rugata, *Sow.* -	..	S								
Cardita deltoidea, *Sow.* -	..	..	..	..	..	..	S			
„ oblonga, *Sow.* -	..	..	..	K	K	..	K			
„ paucicostata, *Sand.* -	..	..	..	..	..	..	W			
„ planicosta, *Lam.* -	..	..	..	S						
„ simplex, *Wood* -	..	..	..	..	..	..	S			
„ sulcata, *Brand.* -	..	..	..	S	S					
Cardium porulosum, *Brand.* -	..	..	..	F	S	..	K			
„ semigranulatum, *Sow.*	..	S	..	..	S	..	S			
„ turgidum, *Brand.* -	..	..	..	..	S					
„ *sp.* -	..	..	..	..	..	..	..	..	..	K
Chama gigantea, *Lowry* -	..	..	..	F						
„ squamosa, *Brand.* -	..	..	..	..	K					
Corbula cuspidata, *Sow.* -	..	..	..	S	..	..	S			
„ ficus, *Brand.* -	..	..	..	..	S					
„ gallica? *Lam.* -	..	..	..	S						

	Reading Beds.	London Clay.	Lower Bagshot Beds.	Bracklesham Beds.	Barton Clay.	Headon Hill Sands.	Headon Beds.	Osborne Beds.	Bembridge Beds.	Hamstead Beds.
Corbula nitida, *Sow.*	..	..	..	..	..	..	S			
„ pisum, *Sow.*	..	..	..	F	S	..	S	..	S	S
„ revoluta, *Sow.* (= costata).	..	..	..	F	S					
„ rugosa, *Lam.*	..	..	..	S						
., vectensis, *Forbes*	..	..	..	..	..	..	..	..	..	S
Crassatella compressa, *Lam.*	..	..	..	F						
„ Sowerbii, *Edw.*	..	..	..	W						
„ subquadrata, *Edw.*	..	..	..	W						
„ sulcata, *Brand.*	..	..	..	S	S					
„ tenuisulcata, *Edw.*	..	..	..	..	F					
Cyclas Bristovii, *Forbes*	..	..	..	..	..	..	..	..	..	S
Cypricardia pectinifera, *Sow.*	..	..	..	..	..	..	K			
„ sp.	..	..	..	..	F					
Cyprina Nysti, *Héb.*	..	..	..	..	..	..	K			
„ planata, *Sow.*	..	S								
Cyrena arenaria, *Forbes*	..	..	..	..	..	..	S			
„ cycladiformis, *Desh.*	..	..	..	..	..	..	S			
„ deperdita, *Lam.*	..	..	..	..	..	..	S			
„ gibbosula, *Morris*	..	..	..	..	..	..	S			
„ obovata, *Sow.*	..	..	..	..	..	..	S	S	S	
„ obtusa, *Forbes*	..	..	..	..	..	..	..	..	S	
„ pulchra, *Sow.* (= *Wrightii.*).	..	..	..	..	..	..	S	..	S	
„ semistriata, *Desh.*	..	..	..	..	..	..	..	..	S	S
„ transversa, *Forbes*	..	..	..	..	..	..	..	..	S	
Cytherea elegans, *Lam.*	..	..	..	..	S					
„ incrassata, *Desh.*	..	..	..	..	S	..	S	..	S	
„ lucida, *Sow.*	..	..	..	F						
„ Lyellii, *Forbes*	..	..	..	..	..	..	..	..	..	S
„ obliqua, *Desh.*	..	..	..	F	S					
„ Solandri, *Sow.*	..	..	..	..	..	..	K			
„ suberycinoides, *Desh.*	..	..	..	F						
„ suessonensis, *Desh.*	..	..	..	..	..	..	K			
„ tenuistriata, *Sow.*	..	S	..	..	K					
„ transversa, *Sow.*	..	..	..	..	S					
„ tellinaria, *Lam.*	..	..	..	..	S					
Diplodonta sp.	..	S								
„ sp.	..	..	..	..	..	..	K			

	Reading Beds.	London Clay.	Lower Bagshot Beds.	Brackleshum Beds.	Barton Clay.	Headon Hill Sands.	Headon Beds.	Osborne Beds.	Bembridge Beds.	Hamstead Beds.
Dreissena Brardii, *Faujas* -	..	..	..	..	..	..	S	..	..	K
Leda minima, *Sow.* -	..	..	..	..	K					
„ partimstriata, *Wood* -	..	W								
„ propinqua, *Wood* -	..	..	..	..	..	..	W			
Leda *sp.* -	..	..	..	F						
Lepton *sp.* -	..	..	..	..	..	..	J			
Limopsis scalaris, *Sow.* -	..	..	..	..	K					
Lithodomus *sp.* -	..	..	..	..	..	..	..	..	..	S
Lucina concava, *Defr.* -	..	..	..	..	..	..	J			
„ gibbosula, *Lam.* -	..	..	..	..	K					
„ inflata, *Lowry* -	..	..	..	..	..	..	K			
„ Thierensi, *Héb.* -	..	..	..	..	..	..	..	..	S	K
„ 4 species -	..	..	..	..	..	..	K			
„ *sp.* -	..	..	..	F						
Mactra fastigata, *Lowry* -	..	..	..	..	..	..	S			
„ *sp.* -	..	..	..	..	..	..	J			
Modiola ? consobrina, *Wood* -	..	..	..	W						
„ elegans, *Sow.* -	..	S								
„ flabellula, *Wood* -	..	..	..	..	..	..	S	..	..	W
„ Nystii, *Kickx.* -	..	..	..	..	..	..	K			
„ Prestwichii, *Morris* -	..	..	..	..	..	..	..	..	..	S
„ simplex, *Sow.* -	..	S								
Mya ? angustata, *Sow.* -	..	..	..	..	..	..	S			
„ (see also Panopæa).										
Mytilus affinis, *Sow.* -	..	..	..	..	..	..	S	..	S	
Necera cochlearella, *Desh.* -	..	..	..	..	S					
Nucula amygdaloides, *Sow.* -	..	S								
„ bisulcata, *Sow.* -	..	..	..	..	S					
„ *deltoidea* (see Trigonocœlia).										
„ Dixoni, *Edw.* -	..	..	..	F						
„ Headonensis, *Forbes* -	..	..	..	..	..	..	S			
„ lissa, *Wood* -	..	..	..	..	..	..	K			
„ nudata, *Wood* -	..	..	..	..	..	..	W			
„ similis, *Sow.* -	..	..	..	..	S	..	S	..	S	
„ sphenoides, *Edw.* -	..	..	..	..	..	..	..	..	..	W
„ subtransversa? *Nyst.* -	..	..	..	F						
Panopæa corrugata, *Sow.* -	..	..	..	F	..	..	S			
„ intermedia, *Sow.* -	..	S	..	..	S					

	Reading Beds.	London Clay.	Lower Bagshot Beds.	Bracklesham Beds.	Barton Clay.	Headon Hill Sands.	Headon Beds.	Osborne Beds.	Bembridge Beds.	Hamstead Beds.
Panopæa minor, *Forbes*.	..	..	..	..	..	..	..	..	S	S
Pectunculus brevirostris, *Sow*.	..	S								
„ deletus, *Brand*.	..	..	..	..	S					
„ pulvinatus, *Lam*.	..	..	..	F						
Pholadomya margaritacea, *Sow*.	..	S								
Potamomya gregaria, *Sow*.	..	..	..	..	..	..	S			
„ plana, *Sow*.	..	..	..	..	..	..	S	S		
Protocardium *sp*.	..	..	..	..	..	..	J			
„ (see also Cardium).										
Psammobia compressa, *Sow*.	..	..	..	..	S	..	S			
„ rudis, *Lam*. (= *solida*).	..	..	..	..	..	..	S			
Sanguinolaria Hollowaysii, *Sow*.	..	..	..	F						
Scintilla *sp*.	..	..	..	..	..	..	J			
Solen affinis, *Sow*.	..	S								
„ obliquus, *Sow*.	..	..	..	F						
Strigilla pulchella, *Ag*.	..	..	..	..	..	..	K			
Sydosmya *sp*.	..	..	..	..	..	..	J			
Tellina ambigua, *Sow*.	..	..	..	..	S	..	S			
„ filosa?, *Sow*.	..	..	..	F						
„ Nystii?, *Desh*.	..	..	..	..	..	..	..	..	..	S
„ plagia, *Edw*.	..	..	..	F						
„ tumescens?, *Edw*.	..	..	..	F						
„ 3 species	..	..	..	..	..	..	K			
Teredo *sp*.	..	..	..	F	S					
Trigonocœlia deltoidea, *Lam*.	..	..	..	..	..	..	S			
Unio Austenii, *Forbes*	..	..	..	..	..	..	..	..	..	S
„ Gibbsii, *Forbes*	..	..	..	..	..	..	S	..	..	S
„ Solandri, *Sow*.	..	..	..	..	..	..	S			
Scaphopoda.										
Dentalium striatum, *Sow*.	..	..	..	?	S					
„ *sp*.	..	..	..	..	..	..	K			
Gasteropoda.										
Achatina costellata, *Sow*.	..	..	..	..	..	..	G	..	S	
Actæon dactylinus, *Desh*.	..	..	..	..	..	..	K			
„ limnæiformis, *Sandb*.	..	..	..	..	..	..	K			

	Reading Beds.	London Clay.	Lower Bagshot Beds.	Bracklesham Beds.	Barton Clay.	Headon Hill Sands.	Headon Beds.	Osborne Beds.	Bembridge Beds.	Hamstead Beds.
Actæon simulatum, *Sow.*							K			
Ancillaria buccinoides, *Lam.*				S			S			
„ canalifera, *Lam.*					S					
Ancylus? latus, *Edw.* (= Limax?).									S	
Aporrhais Sowerbyi, *Mant.*		S					K			
„ sp.										S
Borsonia sulcata, *Edw.*							S			
„ sp.							K			
Buccinum Andrei, *Desh.*					K					
„ desertum, *Brand.*				K	S		S			
„ (Pisania) labiatum, *Sow.*							S		?	
„ „ lavatum, *Sow.*					S					
Bulimus convexus, *Edw.*										W
„ ellipticus, *Sow.*							G	S	S	
„ heterostomus, *Edw.*									S	
„ lævolongus, *Boubée*									G	
„ vectensis, *Edw.*									S	
Bulla attenuata, *Sow.*							K			
„ Sowerbyi, *Nyst.*							K			
„ uniplicata, *Sow.*				S						
„ ?sp.							K			
Cæcum *sp.*							K			
Calyptræa obliqua, *Sow.*					F		J			
„ trochiformis, *Lam.*		S		F	S		S			
Cancellaria elongata, *Nyst.*							S			
„ evulsa, *Brand.*					S		K			
„ læviuscula, *Sow.*		S			S					
„ microstoma, *Brand.*					S					
„ quadrata, *Sow.*				K	K					
Capulus squamiformis, *Desh.*					S					
Cassidaria ambigua, *Brand.*					S		K			
„ coronata, *Desh.*				K						
„ nodosa, *Brand.*				F						
„ striata, *Sow.*		S								
Cerithium Austenii, *Morris*									S	
„ concavum, *Sow.*							S			

	Reading Beds.	London Clay.	Lower Bagshot Beds.	Bracklesham Beds.	Barton Clay.	Headon Hill Sands.	Headon Beds.	Osborne Beds.	Bembridge Beds.	Hamstead Beds.
Cerithium contiguum ?, *Desh.*	..	..	..	..	..	..	K			
„ duplex, *Sow.*	..	..	..	..	..	..	S			
„ elegans, *Desh.*	..	..	..	..	..	..	S	..	S	S
„ filosum, *Charlesw.*	..	..	..	..	S					
„ inornatum, *Morris*	..	..	..	..	..	..	..	..	..	S
„ multispiratum, *Desh.*	..	..	..	..	..	..	K			
„ mutabile, *Lam.*	..	..	..	..	..	..	S	..	S	S
„ plicatum, *Lam.*	..	..	..	..	..	..	S	..	..	S
„ pseudo-cinctum, *D'Orb.*	..	..	..	..	..	..	S	..	..	S
„ Sedgwickii, *Morris*	..	..	..	..	..	..	..	..	..	S
„ trizonatum, *Morris*	..	..	..	..	..	..	S	..	..	S
„ trochiforme, *Desh.*	..	..	..	K						
„ variabile, *Desh.*	..	..	..	..	..	..	K			
„ ventricosum, *Sow.*	..	..	..	..	..	..	S			
Clausilia striatula, *Edw.*	..	..	..	..	..	..	..	..	S	
Clavella (see Fusus).										
Cominella flexuosa, *Lowry*	..	..	..	..	..	..	S			
„ Solandri, *Edw.*	..	..	..	..	K					
Conus (Conorbis) dormitor, *Brand.*	..	..	..	..	S	..	K			
„ „ procerus, *Beyr.*	..	..	..	..	..	..	K			
„ „ scabriculus, *Sow.*	..	..	..	..	S					
Craspedopoma Elizabethæ, *Edw.*	..	..	..	..	..	..	..	..	S	
Cuma Charlesworthii, *Edw.*	..	..	..	..	..	..	..	..	..	S
Cyclostoma mumia, *Lam.*	..	..	..	..	..	..	..	..	S	
Cyclotus cinctus, *Edw.*	..	..	..	..	..	..	..	..	S	
„ nudus, *Edw.*	..	..	..	..	..	..	..	..	S	
Cylichna sp.	..	..	..	..	..	..	J			
Cypræa inflata, *Lam.*	..	..	..	S						
„ platystoma, *Edw.*	..	..	..	..	S					
Eulima sp.	..	..	..	..	..	..	J			
Fasciolaria funiculosa, *Desh.*	..	..	..	..	S					
„ sp.	..	..	..	..	..	..	K			
Fusus armatus, *Sow.*	..	..	..	..	..	..	J			
„ canaliculatus, *Sow.*	..	..	..	..	S					
„ carinella, *Sow.*	..	..	..	F	F					

	Reading Beds.	London Clay.	Lower Bagshot Beds.	Bracklesham Beds.	Barton Clay.	Headon Hill Sands.	Headon Beds.	Osborne Beds.	Bembridge Beds.	Hamstead Beds.
Fusus Edwardsii, *Morris* -	..	..	..	..	..	..	..	..	..	S
„ (Chrysodomus) errans, *Brand.*	..	S	..	..	F					
„ Forbesii, *Morris* -	..	..	..	..	..	..	..	..	S	S
„ interruptus, *Sow.* -	..	..	..	..	F					
„ (Clavella) longævus, *Brand.*	..	..	..	S	S	..	K			
„ minax, *Brand.* -	..	..	..	..	S					
„ Noæ, *Lam.* -	..	..	..	..	F					
„ porrectus, *Brand.* -	..	..	..	..	S					
„ pyrus, *Brand.* (= F. bulbus).	..	..	..	S	S					
„ (Chrysodomus) regularis, *Sow.*	..	S	..	..	F					
„ turgidus, *Brand.* -	..	..	..	F	S					
„ unicarinatus, *Desh.* -	..	..	..	..	F					
Helix D'Urbani, *Edw.* -	..	..	..	..	..	..	..	..	S	
„ globosa, *Sow.* -	..	..	..	..	..	..	..	..	S	
„ headonensis, *Edw.*	..	..	..	..	..	..	S	..	S	
„ labyrinthica, *Say.*	..	..	..	..	..	..	S	..	S	
„ Morrisii, *Edw.* -	..	..	..	..	..	..	..	..	W	
„ occlusa, *Edw.* -	..	..	..	..	..	..	G			
„ omphalus, *Edw.* -	..	..	..	..	..	..	..	..	S	
„ sublabyrinthica, *Edw.*	..	..	..	..	..	..	..	..	S	
„ tropifera, *Edw.* -	..	..	..	..	..	..	..	..	S	
„ vectensis, *Edw.* -	..	..	..	..	..	..	G	..	S	
Hydrobia anceps, *Lowry* -	..	..	..	..	..	..	K			
„ conica (= Chasteli var.).										
„ Chasteli, *Nyst*	..	..	..	..	..	..	S	..	S	S
„ Draparnaudi, *Linn.*	..	..	..	..	..	..	..	..	S	S
„ ? polita, *Edw.* -	..	..	..	..	..	..	S			
„ sp. -	..	..	..	..	..	..	K			
Limnæa angusta, *Edw.* -	..	..	..	..	..	..	S			
„ arenularia, *Brand.* -	..	..	..	..	..	..	S			
„ caudata, *Edw.* -	..	..	..	..	..	..	S			
„ cincta, *Edw.* -	..	..	..	..	..	..	S	..	S	
„ columellaris, *Sow.* -	..	..	..	..	..	..	S			
„ convexa, *Edw.* -	..	..	..	..	..	..	S			
„ costellata, *Edw.* -	..	..	..	..	..	..	S			

	Reading Beds.	London Clay.	Lower Bagshot Beds.	Bracklesham Beds.	Barton Clay.	Headon Hill Sands.	Headon Beds.	Osborne Beds.	Bembridge Beds.	Hamstead Beds.
Limnæa fabula, *Brong.*	..	..	..	..	..	..	S			
„ fusiformis, *Sow.*	..	..	..	..	..	..	S	..	S	
„ gibbosula, *Edw.*	..	..	..	..	..	..	S			
„ longiscata, *Sow.*	..	..	..	..	..	..	S	S	S	
„ minima, *Sow.*	..	..	..	..	..	..	S			
„ mixta, *Edw.*	..	..	..	..	..	..	S	..	S	
„ ovum?, *Brong.*	..	..	..	..	..	..	S			
„ pyramidalis, *Desh.*	..	..	..	..	..	..	S	..	S	
„ recta, *Edw.*	..	..	..	..	..	..	S			
„ sublata, *Edw.*	..	..	..	..	..	..	S			
„ subquadrata, *Edw.*	..	..	..	..	..	..	S			
„ sulcata, *Edw.*	..	..	..	..	..	..	S			
„ tenuis, *Edw.*	..	..	..	..	..	..	S			
„ tumida, *Edw.*	..	..	..	..	..	..	S			
„ sp.	..	..	..	..	..	..	..	S	S	S
Marginella æstuarina, *Edw.*	..	..	..	..	..	..	J			
„ bifido-plicata, *Charlesw.*		S	..	..	S					
„ pusilla, *Edw.*	..	..	..	..	..	..	S			
„ simplex, *Edw.*	..	..	..	..	..	..	K			
„ vittata, *Edw.*	..	..	..	..	..	..	S			
Melania fasciata, *Sow.*	..	..	..	..	..	..	S	..	S	S
„ Forbesii, *Morris*	..	..	..	..	..	..	..	..	S	S
„ inflata, *Morris*	..	..	..	..	..	..	S	..	..	S
„ minima, *Sow.*	..	..	..	..	..	..	S			
„ muricata, *Wood*	..	..	..	..	..	..	S	S	S	S
„ peracuminata, *Charlesw.*	..	..	..	..	..	..	S			
„ turritissima, *Forbes*	..	..	..	..	..	..	..	..	S	S
Melanopsis brevis, *Sow.*	..	..	..	..	..	..	S	S	S	
„ carinata, *Sow.*	..	..	..	..	..	..	S	S	S	S
„ fusiformis, *Sow.*	..	..	..	..	..	..	S	..	S	
„ subcarinata, *Morris*	..	..	..	..	..	..	S	..	S	S
„ subfusiformis, *Morris*	..	..	..	..	..	..	S	..	S	..
„ subulata, *Sow.*	..	..	..	..	..	..	S	..	S	S
Motula juncea, *Sow.*	..	..	..	F	S					
Mitra labratula, *Lam.*	..	..	..	F						
„ parva, *Sow.*	..	..	..	F	S					

	Reading Beds.	London Clay.	Lower Bagshot Beds.	Bracklesham Beds.	Barton Clay.	Headon Hill Sands.	Headon Beds.	Osborne Beds.	Bembridge Beds.	Hamstead Beds.
Mitra porrecta, *Edw.* -	..	..	..	..	S					
„ *sp.* -	..	..	..	..	..	..	K			
Murex asper, *Brander* -	..	..	..	F	S					
„ *Forbesii* (see Fusus).										
„ hantonensis, *Lowry* -	..	..	..	..	..	..	K			
„ minax, *Brand.* -	..	..	..	F	S	..	K			
„ sexdentatus, *Sow.* -	..	..	..	..	..	..	S			
„ *sp.* -	..	..	..	..	..	..	K			
Natica ambulacrum, *Sow.* -	..	..	..	F						
„ depressa, *Sow.* -	..	..	..	..	..	..	S			
„ epiglottina, *Lam.* -	..	..	..	S	S	..	S			
„ hantoniensis, *Sow.* -	..	..	..	..	..	..	S			
„ labellata, *Lam.* -	..	S	..	S	S	..	S	..	S	S
„ mutabilis, *Brand.* (=acuta).	..	..	..	S	..	..	S			
„ sigaretina, *Sow.* -	..	S								
„ Studeri, *Bronn.* -	..	..	..	..	..	..	J			
Nematura parvula, *Desh.* -	..	..	..	..	..	..	S	..	..	S
„ pupa, *Nyst.* -	..	..	..	..	..	..	..	..	..	S
„ *sp.* -	..	..	..	..	..	..	K			
Nerita aperta, *Sow.* -	..	..	..	..	..	..	S			
Neritina concava, *Sow.*	..	..	..	..	..	..	S	..	S	S
„ planulata, *Edw.* -	..	..	..	..	..	..	W			
„ tristis, *Forbes* -	..	..	..	..	..	..	..	..	..	S
„ zonula, *Wood* -	..	..	..	..	..	..	W			
Odostomia 5 species -	..	..	..	..	..	..	J			
Oliva Branderi, *Sow.* -	..	..	..	..	S					
Orthostoma *sp.* -	..	..	..	..	..	..	J			
Paludina angulosa, *Sow.* (= orbicularis).	..	..	..	..	..	..	S	S	S	S
„ lenta, *Sow.* -	..	..	..	..	..	..	S	S	S	S
„ minuta, *Sow.*(= globuloides).	..	..	..	..	..	..	..	S	S	
Phorus agglutinans, *Lam.* -	..	..	..	F	S					
„ *sp.* -	..	..	..	..	..	..	K			
Pisania (see Fusus).										
Planorbis biangulatus, *Edw.* -	..	..	..	..	..	..	S			
„ discus, *Edw.* -	..	..	..	..	..	..	S	?	S	
„ elegans, *Edw.* -	..	..	..	..	..	..	S			

	Reading Beds.	London Clay.	Lower Bagshot Beds.	Bracklesham Beds.	Barton Clay.	Headon Hill Sands.	Headon Beds.	Osborne Beds.	Bembridge Beds.	Hamstead Beds.
Planorbis euomphalus, *Sow.* -	..	..	..	..	..	..	S	S	S	
„ lens, *Brong.* -	..	..	..	..	..	..	S	..	..	S
„ obtusus, *Sow.* -	..	..	..	..	..	..	S	S	S	S
„ oligyratus, *Edw.* -	..	..	..	..	..	..	..	S	S	
„ platystoma, *Wood* -	..	..	..	..	..	..	S	S	S	S
„ rotundatus, *Brard.*	..	..	..	..	..	..	S	S	S	
„ Sowerbyi, *Brong.* -	..	..	..	..	..	..	..	..	S	S
Pleurotoma aspera, *Edw.* -	..	..	..	..	S					
„ attenuata, *Sow.* -	..	..	..	S						
„ comma, *Sow.* -	..	S								
„ conoides, *Brand.* -	..	..	..	..	E					
„ crassa, *Edw.* -	..	S								
„ curta, *Edw.* -	..	..	..	..	S					
„ cymæa, *Edw.* -	..	..	..	..	..	..	S			
„ dentata, *Sow.* -	..	..	..	S						
„ denticula, *Bast.* -	..	S	..	S	E	..	S			
„ exorta, *Brand.* -	..	S	..	..	S					
„ Fisheri, *Edw.* -	..	..	..	F						
„ granulata, *Lam.* -	..	..	..	..	S					
„ headonensis, *Edw.*	..	..	..	..	..	..	S			
„ inflexa, *Lam.* -	..	..	..	F						
„ innexa, *Brand.* -	..	..	..	..	S	..	S			
„ lanceolata, *Edw.* -	..	..	..	..	S					
„ macilenta, *Brand.*	..	..	..	..	S					
„ mixta, *Edw.* -	..	..	..	..	S					
„ plicata, *Lam.* -	..	..	..	S						
„ prisa, *Brand.* -	..	..	..	..	S					
„ rostrata, *Brand.* -	..	..	..	..	S					
„ scalarata, *Edw.* -	..	..	..	F						
„ Sclysii, *De Kon.* -	..	S								
„ subdenticulata, *Goldf.* (= *hantonensis*).	..	..	..	..	..	..	S			
„ transversaria, *Lam.*	..	..	..	..	..	..	J			
„ turbida, *Brand.* -	..	..	..	..	S					
„ turgidula, *Edw.* -	..	..	..	..	?					
„ Woodi, *Edw.* -	..	..	..	..	..	..	E			
„ zonulata, *Edw.* -	..	S	..	..	S					

Potamides (see Cerithium).

	Reading Beds.	London Clay.	Lower Bagshot Beds.	Bracklesham Beds.	Barton Clay.	Headon Hill Sands.	Headon Beds.	Osborne Beds.	Bembridge Beds.	Hamstead Beds.
Pseudoliva obtusa, *Sow.*	..	..	..	S	.					
„ ovalis, *Sow.*	..	..	..	K						
Pupa oryza, *Edw.*	..	..	..	..	..	..	..	..	S	
„ perdentata, *Edw.*	..	..	..	..	..	..	..	..	S	
Pyramidella (Turbonilla) *sp.*	..	..	..	..	..	..	J			
Pyrula nexilis, *Lam.*	..	..	..	F	S					
„ tricostata ?, *Desh.*	..	S					.			
Rissoina cochlearella, *Lowry*	..	..	..	F						
Rostellaria ampla, *Brand.*	..	..	..	..	S	..	K			
„ rimosa, *Brand.*	..	..	..	S	S	..	J			
„ sublucida, *D'Orb.*	..	..	..	F						
Scalaria acuta ?, *Sow.*	..	..	..	..	K					
„ interrupta, *Sow.*	..	..	..	..	K					
„ lævis, *Morris*	..	..	..	..	..	..	S			
„ reticulata, *Brand.*	..	..	..	..	S					
„ undosa, *Sow.*	..	..	..	..	P	..	S			
„ *sp.*	..	..	..	..	..	..	J			
Succinea Edwardsi, *Forbes*	..	..	..	..	..	..	..	..	S	
„ imperspicua, *Wood*	..	..	..	..	..	..	S			
„ sparnacensis ?, *Desh.*	..	..	..	..	..	..	E			
Tcinostoma, 2 *sp.*	..	..	..	..	..	..	J			
Terebellum sopitum, *Brand.*	..	..	..	..	F					
Tornatella (see Actæon).										
Triton argutus, *Sow.*	..	..	..	..	S		.			
Turbonilla, 5 *sp.*	..	..	..	..	..	..	K			
Turritella granulosa, *Desh.*	..	S								
„ imbricataria, *Lam.*	..	S	..	S	S					
„ sulcata, *Lam.*	..	..	..	F						
„ sulcifera, *Desh.*	..	S	..	F						
„ terebellata, *Lam.*	..	:.	..	F						
Typhis fistulosus, *Sow.*	..	..	..	..	S		.			
„ pungens, *Brand.*	..	..	..	..	S	..	K		.	
Vicarya (see Cerithium).										
Voluta ambigua, *Brand.*	..	..	..	..	S					
„ athleta, *Brand.*	..	..	..	..	S					
„ depauperata, *Sow.*	..	..	..	..	S	..	S			
„ digitilina, *Lam.*	..	..	..	..	S					
„ geminata, *Sow.*	..	..	..	..	..	..	G			

	Reading Beds.	London Clay.	Lower Bagshot Beds.	Bracklesham Beds.	Barton Clay.	Headon Hill Sands.	Headon Beds.	Osborne Beds.	Bembridge Beds.	Hamstead Beds.
Voluta humerosa, *Edw.*	..	..	..	..	K					
„ luctatrix, *Brand.*	..	..	..	..	S					
„ maga, *Edw.*	..	..	..	..	S	..	K			
„ nodosa, *Sow.*	..	..	..	F	S					
„ Rathieri, *Héb.* (= Forbesii).	..	..	..	..	..	..	..	..	..	S
„ scalaris, *Sow.*	..	..	..	..	S					
„ selseiensis, *Edw.*	..	..	..	F						
„ Solandri, *Edw.*	..	..	..	..	S					
„ spinosa, *Linn.*	..	..	..	S	S	..	J			
„ suturalis, *Nyst.*	..	..	..	..	..	..	K			
Volvaria acutiuscula, *Sow.*	..	..	..	..	..	..	S			
Pisces.										
Clupea vectensis, *Newt.*	..	..	..	..	..	..	..	S		
Lamna acutissima, *Ag.*	..	S								
„ compressa?, *Ag.*	..	..	..	..	S					
„ contortidens, *Ag.*	..	S	..	..	..	..	K			
„ dubia, *Ag.*	..	..	..	..	S					
„ elegans, *Ag.*	..	S								
„ Hopei, *Ag.*	..	S								
„ verticalis, *Ag.*	..	S								
Lepidosteus sp.	..	..	..	..	..	..	..	S		
Myliobatis sp.	..	..	..	S	S	..		S		
Otodus obliquus, *Ag.*	..	S								
Reptilia.										
Diplocynodon (Crocodilus) sp.	..	..	..	..	..	..	..	S	S	S
Emys sp.	..	..	..	..	..	..	..	..	S	
Ophis sp.	..	..	..	..	..	..	..	..	..	S
Paleryx sp.	..	..	..	..	..	..	..	..	S	
Trionyx incrassatus, *Owen*	..	..	..	..	..	..	S	S	S	?
Aves.										
Ptenornis sp.	..	..	..	..	..	..	..	..	..	S
Bird phalanx	..	..	..	..	..	..	..	..	S	

	Reading Beds.	London Clay.	Lower Bagshot Beds.	Bracklesham Beds.	Barton Clay.	Headon Hill Sands.	Headon Beds.	Osborne Beds.	Bembridge Beds.	Hamstead Beds.
Mammalia.										
Acotherulum saturninum, *Gerv.*	..	..	..	..	..	..	..	..	L	
Anchilophus Desmaresti, *Gerv.*	..	..	..	..	..	..	..	..	L	
Anoplotherium commune, *Cuv.*	..	..	..	..	..	..	..	..	S	
,, minus, *Filhol* -	..	..	..	..	..	..	..	..	?	
,, secundarium, *Cuv.*	..	..	..	..	..	..	..	..	S	
Anthracotherium alsaticum, *Cuv.*	..	..	..	..	..	..	..	..	..	?
,, Gresslyi, *H. von Meyer.*	..	..	..	..	..	..	..	..	L	
,, minus, *Cuv.*	..	..	..	..	..	..	..	..	..	L
Chœropotamus gypsorum, *Desmar.*	..	..	..	.	..	..	..	..	L	
Coryphodon *sp.* -	..	..	..	..	..	..	..	..	..	S
Dacrytherium ovinum, *Owen* -	..	..	..	..	..	..	L			
Dichobune cervinum (see Dichodon).										
Dichodon cervinus, *Owen* -	..	..	..	..	..	..	L	..	L	
,, cuspidatus, *Owen* -	..	..	..	..	..	..	S			
Elotherium magnum, *Pomel* -	..	..	..	..	..	..	..	..	..	L
Hyænodon minor, *Gerv.* -	..	..	..	..	..	..	L			
Hyopotamus bovinus, *Owen* -	..	..	..	..	..	..	..	..	..	L
,, porcinus, *Gerv.* -	..	..	..	..	..	..	..	..	..	L
,, *vectianus, Owen* (see bovinus and velaunus).										
,, velaunus, *Cuv.* -	..	..	..	..	..	..	..	..	?	L
Lophiodon sp. (see Coryphodon).										
Palæotherium annectans, *Owen*	..	..	..	..	..	..	?	..	L	
,, crassum, *Cuv.* -	..	..	..	..	..	..	..	..	L	
,, ?curtum, *Cuv.* -	..	..	..	..	..	..	..	..	?	
,, magnum, *Cuv.*	..	..	..	..	..	..	..	..	L	
,, medium, *Cuv.* -	..	..	..	..	..	..	..	..	S	
,, minus, *Cuv.* -	..	..	..	..	..	..	..	S	S	
Pterodon dasyuroides, *Blainv.*	..	..	..	..	..	..	..	..	L	
Theridomys aquatilis, *Aymard*	..	..	..	..	..	..	L	S	?	?
Xiphodon gracilis, *Cuv.* -	..	..	..	..	..	..	..	..	L	

TABLE V., by Prof. T. R. Jones, F.R.S.

FOSSIL OSTRACODA OF THE ISLE OF WIGHT.

Those marked thus O are known to have been found in the Island; those marked X occur also at localities not in the Island.

	Post-Pliocene.	Hamstead Beds.	Bembridge Beds.	Osborne Beds.	Headon Beds.	Upper Bagshot Beds.	Barton Clay.	Middle Bagshot. { Bracklesham. }	Lower Bagshot Beds.	London Clay.	Woolwich Beds.	Chalk.	Gault.	Wealden.
Cypris gibba, *Ramdohr*.	X	O	..											
„ cornigera, *Jones*													..	O
Potamocypris Brodici, *Jones & Sherborn*			O											
Candona Forbesii, *Jones*		O	..	O										
„ Mantelli, *Jones*													..	O
Cypridea valdensis (*Sow.*)													..	O
„ spinigera (*Sow.*)		O											..	O
„ Austeni, *Jones*													..	?
„ Dunkeri, *Jones*													..	O
„ tuberculata, *Sow.*													..	O
? Pontocypris, sp.					O									
Darwinula leguminella (*Forbes*)													..	O
Cyprione Bristovii, *Jones*													..	O
Metacypris Fittoni (*Mant.*)													..	O
? „ unisulcata, *Jones*				O										
Cythere striatopunctata, *Jones*						O	O	X						
„ Wetherellii, *Jones*						O	..	X						
„ Bosquetiana, *J. & S.*						O	..	X						
„ delirata, *J. & S.*						O								
„ plicata, *Münster*						O	X	X	..	X				
„ transenna, *J. & S.*						O	..	..	..	X				
„ Forbesii, *J. & S.*						O								
Cythereis corrugata (*Reuss.*) var.						O								
„ Prestwichiana, *J. & S.*										O				
„ Bowerbankiana, *J. & S.*										O				
„ cornuta (*Roemer*)						O		X				?		
Cytheridea Muelleri (*Münster*)	X	O	O	O	O						O			
„ „ var. torosa, *Jones*		O									X			
„ montosa, *J. & S.*		O												

The Ostracoda of the Chalk are much the same as those of Sussex and Kent.

	Post-Pliocene.	Hamstead Beds.	Bembridge Beds.	Osborne Beds.	Headon Beds.	Upper Bagshot Beds.	Barton Clay.	Bracklesham.	Lower Bagshot Beds.	London Clay.	Woolwich Beds.	Chalk.	Gault.	Wealden.
							Middle Bagshot.							
Cytheridea debilis, *Jones*	..	..	..	..	O	..	..	X			'			
„ perforata (*Roemer*)	..	..	..	..	O	..	X	..	..	X	..	?	X	
Xestoleberis colwellensis, *J. & S.*	..	..	..	..	O									
„ aurantia (*Baird*), var.	..	..	..	..	O									
Pseudocythere Bristovii, *J. & S.*	..	..	O											
„ sp.	..	..	O											
„ attenuata, *Jones*	..	..	..	..	..	..	O							
Cytherideis colwellensis, *Jones*	..	..	..	..	O									
„ sp.	..	..	..	..	O									
„ gracilis (*Reuss.*)	..	..	..	..	O									
Cytherella Muensteri (*Roemer*)	..	..	..	..	O	..	X	X						
„ sp.	..	..	..	..	O									

APPENDIX III.
WELL SECTIONS AND WATER SUPPLY.

Composition of Water from Springs and Wells in the Chalk and Upper Greensand of the Isle of Wight and from the River Yar.

[Extracted from the 6th Report of the Rivers Pollution Commission (1868).]

| | Temperature, Fahrenheit. | Total Solid Impurity. | Organic Carbon. | Organic Nitrogen. | Ammonia. | Nitrogen as Nitrates and Nitrites. | Total combined Nitrogen. | Previous Sewage or Animal Contamination. | Chlorine. | Hardness. | | | Remarks. | |
										Temporary.	Permanent.	Total.		
VENTNOR:—														
Waterworks Well, near railway station, November 16, 1872.	50·9	34·72	·056	·011	0	·061	·072	290	3·10	21·7	4·6	26·3	Slightly turbid, palatable	Upper Greensand.
Water supply from springs September 12, 1872.	63·5	34·38	·031	·004	0	·187	·191	1,550	3·00	21·0	4·4	25·4	Do.	Do.
Spring in railway tunnel, November 16, 1872.	50·7	32·80	·048	·006	0	·189	·175	1,570	3·15	21·0	4·7	25·7	Do.	Do.
The Wishing Well, St. Boniface Down, November 16, 1872.	51·6	26·40	·007	·018	·006	·078	·101	510	6·40	6·8	5·6	12·4	Turbid, palatable.	Chalk.
CARISBROOK:—														
Spring, November 4, 1871	47·5	28·50	·008	·005	·001	·369	·375	3,380	3·30	23·4	6·0	29·4	Clear and palatable.	Chalk.
Well in Castle,* March 8, 1873	52·3	43·28	·169	·043	·002	1·363	1·410	13,340	6·40	13·9	10·0	23·9	Do.	Do.
SURFACE WATER:—														
The Yar,† above Sandown, September 29, 1873	—	22·52	·252	·064	0	·158	·222	1,200	·385	5·1	6·4	11·5	Slightly turbid	River water.
Do, April 8, 1874	—	20·20	·254	·039	·008	0	·046	0	3·80	3·8	7·1	10·9	Very turbid	Do.

* An old polluted well, 240 feet deep.

† The water of the river is supplied, after filtering, to Sandown.

BEMBRIDGE.   At the Bembridge Hotel.

R. F. GRANTHAM.   *Trans. Surveyors' Inst.*, vol. xx., pt. v., p. 144, plate. (1888.)

23¾ feet above Ordnance Datum.

Shaft 70 feet, the rest bored.

Water-level 24½ feet down.   Yield 2,200 gallons in 12 hours.

		THICKNESS.	DEPTH.
		FEET.	FEET.
[Bembridge and Osborne Series.]	Brown and blue clay [no details]	70	70
	Clay - - - -	5	75
	Stone - - -	2	77
	Mixture of sand - - -	12	89
	Light [-coloured] sand - -	4	93
	Stone - - - -	2½	95½
	Dead grey sand - - -	4½	100
	Coloured [mottled] clay - -	36	136
	Stone - - - -	1	137
	Blue clay with shells - -	10	147
	Blue clay with sand - -	3	150
[Headon Beds.]	Rock - - - -	2½	152½
	Green sand - - -	3½	156
	Clay and stone - - -	5½	161½
	Green sand - - -	1	162½
	Sandstone - - -	2½	165
	Green sand - - -	½	165½
	White marl - - -	3½	169
	Green sand with clay - -	6	175
	Purple clay - - -	23	198
	Clay and shells - - -	22	220
	Green clay - - -	3	223
	Small shells - - -	7	230
	Dark green clay - - -	6	236
	Light [-coloured] sand - -	6	242
	Hard rock - - -	2¾	244¾
	Sand - - - -	2¼	247
	Brown clay - - -	2½	249½
	Hard rock - - -	3	252½
	Black clay and shells - -	4	256½
	Mixture of sand - - -	2½	259
	Light [-coloured] sand - -	4	263
	Rock - - - -	1	264

The Bembridge Limestone was probably reached at about 35 feet, but no record has been kept of the beds passed through in the shaft.

CARISBROOK.   Newport Waterworks.   Height above Ordnance Datum about 58 feet.

From information obtained by MR. WHITAKER on the spot.

Shaft 25 feet, bore of 20 inches diameter, 30 feet.   Water pumped down 10 feet, but soon rises (to the surface) on cessation of pumping. Supply abundant.   Chalk.

FRESHWATER, Golden Hill Fort.   For H.M. Government.

(Communicated by MESSRS. DOCWRA.)

			FEET.
	Light red clay -	-	
	,, coloured clay	-	
	Dark red clay -	-	
	Yellow clay -	-	All thin beds - 6
	Red clay and shells	-	
	Light stone -	-	

		FEET.
[Osborne Beds? 74 feet.]	Light loam Brown clay Light loam ,, blue clay Brown loam Light blue clay and shells Blue mottled clay Rock and shells Shells Black sand and shells Light red clay Dark blue clay Light blue clay } Thicker beds	44
	,, red clay ,, blue clay Red mottled clay } Thick beds	24
	Brown clay and shells Light rock - ,, loam ,, blue clay and shells Blue clay - ,, mottled clay (dark) Light loam Shells - Blue clay and shells Rock and shells Blue clay and shells Shelly stone Light clay - } Thin beds	22
[Headon Beds, 99½ feet.]	Mottled loam Green loam Brown loam Stone Green loam Sand rock - Mottled loam Dark sand - Brown sand and clay ,, clay and sand ,, sand Blue clay and sand Dark sand - Blue loam - } Moderately thick	50
	Black sand Dark sand - Stone Blue clay - Black sand ,, sand ,, sand and shells Blue clay - Black sand Sand Blue clay - Yellow mottled clay (Bed, not named) - } Thinner beds	24
	Black clay - Limestone - Light green clay Dark green clay } Thin beds	3½
	Total	173½

Total depth, 173½ feet.
Water level 95 feet down.
94 feet to bottom of shaft, the rest is bored.

HAVEN STREET. 6 chains north-west of the Church.

From specimens and notes communicated by MR. TOWNEND.

Old well 30 feet, then bored to 378 feet.

No water obtained.

			FEET.
Hamstead Beds and Bembridge Marls.	Sand Clay } old well—no record	-	{ about 20 0 { about 10 0
	Shelly blue slipper -	- -	- at 130 to 208
Bembridge Limestone.	} Hard earthy limestone with *Limnæa*	-	- at 208 to 210
Osborne Beds and Headon Beds.	Blue and black slipper	- -	- to 230
	Sand (?) -	- -	- at 249
	Blue shelly slipper -	- -	- at 264
	Mottled yellow and white marl	-	- at 278
	Stiff red clay	- -	- 280 to 286
	Shaly slipper	- -	- 290 to 320
	Yellow and green slipper	-	- at 330
	Reddish slipper	- -	- at 343
	Reddish marl	- -	- at 350
	Greenish slipper and clay	-	- at 357
	Rock, light blue	- -	- at 366
	Hard green sandy marl	-	- at 368
	Spongy fine-grained grit	-	- at 378

Owing to the destruction of the fossils it is impossible to fix the limits of the different beds in this boring. The "sand" in the old well is the bed at the base of the Middle Hamstead Beds. The "limnæan limestone" is apparently the Bembridge Limestone. The boundary between the Osborne and Headon beds is quite uncertain.

HAVEN STREET. Longford House.

From specimens communicated by MR. TOWNEND.

Old well 100 feet (no record), the rest a 10-inch bore (on Parson's system). At first yielded over 22,000 gals. a day, the water rising 12 feet above the ground. In July 1887 the water rose 9 feet above the ground after several hours pumping. In October 1887 the supply had fallen off greatly, the water not rising above the surface and being greatly lowered by pumping. The water is unpalatable and ferruginous. Temperature 55°.

		THICKNESS.		DEPTH.	
		FT.	IN.	FT.	IN.
Hamstead Beds (perhaps 40 feet).	Old Well (no record) -	100	0	100	0
	Shelly blue and green clay	42	0	142	0
	Whitish marl -	5	0	147	0
	Green clay -	2	0	149	0
	White granular marl	1	0	150	0
Bembridge Marls (about 120 feet).	Shelly blue clay -	6	6	156	6
	Hard and soft whitish marl	3	6	160	0
	Black and green clay	1	6	161	6
	Bluish white very shelly marl	2	6	164	0
Bembridge Limestone.	Grit and rotten stone, with much water -	0	10	164	10
	Rock, very hard -	2	2	167	0

Analysis of sample of water taken 13th August 1887.

Total Solids	-	-	-	-	25·0	Grains per Gallon.
Chlorine	-	-	-	-	2·5	,, ,, ,,
Free Ammonia	-	-	-	-	·063	,, ,, ,,
Albuminoid Ammonia	-	-	-	·0014	,, ,, ,,
Nitrogen as Nitrates and Nitrites	-	-	·03228	,, ,, ,,

KNIGHTON. South-east part of the Pumping Station of the Ryde Corpora-
tion Waterworks, about 130 yards south of Knighton Mill. 1885. About
46 feet above Ordnance Datum.

From information and specimens communicated by Mn. F. NEWMAN,
Borough Engineer, to Mr. Whitaker.

Shaft 15 feet, the rest bored. Water at 53 feet, rose above the surface, but
the tubes soon filled with sand. Water was again met with at 66 feet, and from
this downward the sand was all wet. The greatest quantity was at 53 feet.

		DEPTH OF SPECIMENS IN FEET.
[Alluvial Beds, about 12½ feet.]	Dark-grey (blackish) sand, with plant-remains	9
	Grey and brown dirty sand - - -	10
	Dry. Pieces of chalk, a little 'grey clay and pieces of flint - - - -	11
	Moist. Grey and brownish sandy clay, with green sand, plant-remains and bits of flint -	12
[Carstone. Base uncertain, about 40 feet.]	Brown gritty sand - - - - -	12½
	Dry. Brownish grey firm clayey sand -	22
	Moist. Brownish grey firm clayey sand. This and the above with small pieces of a more clayey character - - -	40
	Moist. Brownish grey clayey sand -	44
	Brown clayey sand, with quartz grains and small pebbles; only slight differences in the specimens - - - -	45 46 49 50
	Brown and grey clayey sand, like the above but finer, partly hard, with a trace of plant-remains - - - - -	51
	Described as stony and with water at great pressure. Specimen brown firm clayey sand with quartz grains - - - -	53
[Sandrock Series, about 57 feet +]	Dry. Grey and greenish-grey firm clayey sand - - - - -	56
	Described as moist Greensand, as also are the beds below. Specimen grey and blackish firm clayey sand - - - -	66
	Grey firm clayey sand, with quartz grains and pebbles - - - - -	74
	Greenish sand - - - - -	78
	Green clayey sand - - - -	82
	Fine grey sand - - - - -	91
	Loose light-grey fine sand - - -	101
	Fine grey sand - - - -	110

KNIGHTON. Ryde Waterworks. Just north of the Engine House, 1885.
About 45½ feet above Ordnance Datum.

Communicated by MR. F. NEWMAN, Borough Engineer, to Mr. Whitaker.

Gault, to Lower Greensand, with water, 46 feet.

The boring at the Mill, of which a note follows this, is 185 feet to the north.
The difference of level of the bottom of the Gault in the two borings shows a
northerly dip of between 16° and 17°, supposing that the inclination is uniform:
it probably increases northwards.

KNIGHTON. Ryde Waterworks. Boring in the Mill, 1885. Floor of Mill
48 feet above Ordnance Datum.

Communicated by MR. F. NEWMAN, Borough Engineer, to Mr. Whitaker.

Gault, mixed with sand at 101 feet below the floor of the Mill. At
120½ feet a specimen of clayey sand, with clay and small pebbles [? junction of
Gault and Lower Greensand].

. Water flowed up from the bottom, and, at the surface, seemed to have some head.

NEWPORT. Messrs. Mew & Co.'s Brewery.

(From information and samples communicated by ARTHUR KINDER, ESQ.)
Surface 12 feet above O.D. Well sunk 138 feet; boring carried to
160 feet. Temperature of the water 61·5°.

		THICKNESS, FEET.	DEPTH. FEET.
Hamstead and Bembridge Beds.	Clay, with thin rock at 26 feet and 90 feet (no samples preserved) - - - -	148	148
Bembridge Limestone.	Limestone - - -	4	152
Osborne Beds 107 feet.	Mottled clays - - -	28	180
	Shell limestone and green marl full of *Cyrena* at about 180 feet. Platy shale full of Ostracoda at about 180 feet. *Cyrena obovata* in green clay at about 200 feet [samples are not marked with the depths] - - -	20	200
	Mottled clays. *Cyrena* at 245 feet	59	259
Upper Headon Beds 82½ feet.	Lead-coloured shelly clays -	8	267
	Mottled green, red, and yellow clays - - -	22	289
	Greenish sand - - -	1	290
	Mottled dark-red and green clays	5	295
	Mottled green, red, and yellow clays - - -	9	304
	Limestone - - -	4	308
	Green clay - - -	4	312
	Limestone. Lignite at 313 feet. Turtle bone at 313 feet -	3	315
	Pale green, red, and yellow clays -	26½	341½
	Brown and green clays - -	2½	344
	Whitish marl and green soapy clay	3	347
	Darker green marl -	1	348
	White marl, with indeterminable shells and fish bones - -	2	350
	Lead coloured clay and shell marl	4	354
	Green clay - - -	½	354½
	Lead coloured clay with *Cyrena* -	1½	356
	White chert [Fragments marked 356 feet]		
	Greenish marl full of *Cyrena* -	3	359
	Pale green marl - - -	2½	361½
	Green marl. *Potamomya, Cyrena, Serpula* - - -	5½	367
	Dark-green and yellow marl. *Melania muricata* - -	3	370
Middle Headon Beds 106½ feet.	White shell-marl with indeterminable bivalves and fish bones -	5	375
	White marl and dark-green clay. *Potamomya* - - -	3	378
	Green clay and ironstone nodule -	1	379
	Lead-coloured shelly marl -	1	380
	Dark-green shelly marl full of *Cyrena* - - -	1	381
	Limestone or hard marl, full of indeterminable shell fragments -	7	388
	Hard shell bed (pyrites) - -	1	389

	THICKNESS.	DEPTH.
	DEPTH.	FEET.
Hard flaggy sandstone with nodule	1	390
Black sandy clay with shells -	13	403
Dark-green shelly clay with iron-stone nodules. *Cyrena obovata, Paludina, Melania, Planorbis, Balanus,* and *Serpula* at 409 feet	6	409
Lead-coloured shelly clays. *Cytherea incrassata, Melania ? Natica,* and *Balanus* - - -	10	419
Greenish and lead-coloured clay -	1	420
Green sandy clay. *Cythereu incrassata, Cyrena, Natica* at 420 feet - - - -	28	448
Green sand and sandy clay. Water at 448 feet - - -	13	461

NEWPORT.   At the Steam Mills in Pyle Street.
Communicated by MR. TAYLOR.

	Ft.	In.
Clay, dry - - - - - - -	70	0
Clay, bored - - - - - -	75	0
Soft marly rock - - - - -	4	6
	149	6

NEWPORT.   At the corner of South Street and Archer Street.
Communicated by MR. LOCK.

	FEET.
To rock - - - - - - -	145

NEWPORT.   At the Round pump.
Communicated by MR. LOCK.

	FEET.
Clay to rock - - - - - - -	140

NEWPORT.   Anchor Brewery, 3 wells.
Communicated by MR. LOCK.

	Ft.	In.
To rock - - - - - - -	150	0
Rock - - - - - - -	7	5

NEWPORT.   West Medina Cement Works.
Sunk and communicated by MR. PARSONS.

		THICKNESS.	DEPTH.
		FEET.	FEET.
Hamstead and Bembridge Beds 173 feet.	Clay with 5 beds of shaly rock { old	153	153
	Stone, with water - - { well	5	158
	Black and green clays - -	12¼	170½
	Yellow and white marl - -	1½	172
	Limestone and marl - -	1	173
Bembridge Limestone 6 feet.	Limestone - - -	6	179

		THICKNESS.	DEPTH.
		FEET.	FEET.
Osborne Beds 113 feet.	Green and carbonaceous clays -	8	187
	Mottled red, green and yellow clays	26	213
	Hard fine-grained grit (concretion?)	—	213
	White and green clays - -	2	215
	Green and red clays - -	1	216
	Mottled green, yellow and carbonaceous clays - - -	17	233
	Black clay - - -	6	239
	Mottled clays—green, black, yellow, and brown - - -	26	265
	Hard green clay with *Paludina* -	½	265½
	Green clayey sand - -	1½	267
	Limestone - - -	1½	268½
	Sand - - - -	½	269
	Rock - - - -	3	272
	Green clay - - -	1½	273½
	Red clay - - -	4½	278
	Red and green clay - -	4	282
	Green clay - - -	½	282½
	Sand rock - - -	1½	284
	Light-green clay - - -	2	286
	Blue clay - - - -	1	287
	Rock 1 foot 4 inches (sandy limestone) - - - ; Blue clay - - - ; Hard detrital limestone 3 feet 4 inches - - -	—	292
Upper Headon Beds 64¼ feet.	Light-green clay - - -	¾	292¾
	Limestone - - -	1¼	294
	Light-green sandy clay - -	2½	296½
	Limestone 2 feet 5 inches -	2½	299
	Dark green clay - - -	1	300
	Black peaty substance - -	1½	301½
	Green clay - - -	4½	306
	Limestone - - -	2	308
	Red, green and mottled clays -	21	329
	Green clay and ½-inch concretionary limestone - -	2	331
	Dark green clay - - -	¾	331¾
	Dark blue clay - - -		
	Black clay full of shells - -	6¼	338
	Light-coloured very fine loam -	1	339
	Dark green shelly clay - -	½	339½
	Dark-coloured shelly clays -	3	342½
	Whitish clays - - -	2½	345
	Very dark shelly clays, black at the base - - -	11¼	356¼
Middle Headon Beds.	Green clays - - -	2¼	358½
	Black clays full of shells, *Cyrena obovata*, *Potamomya gregaria*, *Limnæa*, Fish-bones - -	6¼	364¾
	Dark-green clay - - -	¾	365½
	Black shelly clay - - -	1½	367
	Sandy clay, very shelly - -	1	368
	Do. do. with water, 2,500 gals. per hour - -	8	376
	Dark sandy clay - - -	½	376½
	Deep black clay - - -	13½	390
	Dark-green sandy clay, with *Cyrena obovata*, *C. deperdita*, *Melania muricata*, *Buccinum*		

	THICKNESS.	DEPTH.
	FEET.	FEET.
labiatum, Nematura parvula, Planorbis, and *Cerithium pseudocinctum* - - - -	15	405
Blue clay with *Cytherea incrassata* [Venus Bed ?] - - -	5	410
Blue sandy clay with *Cytherea incrassata* - - -	1	411
Very shelly greenish clay with *Cytherea* - - -	1	412
Blue and brown clay with *Cytherea*	2¼	414½
Greenish clay full of *Cytherea* -	2½	417
Brown sandy clay - -	3	420
Brown very sandy clay - -	6	426
Hard blue clay - - -	3	429
More sandy brown clay - -	7	436
Hard earthy limestone - -	4½	440½
Fine micaceous sandy loam -	7	447½
Micaceous loam and lignite -	3	450½
Brown sandy loam - -		

At the time of going to press this well was still unfinished.

PARKHURST Upper Prison.

Communicated by MR. LOCK.

	FT.	IN.
Clay, &c. - - - - - - -	255	0
Limestone [Bembridge Limestone] - - -	4	6
	259	6

PARKHURST PRISON FARM.

Communicated by MR. LOCK.

	FEET.
To rock - - - - - - - -	about 200

PARKHURST LOWER PRISON.

Communicated by MR. LOCK.

	FEET.
Clay with thin rocks - - - - - -	239
Freestone.	

At the Prison this well is said to be 250 feet deep. It was probably deepened afterwards.

PARKHURST BARRACKS.

	FEET.
Clay, to rock - - - - - - -	236

Water rises to 56 feet below surface, but after pumping sinks much and continuously. Pumping affects the wells at the Cement Works and Prison, as also at High Street, Newport [?]. Now (Aug. 29th, 1887) water stands at 70 feet from the surface.

St. Helens. Nearly half a mile south-east of the Church. Height about 150 feet above the sea. Sunk 15 feet, the rest bored.

Sunk and communicated by Mr. Parsons.

		THICKNESS.	DEPTH.
		FEET.	FEET.
Hamstead Beds? 15 feet (?)	Blue slipper, black at base [no specimens] - - -	15	15
Bembridge Marls 118½ feet (?)	Green and brown clay - -	75	90
	Stone (3 or 4 inches) - -	—	—
	Blue clay (shelly at 100 feet) -	11	101
	Green clay - - -	2	103
	Green clay and marl - -	2	105
	Green clay - - -	3	108
	Brown clay - - -	2	110
	Green clay - - -	1	111
	Mottled brown and green clay -	½	111½
	Green clay - - -	½	112
	Green marl - - -	2	114
	Green clay - - -	4	118
	Green stone - - -	1	119
	Dark marl and black clay -	1	120
	Green clay - - -	½	120½
	Green stone and clay - -	3¼	123¾
	Brown carbonaceous clay -	2¼	126
	Black shelly clay - -	½	126½
	Black clay with *Serpula* - -	½	127
	Dark-green shelly clay with *Cyrena*	3½	130½
	Black clay - - -	1½	132
	Green clay and pyrites - -	2	134
Bembridge Limestone 16 feet.	Freestone - - -	5	139
	Greenish grey clay - -	4	143
	Sandy clay - - -	4	147
	Freestone - - -	3	150
Osborne Beds (St. Helen's Sands) 25½ feet.	Dark green clay - - -	1	151
	Very dark green clay - -	1¾	152¾
	Do. sandy -	1¾	154½
	Dark green and brown clay -	1	155½
	Green sandy clay and sandstone -	3½	159
	Grit - - - -	1	160
	Fine-grained sandstone - -	1¾	161¾
	Blue sandstone - - -	1¼	163
	Buff sandstone - - -	6	169
	Rock - - - -	½	169½
	Buff sandstone - - -	5	174½
	Hard sandstone - - -	¾	175¼

No fossils from the first 15 feet could be found among the waste and no fragments of the Black Band. A thin black seam is said to have been passed through at 15 feet, but samples were only preserved below that depth. Perhaps the first 133½ feet is entirely in Bembridge Marls.

St. Helens. North-east of the Station. Height about 5 feet above high-water.

Sunk and communicated by Mr. Parsons.

		THICKNESS.	DEPTH.
		FEET.	FEET.
Bembridge Marl.	Blue marl with *Ostrea vectensis*, *Cyrena obovata*, *C. obtusa*, *C. semistriata*, *Melania muricata*, *Cerithium mutabile*, *Serpula tenuis* - - - -	28	28
Bembridge Limestone.	Limestone - - -	9	37
Osborne Beds	Blue and various coloured clays -	11	48

ST. HELEN'S FORT.　1867?

Sunk and communicated by MESSRS. DOCWRA AND SON. (The words in brackets from an account communicated by MR. MYLNE.)
Bored throughout.

	THICKNESS.		DEPTH.	
	FT.	IN.	FT.	IN.
Concrete - - - - - -	19	0	19	0
Speckled sand - - - -	3	0	22	0
Shingle and black pebbles - - -	15	0	37	0
Grey clay (Yellow sandy clay, 57) - -	54	0	91	0
Peat (Black earth) - - -	2	0	93	0
Greenish sand (Coarse green sand) -	7	0	100	0
Stones (Flint gravel) - - -	2	0	102	0
Greenish clay and shells - -	15	0	117	0
Pale green shell-marl (Shelly clay) -	13	0	130	0
Green clay and shells (Hard green clay)	10	0	140	0
Claystone - - - - -	0	6	140	6
Grey clay and shells (Brown shelly clay) -	9	0	149	6
Claystone - - - - -	0	4	149	10
Green clay and shells - - -	1	6	151	4
Stones - - - - -	0	4	151	8
Dark green clay and shells - -	2	6	154	2
Claystone - - - - -	0	10	155	0
Green sand - - - - -	7	0	162	0
Green clay and pebbles - - -	2	0	164	0
Grey sand - - - - -	6	0	170	0

MR. MYLNE'S account is as follows, below 149 feet.

	THICKNESS.	DEPTH.
	FT.	FT.
Claystone } Hard blue clay } - - - -	5	154
Limestone } Green clayey sand } - - - -	8	162
Dark blue clay - - - -	6	168
Dark sandy clay - - - -	2	170

SPITHEAD DEFENCES—Horse Sand Fort.

Communicated by CAPT. HEWETT, R.E., to H. W. BRISTOW. The fossils determined by MR. ETHERIDGE.

Surface of shoal 24¼ feet below high-water of ordinary spring-tides. Measurements from the Pump Room Floor, 3½ feet above high-water. 6 foot Cylinder to 83 feet; the rest bored.

		THICKNESS.	DEPTH.
		FEET.	FEET.
	Water, &c. to surface of shoal -	27¾	27¾
	Shingle and a little sand - ⎤		
	" Natural concrete " - ⎬	5	32¾
	Clean shingle - - - ⎦		
	Moderately fine sand and occasional shingle, pieces of bark and branches of trees - -	18	50¾
Recent Marine Deposits 70½ feet.	Shingle, sand and vegetable matter, the latter almost entirely compressing to centre dark band [shown on the drawing sent] -	8	58¾
	Shingle, sand and shells - -	5	63¾
	Blue clay, shingle and sand -	14	77¾
	Pure sand - - - -	¼	78

	THICKNESS.	DEPTH.
	FEET.	FEET.
Blue clay - - -	1	79
Chalk flints - - -	½	79½
Shingle, sand and shells -	2½	82
Rock - - - -	¼	82¼
Flint shingle and clean orange sand - - - -	15¾	98
Greenish-grey clay with slight sand and occasional flint pebbles and stone - - - -	45	143
Greenish-grey clay - -	25	168
Greenish-grey clay and slightly more sand. *Ostrea, Cardita planicosta* - - -	37	205
Greenish-grey clay, less sand, no fossils - - - -	30	235
Greenish-grey sandy clay -	20	255
Greenish-grey clay, no fossils -	18	273
Greenish-grey clay. *Nummulites, Corbula* - - -	15	288
Brownish-grey clay. Nodules of siliceous sandstone full of glauconite at 335 feet. No fossils	50	338
Fine clean greenish-grey and black sand. *Cardita planicosta.* Many nodules of sandstone and iron pyrites - - -	20	358
Grey rock. *Pecten corneus, Cardium semigranulatum* - -	1½	359½
Brownish-grey clay. *Cardium semigranulatum, Pectunculus pulvinatus, Pecten corneus* -	12	371½
Darker brownish-grey clay and flint pebbles. *Pectunculus pulvinatus, Turritella imbricataria*	4½	376 .
Very fine greenish-grey and some orange sand, and flint pebbles. *Cardium semigranulatum, Pectunculus, Voluta, Turritella imbricataria, Fusus longævus* -	10	386
Greenish-grey sandy clay, slightly stratified - - -	34½	420½
Greenish-grey clay, with some sand, slightly stratified. *Cytherea suberycinoides, Pectunculus* - -	24	444½
Greenish-grey sand rock, numerous fossils. *Nummulites* - -	2	446½
Light greenish-grey and black very fine quicksand. *Cardita planicosta* and *Turritella* at 494 feet - - -	58½	505
Rather darker green-grey sand with clay in lumps. *Cytherea lucida, Corbula gallica, Cardita planicosta, Fusus pyrus* -	27¾	532¾
Dark-green band of sandstone and iron pyrites - - -	½	533¼
Light-grey clean sand. Frequent nodules of iron pyrites and pieces of lignite. *Cardita planicosta* - - - -	24¼	557½

Bracklesham Beds 471¼ feet.

	THICKNESS.	DEPTH.
	FEET.	FEET.
Brownish-grey sand and stratified clay with iron pyrites - -	$2\frac{1}{2}$	560
Brownish-grey clay, occasionally stratified and with vegetable impressions and plant remains	7	567
Clean sharp light-grey (almost white) siliceous sand. No fossils - - - -	$2\frac{1}{4}$	$569\frac{1}{4}$

SPITHEAD DEFENCES—Noman Fort.

Communicated by MAJOR E. A. HEWITT, R.E., to H. W. BRISTOW. The fossils determined by MR. ETHERIDGE.

Surface of shoal 34 feet below high-water. Measurements from Powder Magazine floor, $3\frac{1}{2}$ feet above High-water.

		THICKNESS.	DEPTH.
		FEET.	FEET.
	Water &c. to surface of shoal -	$37\frac{1}{2}$	$37\frac{1}{2}$
Recent Marine Deposits.	Hard compact flint shingle, bright sand, chalk stones, Isle of Wight stone, shells &c. Jaw of Red Deer fifty feet down. Large flint shingle, fine pale-yellow sand, shells, &c. - - - Fine flint-shingle, coarse angular pale-yellow sand. Remains of trees, shells, &c. *Nassa reticulata, Trochus ziziphinus* - -	90	$127\frac{1}{2}$
Bracklesham Beds.	Grey sand with slight clay and occasional flint shingle, shells, &c.	28	$155\frac{1}{2}$
	Greenish-grey sandy clay. No fossils	$11\frac{3}{4}$	$167\frac{1}{4}$
	Green-grey clay, fossils. *Cardita acuticosta, Astarte, Cytherea, Ostrea tenera*? fragments -	130	$297\frac{1}{4}$
	Green-grey clay, fossils numerous. Indurated phosphatic nodules with *Plicatulæ* - - - -	$43\frac{1}{4}$	$340\frac{1}{2}$
	Green-grey clay, rather more sand, fossils - - - -	100	$440\frac{1}{2}$
	Brown-grey clay, fossils. *Conus, Turritella sulcifera,* in sandy clay	$25\frac{1}{2}$	466
	Brown-grey clay, slight sand, fossils. *Pinna margaritacea, Corbula, Turritella sulcifera, Nummulites variolarius, Serpula* at 502 feet from surface; at 506, *Cardium semigranulatum;* at 510 *Pecten corneus, Cytherea suberycinoides, Cardium semigranulatum* - - -	47	513
	Darker green-grey sandy clay -	27	540
	Hard grey-green sandstone rock, numerous fossils. *Cardium semigranulatum, Cytherea suberycinoides, Cardita planicosta, Turritella, Fusus* - - -	1	541
	Pale green-grey sand, numerous fossils. *Cardita planicosta, Turritella sulcifera, Seraphs,* sp., *Cardium semigranulatum, Pectunculus pulvinatus* - - -	9	550

	THICKNESS.	DEPTH.
	FEET.	FEET.
Brown-green clay, with slight sand in layers, chalk [weathered flint?] pebbles from 558'–560'. *Turritella imbricataria, Pectunculus pulvinatus*, flint pebbles - -	21½	571½
Green-grey with some orange sand slightly stratified in places, fossils. *Cerithium giganteum, Turritella imbricataria* - - -	8⅓	579
Bottom of bore at 571 feet.		

Water rises to 4' 4" below Powder Magazine, *i.e.* 10" below high-water ordinary spring tides. The supply of water

at 50' below Powder Magazine Floor is 10,800 gallons per diem.
,, 100' ,,   ,,   ,,   23,000 ,,   ,,

STAPLERS. Farm west of the Gravel Pits. Height about 257 feet above the sea.

From information supplied by the farmer.

		THICKNESS.		DEPTH.	
		FT.	IN.	FT.	IN.
Drift	Gravel - - - -	1	6	1	6
Middle ? Hamstead Beds.	Blue clay - - -	} 64	0	65	6
	Mottled clay - - -				
	Fine-grained hard concretionary sandstone - - -	2	6	68	0
	Clay - - - -	5	0	73	0

WEST COWES WATERWORKS. 7 chains east of Broadfield. Height 165 feet above Ordnance Datum.

From samples and measurements communicated by MR. ATKEY and MESSRS. TILLEY & SONS.

		THICKNESS.		DEPTH.	
		FT.	IN.	FT.	IN.
Drift 10 feet -	Gravel - - - -	10	0	10	0
Lower Hamstead Beds 29 feet.	Greenish clay (disturbed on one side of the well and containing a drain at 23 feet) -	13	0	23	0
	Blue clay (a 2 inch seam of shells at 30 feet) - - -	9	0	32	0
	Flat cement stone - -	0	6	32	6
	Blue clay - - -	5	6	38	0
	Black shaly clay - - -	1	0	39	0
Bembridge Marls 116 feet.	Blue and green clay. A shell bed with *Melania muricata* at 40 feet. Rock with *Melanopsis* and *Paludina lenta* at 61 feet -	33	6	72	6
	Stone and a little water - -	0	6	73	0
	Blue and green clay - -	31	0	104	0
	Cement stone - - -	1	0	105	0
	Green shelly clay and shale with nodular stone at 110 feet. Very shelly at 115 feet	14	0	119	0
	Green clay and stone - -	5	0	124	0
	Blue clay (pyrites at 127 feet)	31	0	155	0
Bembridge Limestone 9 feet.	Very hard freestone - -	5	0	160	0
	White bed - - -	2	0	162	0
	Black brown and white clay -	2	0	164	0

		THICKNESS.		DEPTH.	
		Ft.	In.	Ft.	In.
Osborne Beds 103½ feet.	Red and green mottled clays	36	0	200	0
	Blue shell marl	30	0	230	0
	Green clay	21	0	251	0
	Do.　rather sandy	0	6	251	6
	Stone, and a little water	1	0	252	6
	Dark green and brown mottled clay	14	0	266	6
	Stone	1	0	267	6
Upper Headon Beds 53½ feet.	Mottled clay, with veins of sand and a little water	21	0	288	6
	Stone and a little water	1	0	289	6
	Blue clay	3	0	292	6
	Stone and a little water	0	6	293	0
	Blue clay (fragments of shell at 320 feet)	28	0	321	0
Middle Headon Beds 116 feet.	Sand with shells and water (pumping from this spring dried the well at Woodvale)	7	0	328	0
	Green sandy clay and blue clay, full of *Cyrena oborata* and *Melania muricata* at 331 feet; green and carbonaceous at 341; at 365 blue and very shelly, with *Cytherea incrassata*, *Cyrena*, sp., *Natica labellata*, *Nematura parvula*, *Buccinum labiatum*, Fish otolith (VENUS BED); at 375 green clay with *Natica*, *Cerithium* &c.; at 385 blue shelly clay; at 400 hard clay; at 414 green sandy clay full of fossils (the following species were found in the spoil heap, but the exact depth to which they belong is uncertain, but lies between 414 and 420 feet— *Ostrea ventilabrum*, *Cardita simplex*, *Cytherea incrassata*, *Cyrena oborata* *C. deperdita*, *Corbula cuspidata*, *C. pisum*, *Cancellaria elongata*, *Buccinum labiatum*, *Voluta geminata*, *Pleurotoma plebia*, *Rostellaria*, sp., *Cerithium elegans*, *Natica labellata*, *Bulla*, sp. (BROCKENHURST BED?)	92	0	420	0
	Grey shelly sand, *Natica*, *Pleurotoma*, *Nematura parvula*, *Planorbis*, *Cyrena*, *Potamomya*	14	0	434	0
	Clay	3	0	437	0

An Analysis by PROFESSOR J. ATTFIELD, F.R.S. (November 1887) of the spring at 320 feet gave the following results:—

	Grains per Gallon.
Total suspended solid matter, dried at 250° F.	None after subsidence.
Total dissolved solid matter, dried at 250° F.	17·00
Ammonia	0·07
(Equal to ammonia per million 1·00).	
Albumenoid organic matter, yielding 10 per cent. of nitrogen	0·01
(Equal to ammonia per million, 0·02).	
Nitrites	None.

Grains per Gallon.

Nitrates containing 17 per cent. of nitrogen - 0·35
(Equal to grains of nitrogen per gallon, 0·06).
Chlorides containing 60 per cent. of chlorine - 3·20
(Equal to grains of chlorine per gallon 1·9).
Hardness, reckoned as chalk grains or " degrees ";
removed by ebullition - - 10·
unaffected by ebullition - - 0·
Total hardness - - - - - 10·00
Lead or Copper - - - - - None.
Physical examination after subsidence - - Satisfactory
Oxygen absorbed in three hours - - 0·02

WEST COWES. Boring at Egypt Point by MR. VIGNOLLES.
Height 8 feet above datum line.

FEET.
Clays of various colours [Osborne Beds] - - - 94

WOODVALE. West of West Cowes, Isle of Wight.

Communicated by the MESSRS. ADDIE, of Preston, to Mr. Whitaker.
The Fossils determined by MR. J. W. ELWES.

		THICKNESS.		DEPTH.	
		FT.	IN.	FT.	IN.
	Whitish earth, calcareous -	2	0	2	0
	Grey shelly clay. *Melania muricata* - -	10	6	12	6
	Light-buff calcareous earth -	1	6	14	0
	Grey clay with shells. *Melania muricata* - -	30	0	44	0
Bembridge Beds.	Light greenish-grey clay with some broken shells -	2	0	46	0
	Grey and brownish clay - -	15	0	61	0
	Dark-grey or blackish clay with some shells - -	2	0	63	0
	Dark-grey and brown clay. *Melania muricata* - -	3	0	66	0
	Grey clay with some broken shells	9	0	75	0
	Cream-coloured limestone -	4	3	79	3
	Light-grey clay, mottled brownish	8	0	87	3
	Puce and grey mottled clay -	2	0	89	3
	Pale-grey clay - -	2	0	91	3
	Crimson and grey mottled clay -	42	0	133	3
Osborne Beds 109 feet.	Light-greenish and brownish clay	2	0	135	3
	Grey clay with some very fine and soft sand? - -	2	0	137	3
	Grey clay, partly brownish (specimen from 138 feet) - -	24	0	161	3
	Crimson, grey and brown mottled clay - - -	27	0	188	3
	Limestone ? nodular -	3	6	191	9
	Pale greenish-grey clay - -	18	3	210	0
	Greenish-grey and puce mottled clay - - -	30	0	240	0
Upper Headon Beds 73 feet.	Light-grey clay with some broken shells - - -	10	3	250	3
	Grey clay with crushed shells ? *Potamomya* - -	1	0	251	3
	Grey clay with some broken shells	3	0	254	3
	Calcareous nodule (?) - -	0	6	254	9
	Greenish-grey clay with broken shells - - -	6	0	260	9
	Stone (no specimen) - -	0	6	261	3

		THICKNESS.	DEPTH.
		FT. IN.	FT. IN.
Middle Headon Beds 13½ feet.	Fine grey sand, with shells: *Cerithium concavum* (many), *C. trizonatum?*, *Melania muricata*, *Cyrena obovata*, *Potamomya gregaria* (? 2 vars.) *Ostrea* with *Serpula* - - - -	9 0	270 3
	Firm grey clayey sand with some shells - - - -	4 6	274 9

This well was subsequently deepened to 437 feet, at which depth shelly sand occurred—perhaps representing the Headon Hill Sands—but no further details can be obtained.

WOOTTON. In Beech Lane. 6 chains north of the Station.

From specimens communicated by MR. NEWBURY of Wootton, and notes and specimens communicated by MR. BROWN of Tottenham.
Water at 370 feet, rose to 100 feet from surface.

		THICKNESS.	DEPTH.
		FEET.	FEET.
Lower Hamstead Beds about 110 feet.	Light-blue clay [no specimens] -	40	40
	Clay [no specimens] about	50	90
	Dark-blue and carbonaceous clay, full of fossils. *Paludina lenta, Hydrobia pupa, H. Chasteli, Neritina tristis, Melania Forbesii, M. muricata, Melanopsis carinata, M. subulata, Planorbis,* small sp., *Cyrena semistriata, Modiola Prestwichii* - -	5	95
	Clay - - - -	15	110
Bembridge Marls about 115 feet.	Green clay &c. [specimens preserved are green clay at 114', Cement stone and pyrites with *Paludina* and *Melania turritissima* at 140', Grey clay at 143', Green clay at 155', 160', 166', Green and black clay at 169', 170', Green clay at 175', 180', 185', Bright-green clay at 206'] - - -	115	225
Bembridge Limestone.	Limestone [Black clay and limestone at 208'.] - - -	3	228
Osborne Beds about 117 feet.	[Red clay at 254', Red and green clay at 260', 265', 278', 285', Bright-green clay at 290' and 340', Red and green clay at 341' and 345'] - - -	117	345
Upper Headon Beds 59 feet.	Clay - - - -	11	356
	Rock - - - -	3½	359½
	White sand, very sharp. [Very fine brown sand at 370']	10½	370
	? [no record kept] - -	34	404

The thicknesses are only approximate, as no complete record or series of specimens is available. Another memorandum gives 385 feet to sand with water, and a total depth of 420 feet.

WOOTTON. At Briddlesford Lodge, in the middle of the Farm Buildings. Height 181 feet above the sea.

From notes made during the excavations.

		THICKNESS. Ft. In.	DEPTH. Ft. In.
Drift -	- Clayey gravel - - -	4 6	4 6
Upper Hamstead Beds.	Yellow clay, much weathered -	5 0	9 6
	Dark-blue shelly clay, full of *Cerithium plicatum* and *Melania inflata* - - - -	1 0	10 6
Middle Hamstead Beds.	Grey loamy clay - - -	1 0	11 6
	Green clay - - -	8 6	20 0
	Green clay with faint red mottling	3 6	23 6

WOOTTON. At Briddlesford Lodge. At the south-east corner of the farm buildings. Height 190 feet above the sea.

From notes made during the excavations.

		THICKNESS. Ft. In.	DEPTH. Ft. In.
Middle ? Hamstead Beds.	Mottled light-grey and dark-red clay - - - -	8 0	8 0
	Yellow and brown mixed clay, perhaps reconstructed shaly clay - - - -	2 0	10 0
	Greenish-blue clay - -	1 0	11 0
	Tenaceous blue clay - -	6 0	17 0
	Sand parting - - -	0 1	17 1
	Reconstructed clay - -	0 11	18 0
	Mottled green and red clay, slightly carbonaceous - -	7 0	25 0
	Blue carbonaceous clay, full of *Unio* - - - -	5 0	30 0

Though these two wells are only 2 chains apart the sections are quite different. No trace of the bed with *Cerithium plicatum* could be found in the second.

WOOTTON. A quarter of a mile north of Beech.

		THICKNESS. Ft. In.	DEPTH. Ft. In.
Drift -	Gravel - - - -	15 0	15 0
	Sand - - -	[?] 15 0	30 0
	Loam and ironstone - -	4 0	34 0

A good supply of water.

WOOTTON. 5 chains west of Fernhill.

		THICKNESS. Ft. In.	DEPTH. Ft. In.
Drift -	Gravel - - - -	13 0	13 0
	Sand - - -	3 0	16 0
Hamstead Beds.	Clay - - - -	2 0	18 0

WOOTTON.   Close to Brannon's Cottage.   Height about 170 feet above
the sea.

From notes made during the excavation.

		THICKNESS.		DEPTH.	
		FT.	IN.	FT.	IN.
	Red and green clay - -	10	0	10	0
Middle and	Sand - - - -	4	0	14	0
Lower (?)	Green clay - - -	10	0	24	0
Hamstead	Concretionary sandstone -	0	8	24	8
Beds.	Hard blue and green loamy clay -	20	0	44	8
	Ironstone with casts of *Limnæa* -	0	4	45	0
	Harder green and purple clay -	12	0	57	0

WOOTTON.   At Beech.

		THICKNESS.		DEPTH.	
		FT.	IN.	FT.	IN.
Middle	Clay - - - -	36	0	36	0
Hamstead	Sand - - - -	5	0	41	0
Beds.	Clay - - - -	3	0	44	0

The bed of sand corresponds with the one seen at Brannon's Cottage, and
in the cutting above the Station.

WOOTTON.   At Whitehayes.

From notes made during the excavation.

		THICKNESS.		DEPTH.	
		FT.	IN.	FT.	IN.
	Yellow clay - - -	10	0	10	0
Middle	Blue clay - - - -	3	0	13	0
Hamstead	Red clay - - - -	1	6	14	6
Beds.	Blue and yellow clay with turtle bones - - - -	3	6	18	0

This well was still unfinished at the time of the completion of the Survey.

APPENDIX IV.

GEOLOGICAL BIBLIOGRAPHY OF THE ISLE OF WIGHT.

1. Publications of the Geological Survey.

Maps and Sections.

Sheet 10 of the Map. Originally geologically surveyed on the One-Inch Scale, by H. W. Bristow and W. T. Aveline (1856). The Isle of Wight re-surveyed on the Six-Inch Scale, by Clement Reid (Tertiary) and Aubrey Strahan, M.A. (Cretaceous) 1888.

Geological Map of the Isle of Wight (in MS.), surveyed by Clement Reid (Tertiary area) and Aubrey Strahan (Secondary area), on a scale of 6 inches = 1 mile. Exhibited at the fourth meeting of the International Geological Congress in September 1888, and subsequently hung in the Museum of Practical Geology.

Horizontal Sections, Sheet 47. By H. W. Bristow, 1858. Revised Edition in 1870. [Under revision in 1889.] No. 1, from Totlands Bay, across the western extremity of Headon Hill to the Sea near the Main Bench. No. 2. Section from the Solent, near Worsley's Tower, to the Sea under High Down Beacon. No. 3. Section from Hempstead Cliff to Hanover Point. No. 4. Section from Norris to Rocken End. No. 5. Section from Binstead to Ventnor Cove.

Vertical Sections, Sheet 25. By H. W. Bristow in 1858. [Under revision in 1889.] Illustrative of the Upper, Middle, and Lower Eocene strata of Hempstead, St. Helens, Colwell and Totland Bays, Headon Hill, Alum Bay, and Whitecliff Bay.

Memoirs.

On the Tertiary Fluvio-marine Formation of the Isle of Wight, by Prof. E. Forbes. 8vo. 1856. (Edited by R. A. C. Godwin-Austen. With notes by H. W. Bristow, and Descriptions of Fossils by Prof. J. Morris, J. W. Salter, and T. R. Jones.)

Description of Horizontal Section, Sheet 47. By H. W. Bristow. 8vo. 1859. (Pamphlet.)

Description of Vertical Section, Sheet 25. By H. W. Bristow. 8vo. 1859. (Pamphlet.)

A Descriptive Catalogue of the Rock Specimens in the Museum of Practical Geology. By Prof. A. C. Ramsay, H. W. Bristow (and others). 8vo. 1862. (3rd Ed.), pp. 154, 158, 160, 166, 167, 170–173.

A Catalogue of the Collection of Fossils in the Museum of Practical Geology. By Prof. T. H. Huxley and R. Etheridge. 8vo. 1865.

A Catalogue of the Cretaceous Fossils in the Museum of Practical Geology. By E. T. Newton. 8vo. 1878.

A Catalogue of the Tertiary and Post-Tertiary Fossils in the Museum of Practical Geology. By E. T. Newton. 8vo. 1878.

2. List of Works, other than those of The Geological Survey, by H. W. Bristow, F.R.S., F.G.S.*

[This list is arranged in chronological order. For Index of Authors, see p. 336.]

---

* In the compilation of this List much assistance has been derived from the excellent "List of Works on the Geology, Mineralogy, and Palæontology of the Hampshire Basin," by W. Whitaker, published in the *Proc. Winchester and Hampshire Sci. Soc.* for 1873, pp. 108–127.

1605.

VERSTEGAN, R.—Restitution of Decayed Intelligence in Antiquities Concerning the most noble and renowned English Nation. 4to. *Antwerp.* (Other Editions in 1628, 1634, 1655, 1673, 1723.)

1738.

COOKE, B.—An Observation of an extraordinary damp in a well in the Isle of Wight. (Letter dated 1736.) *Phil. Trans.*, vol. xl., p. 379 (vol. viii. of Abridgment, pp. 244 and 658).

1749–50.

COOKE, B.—Account of an earthquake felt in the Isle of Wight, March 18, 19. *Phil. Trans.*, vol. xlvi. p. 651 (vol. x. of Abridgment, p. 508).

1755.

FIELDING, H.—Journal of a Voyage to Lisbon (1753). 12mo. *London.* (Gives an account of the shore at Ryde, I.W.)

1782.

ANON.—The Isle of Wight : A Poem, with Plate of Needle Rocks before the Fall of the Pointed Rock, from which the group takes its name. 12mo.

1794.

DRIVER, A. & W.—General View of the Agriculture of the County of Hants, with View of the Isle of Wight.
Agriculture by Rev. R. WARNER, and a Postcript by A. YOUNG. 4to.
WARNER, REV. R.—General View of the Agriculture of the Isle of Wight (forming a part of Hampshire), with observations on the means of its Improvements. 4to. *London.*

1795.

WARNER, REV. R.—The History of the Isle of Wight : Military, Ecclesiastical, Civil, and Natural : to which is added a View of its Agriculture (folding Map). 8vo. *Southampton.*

1798.

MARSHALL, W.—The Rural Economy of the Southern Counties of England, comprehending . . . . . the Isle of Wight, &c. . . . . . &c. 8vo. 2 vols. *London.* (Second edition in 1799.)

1801.

PENNANT, T.—A Journey from London to the Isle of Wight. (Plates and Maps.) 2 vols. 4to. *London.*

1802.

ENGLEFIELD, SIR H. C.—Observations on some remarkable Strata of Flint, in a Chalk-pit in the Isle of Wight. *Trans. Linn. Soc.*, vol. vi. p. 103. Additional Observations, p. 303.

1805.

BRAYLEY, E. W., and BRITTON, J.—The beauties of England and Wales : or delineations, topographical, historical, and descriptive of each county. 8vo. *London.*
[Vol. vi. : Topographical and Historical Description of Hampshire and the Isle of Wight.]
CAMDEN, W.—History of Hampshire and the Isle of Wight; with additions by R. GOUGH. Fol.

1806.

CAMDEN, W.—*Britannia:* or, a Chorographical Description of the Flourishing Kingdoms of England, Scotland, and Ireland, and the Islands adjacent; from the earliest antiquity. Translated from the edition published by the Author in MDCVII. Enlarged by the latest discoveries, by RICHARD GOUGH. (Vol. i., pp. 174, 208.)

1808.

ALBIN, J.—Vectiana, or a Companion to the Isle of Wight. Seventh Edition. 12mo. *London.* (Twelfth Edition, 1831).

1809.

COOKE, W.—A New Picture of the Isle of Wight, illustrated with 36 plates, and a voyage round the Coast. 8vo. and 4to. *London.* (Second Edition, *Southampton*, 1813).

1810.

VANCOUVER, C.—General View of the Agriculture of Hampshire, including the Isle of Wight. [Map and Account of Soils and Minerals.] 8vo. *London.* (Another Edition in 1813.)

1811.

BERGER, DR. J. F.—A Sketch of the Geology of some parts of Hampshire and Dorsetshire. *Trans. Geol. Soc.,* vol. i., p. 249. DE LUC, J. A.—Geological Travels. Translated from the French MS. 8vo. *London.* (Vol. ii., p. 124.) MARCET, DR. A.—A Chemical Account of an Aluminous Chalybeate Spring in the Isle of Wight. *Trans. Geol. Soc.,* vol. i., pp. 213-248., and *Journ. Nat. Phil. Chem. and Arts,* ser. 2, vol. xxxii., pp. 52, 85.

1812.

LEMPRIERE, W.—On the medicinal effects of an Aluminous Chalybeate Water lately discovered at Sand Rocks, in the Isle of Wight. 8vo. *London,* Other editions 1820 and 1827). See also Nicholson, *Journ. Nat. Phil.,* p. 52-66 and 85-100. MIDDLETON, J.—Outlines of the Mineral Strata of Great Britain. *Monthly Mag.,* vol. xxxiv., No. 233, p. 310, and No. 234, p. 393.

1813.

TOWNSEND, REV. JOSEPH.—The character of Moses established for veracity as an Historian, recording events from the Creation to the Deluge (pp. 190, 231, 310). 4°. *Bath and London.* WATERWORTH, DR. T. L.—Account of a Chalybeate Spring in the Isle of Wight. *Thom. Ann. Phil.,* Ser. i., vol. i., p. 447. WEBSTER, T.—On the Freshwater Formations in the Isle of Wight. (MS. in the Library of the *Geol. Soc.*) ———, ———.—On the Isle of Wight and the discovery of Freshwater Shells. (MS. in Library of the *Geol. Soc.*)

1814.

WEBSTER, T.—On the Freshwater Formations in the Isle   Wight, with Some Observations on the Strata over the Chalk in the South-east of England. *Trans. Geol. Soc.,* 1st. ser., vol. ii., p. 161, pl. 9-1 . ———, ———.—On some new varieties of Fossil Alcyonia. *Ibid.,* vol. iii., p. 377, pp. 27-30.

E 56786.   X

1816.

ENGLEFIELD, SIR H. C.—A Description of the Principal Picturesque
Beauties, Antiquities, and Geological Phænomena of the Isle of Wight, with
Additional Observations on the Strata of this Island, and their continuation
in the Adjacent Parts of Dorsetshire, by THOMAS WEBSTER. Fol. *London.*
50 plates.

1816–1829.

SOWERBY, J. DE C.—The Mineral Conchology of Great Britain. 8vo.
London. Vols. ii. to vi. (Vol. i. in 1812.)

1818.

ANON.—Animal Remains (Bones at Motteston and Northwood, Isle of
Wight). *Phil. Mag.*, vol. lii., p. 68.
FAREY, J.—An Alphabetical Arrangement of the Places from whence
Fossil Shells have been obtained by MR. JAMES SOWERBY, and drawn and
described in vol. ii. of his "Mineral Conchology . . . . . ." *Phil.
Mag.*, vol. li., p. 348.

1820.

SCUDAMORE, C.—A Chemical and Medical Report on the properties of the
Mineral Waters of . . . . the Isle of Wight. 8vo. *London.* (Another
edition in 1833.)

1821.

SOWERBY, G. B.—On the Geological Formations of Headen Hill in the
Isle of Wight. *Ann. of Phil.*, vol. xviii. (ser. 2, vol. ii.), p. 216.

1822.

CONYBEARE, REV. W. D., and PHILLIPS, W.—Outlines of the Geology of
England and Wales. 8vo. *London.*
SEDGWICK, REV. PROF. A.—On the Geology of the Isle of Wight, &c.
Ann. of Phil., vol. xix. (ser. 2, vol. iii.), pp. 329 to 355.

1824.

FITTON, DR. W. H.—Inquiries respecting the Geological Relations of the Beds
between the Chalk and the Purbeck Limestone in the South-east of England.
Ann. of Phil., vol. xxiv. (new series, vol. viii.), pp. 365 and 458. (Reprinted
in 4°. in 1833.)
WEBSTER, T.—On a Freshwater Formation in Hordwell Cliff, Hampshire,
and on the subjacent Beds from Hordwell to Muddiford. *Trans. Geol. Soc.*,
ser. 2, vol. i., p. 90, pl. xii.

1825.

BUCKLAND, REV. PROF. W.—On the Discovery of the *Anoplotherium
commune* in the Isle of Wight. *Ann. of Phil.*, ser. 2, vol. x., p. 360.
SEDGWICK, REV. PROF. A.—On the Origin of Alluvial and Diluvial Formations. *Ann. of Phil.*, ser. 2, vol. x., p. 18. (Isle of Wight, p. 20.)
WEBSTER, T.—Reply to Dr. Fitton's paper entitled "Inquiries respecting
the Geological Relations of the Beds between the Chalk and the Purbeck
Limestone in the south-east of England." *Ann. of Phil.*, vol. xxv. (vol. ix. of
new series), p. 33.

1831.

SEDGWICK, REV. PROF. A.—Address to the Geological Society, delivered
18th February 1831. On the deposits of the Isle of Wight above the London
Clay. *Proc. Geol. Soc.*, vol. i., p. 294. See also p. 289.

1832.

CONYBEARE, REV. W. D.—Inquiry how far the Theory of M. E. de Beaumont concerning the Parallelism of Lines of Elevation of the same Geological Æra, is agreeable to the Phœnomena as exhibited in Great Britain. *Phil. Mag.*, ser. 3, vol. i., p. 118.

BROWNE, H.—The Geology of Scripture. 8vo. *Frome.* Elevation of the Isle of Wight, p. 23. Formation of Haden [Headon] Hill, p. 30.

1833.

DE LA BECHE, SIR H. T.—A geological Manual. 3rd Edition, considerably enlarged. 8vo. *London.* (Supracretaceous Rocks of the Isle of Wight, pp. 260–264.)

1835.

BUCKLAND, REV. W.—On the Discovery of Fossil Bones of the Iguanodon in the Iron Sand of the Wealden Formation in the Isle of Wight, and in the Isle of Purbeck. *Trans. Geol. Soc.*, ser. 2, vol. iii., p. 425.

MORRIS [PROF.] J.—Fact and situation of the Occurrence of Seeds and certain Species of Shells in the Lower Freshwater Formation of the Isle of Wight. *Mag. Nat. Hist.*, vol. viii., p. 391.

PRATT, S. P.—Remarks on the Existence of the *Anoplotherium* and *Palœotherium* in the Lower Freshwater Formation at Binstead, near Ryde, in the Isle of Wight. *Trans. Geol. Soc.*, ser. 2, vol. iii., pp. 451–453.

1836.

FITTON, DR. W. H.—Observations on some of the Strata between the Chalk and the Oxford Oolite, in the South-east of England. *Trans. Geol. Soc.*, Ser. 2, vol. iv., p. 103.

1837.

FAIRHOLME, G.—Description of the Isle of Wight and its coasts, together with the evidences which they present of the recent origin of the island as a dry land, forming chapter 7 of "New and Conclusive Physical Demonstrations both of the Fact and Period of the Mosaic Deluge &c." 8vo. *London.*

PRÉVOST, C.—Coupe d'Alum Bay et d'Headen-Hill, dans l'île de Wight. *Bull. Soc. Géol. France*, t. viii., p. 76.

SOWERBY, J. DE C.—On his new genus of fossil shells, *Tropœum. Proc. Geol. Soc.*, vol. ii., p. 535.

TOOKE, A. W.—The Mineral Topography of Great Britain. *Mining Review*, No. 9, p. 39. (Isle of Wight, p. 45.)

1838.

BOWERBANK, DR. J. S.—An Account of a deposit containing Land Shells at Gore Cliff, Isle of Wight. *Proc. Geol. Soc.*, vol. ii., p. 449.

———, —.—Lower freshwater formation in the Isle of Wight. *Mag. Nat. Hist.*, ser. 2, vol. ii., p. 674.

D'ARCHIAC, VICOMTE.—Note sur les Sables et Grès Moyens Tertiares. *Bull. Soc. Géol. France*, vol. ix., p. 54. "En Angleterre," pp. 65–67.

IBBETSON, CAPT. L. L. B.—Typorama, a Modelled View of the Undercliff in the Isle of Wight. (Descriptive Letterpress to the above.) 8vo. *London.*

———., —.—A Trigonometrical Model of the Undercliff on the scale of 3 feet to one mile (coloured geologically).

MANTELL, DR. G. A.—The Wonders of Geology, or a Familiar Exposition of Geological Phenomena. 2 vols. 8vo. *London.*

(Other Editions with additions published in 1839, 40, 42, 44, 48. Seventh edition, revised by PROF. T. R JONES, in 1858.)

OWEN, PROF. [SIR] R.—On some Fossil Remains of Palœotherium, Anoplotherium and Chæropotamus, from the freshwater beds of the Isle of Wight. *Proc. Geol. Soc.*, vol. iii., p. 1.

1839.

CLARKE, REV. W. B.—Illustrations of the Geology of the South East of Dorsetshire. *Mag. Nat. Hist.*, vol iii., New series, pp. 390, 432, 483.

D'ARCHIAC, VICOMTE.—Note sur la coordination des terrains tertiaires du nord de la France, de la Belgique, et de l'Angleterre. *Bull. Soc. Géol. France*, vol. x., p. 159.

OGILBY, W.—Description of the Frontal Spine of a second species of *Hybodus;* from the Wealden Clay, Isle of Wight. *Mag. Nat. Hist.*, vol. iii., series 2, p. 279.

1840.

RICKMAN, W.—Earth Falls at the Undercliff in the Isle of Wight. *Inst. Civ. Eng.*, vol. i., p. 35.

SOWERBY, J. DE C.—Letter on the Genus *Crioceratites* and on *Scaphites gigas. Trans. Geol. Soc.*, ser. 2, vol. v., p. 409.

1841.

BOWERBANK, DR. J. S.—On the London and Plastic Clay Formations of the Isle of Wight. *Trans. Geol. Soc.*, ser. 2, vol. vi., p. 169.

GRANVILLE, DR. A. B.—The Spas of England and principal Sea-bathing places. 8vo. *London.* (Isle of Wight, pp. 537–549.)

MUDIE, R.—The Isle of Wight: its Past and Present Condition, and Future Prospects. 8vo. *London and Winchester.* (Vol. iii., chap. 3, Geology of the Isle of Wight.)

OWEN, PROF. [SIR] R.—Description of some Fossil Remains of Chæropotamus, Palæotherium, Anoplotherium, and Dichobunes, from the Eocene Formation, Isle of Wight. *Trans. Geol. Soc.*, ser. 2, vol. vi., p. 41.

TRIMMER, J.—Practical Geology and Mineralogy. 8vo. *London.* (Freshwater Formations of the Isle of Wight, pp. 359–61.)

1842.

OWEN, PROF. [SIR] R.—Report on British Fossil Reptiles (Part II.). *Rep. Brit. Assoc.* for 1841, pp. 87, 91 to 95, 128, 168.

1843.

FITTON, DR. W. H.—Observations on part of the Section of the Lower Greensand at Atherfield, on the coast of the Isle of Wight. *Proc. Geol. Soc.*, vol. iv., p. 198.

———, ——.—Comparative Remarks on the Lower Greensand of Kent and the Isle of Wight. *Ibid.*, p. 208.

LEE, J. E.—Notice of Saurian Dermal Plates from the Wealden of the Isle of Wight. *Ann. and Mag. Nat. Hist.*, vol. xi., p. 5.

MURCHISON, SIR R. I.—Observations on the Occurrence of Freshwater Beds in the Oolitic Deposits of Brora, Sutherlandshire; and on the British Equivalents of the Neocomian System of Foreign Geologists. *Proc. Geol. Soc.*, vol. iv., p. 174.

1844.

FITTON, DR. W. H.—Observations sur le lower greensand de l'île de Wight. *Bull. Soc. Géol. France*, vol i., 2ᵉ série, p. 438.

LEYMERIE, A.—Observations sur la communication faite sur le lower-green-sand de l'île de Wight, par M. Fitton, dans la séance du 20 mai 1844. *Bull. Soc. Géol. France*, vol. ii., 2ᵉ série, p. 41–47 (1844 à 1845).

MANTELL, DR. G. A.—Medals of Creation or First Lessons in Geology and in the study of Organic Remains. 12mo. 2 vols.

OWEN, PROF. [SIR] R.—Report on the British Fossil Mammalia. (Part II. Ungulata.) *Rep. Brit. Assoc.* for 1843, pp. 224–226.

1845.

EGERTON, SIR P. DE M. G.—Description of the mouth of a *Hybodus* (*H. basanus*) found in the Isle of Wight. *Quart. Journ. Geol. Soc.*, vol. i., p. 197.

FITTON, DR. W. H.—Comparative Remarks on the Sections below the Chalk on the Coast near Hythe, in Kent, and Atherfield, in the Isle of Wight. *Quart. Journ. Geol. Soc.*, vol. i., p. 179.

FORBES, PROF. E.—Catalogue of Lower Greensand Fossils in the Museum of the Geological Society. *Quart. Journ. Geol. Soc.*, vol. i., pp. 237 and 345.

FORBES, PROF. E., and CAPT. L. L. B. IBBETSON.—On the Section between Black-Gang-Chine and Atherfield Point, Isle of Wight. *Quart. Journ. Geol. Soc.*, vol. i., p. 190.

———, ———.—On the Tertiary and Cretaceous Formations of the Isle of Wight. *Rep. Brit. Assoc.* for 1844, trans. of sections, p. 43.

LEYMERIE, PROF. A.—Observations on a Communication made by Dr. Fitton to the Geol. Soc. France at the Meeting of May 20, 1844, on the Lower Greensand of the Isle of Wight. *Phil. Mag.*, ser. 3, vol. xxvi., p. 281.

SIMMS, F. W.—On the Thickness of the Lower Greensand Beds of the South-east coast of the Isle of Wight. *Quart. Journ. Geol. Soc.*, vol. i., p. 76.

1846.

FITTON, DR. W. H.—Stratigraphical account of the Section from Atherfield to Rocken-end in the Isle of Wight. *Quart. Journ. Geol. Soc.*, vol. ii., p. 55.

MANTELL, DR. G. A.—Notes on the Wealden Strata of the Isle of Wight, with an account of the Bones of Iguanodons and other Reptiles, discovered at Brook Point and Sandown Bay. *Quart. Journ. Geol. Soc.*, vol. ii., pp. 91–96.

OWEN, PROF. [SIR] R.—Description of an Upper Molar Tooth of *Dichobune cervinum*, from the Eocene Marl at Binstead, Isle of Wight. *Quart. Journ. Geol. Soc.*, vol. ii., p. 420.

———, ———.—A History of British Fossil Mammals and Birds. 8°. *London.*

PRESTWICH [PROF.] J.—On the Tertiary or Supracretaceous Formations of the Isle of Wight, as exhibited in the Sections at Alum Bay and White-cliff Bay. *Quart. Journ. Geol. Soc.*, vol. ii., pp. 223, 259. Pl. ix.

SAXBY, S. M.—On the Discovery of Footmarks in the Greensand of the Isle of Wight. *Phil. Mag.*, ser. 3, vol. xxix., p. 310.

1847.

FITTON, DR. W. H.—A Stratigraphical Account of the Section from Atherfield to Rocken-end, on the South-west Coast of the Isle of Wight. *Quart. Journ. Geol. Soc.*, vol. iii., p. 289. (Plate xii., comparative sections of the Lower Greensand in England.)

———, ———.—On the Arrangement and Nomenclature of some of the sub-cretaceous Strata. *Rep. Brit. Assoc.* for 1846. Trans. of sections, p. 58.

MANTELL, DR. G. A.—Geological Excursions round the Isle of Wight, &c. 8°. *London.* 2nd Edition, 1851. 3rd Edition, 1854.

———.—Fossil Remains of the Reindeer in the Isle of Wight. *London Geol. Journ.*, vol. i., p. 36.

———.—On the occurrence of a large Species of Unio in the Wealden Strata of the Isle of Wight. (*Brit. Assoc.* 1844.) *Ibid.*, p. 41, and Plate 14.

PRESTWICH, [PROF.] J.—On the probable Age of the London Clay, and its Relations to the Hampshire and Paris Tertiary Systems. *Quart. Journ. Geol. Soc.*, vol. iii., p. 354.

———.—On the main points of structure, and on the probable Age of the Bagshot Sands, and on their presumed equivalents in Hampshire and France. *Ibid.*, vol. iii., p. 378.

———.—On the Occurrence of Cypris in a part of the Tertiary Freshwater Strata of the Isle of Wight. *Rep. Brit. Assoc.* for 1846. Trans. of sections, pp. 56 and 58.

1848.

CHAMBERS, R.—Ancient Sea-Margins, as Memorials of Changes in the Relative Level of Sea and Land. 8°. *Edinburgh and London.* Pp. 241–3.

MORRIS, [PROF.] J.—A description of a New Species of *Nautilus* (N. Saxbii) from the Lower Greensand of the Isle of Wight. *Quart. Journ. Geol. Soc.,* vol. iv., p. 193. *Ann. Nat. Hist.,* vol. i., pp. 106–107.

NESBIT, J. C.—On the presence of Phosphoric Acid in the subordinate members of the Chalk Formation. *Quart. Journ. Geol. Soc.,* vol. iv., p. 262.

OWEN, PROF. [SIR] R.—Description of Teeth and portions of Jaws of two extinct Anthracotherioid Quadrupeds—(*Hyopotamus Vectianus* and *H. bovinus,*) discovered in the Eocene deposits on the N.W. Coast of the Isle of Wight, &c. *Ibid.,* vol. iv., p. 103–141.

1849.

IBBETSON, CAPT. L. L. B.—Notes on the Geology and Chemical Composition of the various Strata in the Isle of Wight. Map, in relief, coloured geologically. 8vo. *London.*

——, ——.—On the Position of the Chloritic Marl or Phosphate of Lime Bed in the Isle of Wight. *Rep. Brit. Assoc.,* trans. of sections, p. 69.

LONSDALE, W.—Notes on Fossil Zoophytes found in the Deposits described by Dr. Fitton in his Memoir entitled "A Stratigraphical Account of the Section from Atherfield to Rocken End." *Quart. Journ. Geol. Soc.,* vol. v., p. 55.

McCoY, F.—On the Classification of some British Fossil Crustacea . . . . *Ann. Nat. Hist.,* ser. 2, vol. iv., p. 330.

MANTELL, DR. G. A.—A brief Notice of Organic Remains recently discovered in the Wealden Formation. *Quart. Journ. Geol. Soc.,* vol. v., p. 37.

PAINE, J. M., and WAY, J. T.—On the Phosphoric Strata of the Chalk Formation. *Journ. Roy. Agric. Soc.,* vol. ix., pp. 56–84.

PRESTWICH, [PROF.] J.—On the Position and General Characters of the Strata exhibited in the Coast Section, from Christchurch Harbour to Poole Harbour. *Quart. Journ. Geol. Soc.,* vol. v., p. 43.

1850.

GODWIN-AUSTEN, R. A. C.—On the Valley of the English Channel. *Quart. Journ. Geol. Soc.,* vol. vi., p. 69.

PRESTWICH, [PROF.] J.—On the Structure of the Strata between the London Clay and the Chalk in the London and Hampshire Tertiary Systems. Part I. The Basement-bed of the London Clay. *Quart. Journ. Geol. Soc.,* vol. vi., p. 252.

1851.

DUMONT, A.—Note sur la position géologique de l'argile rupelienne et sur la synchronisme des formations tertiaires de la Belgique, de l'Angleterre et du nord de la France. *Bull. Acad. Roy. Sciences Belgique,* t. xviii., IIe. partie, p. 179.

WRIGHT, DR. T.—A Stratigraphical Account of the Section from Round Tower Point to Alum Bay, on the North-west Coast of the Isle of Wight. *Ann. and Mag. Nat. Hist.,* ser. 2, vol. vii., p. 14. *Proc. Cotteswold Nat. Club.,* vol. i., p. 87.

1852.

DUMONT, A.—Observations sur la Constitution Géologique des terrains tertiaires de l'Angleterre, comparés à ceux de la Belgique, faites en Octobre 1851. *Bull Acad. Roy. Sciences Belgique,* t. xix., IIe. partie, p. 344.

HÉBERT, E.—Comparaison des couches tertiares inférieures de la France et de l'Angleterre. *Bull. Soc. Géol. Fr.,* sér. 2, t. ix., p. 350.

WETHERELL, N. T.—Note on a new species of Clionites. *Ann. and Mag. Nat. Htist.,* ser. 2, vol. x., p. 354.

WRIGHT, DR. T.—Contributions to the Palæontology of the Isle of Wight. *Ann. and. May. Nat. Hist.*, ser. 2, vol. x., p. 87.   *Proc. Cotteswold Nat. Club*, vol. i., p. 229.

———, ————.—A Stratigraphical Account of the Section of Hordwell, Beacon and Barton Cliffs, on the Coast of Hampshire.   *Proc. Cotteswold Club*, vol. i., p. 120.

1853.

FORBES, PROF. E.—On the Fluvio-marine Tertiaries of the Isle of Wight. *Quart. Journ. Geol. Soc.*, vol. ix., p. 259.
FORBES, PROF. E.—On some New Points in British Geology.   *Edin. New Phil. Journ.*, vol. 55, p. 263.
SORBY, H. C.—On the Microscopical Structure of some British Tertiary and Post-Tertiary Freshwater Marls and Limestones.   *Quart. Journ. Geol. Soc.*, vol. ix., pp. 344–346.
WAY, J. T., and PAINE, J. M.—On the Silica Strata of the Lower Chalk. Isle of Wight, p. 235.   *Journ. Roy. Agric. Soc.*, vol. xiv., p. 225.

1854.

McCOY, F.—On some New Cretaceous Crustacea.   *Ann. Nat. Hist.*, ser. 2, vol. xiv., pp. 116–122.
MANTELL, DR. G. A.—See 1847.
MORRIS, PROF. J.—A Catalogue of British Fossils.   8vo.   London. 2nd Edition.
PRESTWICH, [PROF.] J.—On the Correlation of the Lower Tertiaries of England with those of France and Belgium.   *Quart. Journ. Geol. Soc.*, vol. x., p. 454–456.
———, ———.—On the Structure of the Strata between the London Clay and the Chalk in the London and Hampshire Tertiary Systems.   Part II. The Woolwich and Reading Series.   *Ibid.*, vol. x., pp. 75–170.   Isle of Wight, 78–80, 81.
TRIMMER, J.—On the Superficial Deposits of the Isle of Wight.   *Ibid.*, vol. x., p. 51.

1855.

GODWIN-AUSTEN, R. A. C.—On Land-Surfaces beneath the Drift-Gravel. *Ibid.*, vol. xi., p. 116.
TRIMMER, J.—On the Agricultural Relations of the Western portion of the Hampshire Tertiary District, and on the Agricultural Importance of the Marls of the New Forest.   *Journ. Roy. Agric. Soc.*, vol. xvi., pp. 125–151.

1856.

LA HARPE, DR. P. DE.—Flore Tertiaire de l'Angleterre.   *Bull. Soc. Vaudoise Sci. Nat.*   Vol. v., pp. 133–156 (Alum Bay Leaf-Bed).
RENEVIER, E.—Notes sur quelques points de la géologie de l'Angleterre *Ibid.*, vol. v., pp. 51, 52, and *Quart. Journ. Geol. Soc.*, vol. xii., p. 2.
WHITLEY, N.—The Physical Geography of the South-Western Counties of England.   *Journ. Bath and W. of England Agric. Soc.*, N. Ser., vol. iv., p. 227.

1857.

GODWIN-AUSTEN, R. A. C.—On the Newer Tertiary Deposits of the Sussex Coast.   *Quart. Journ. Geol. Soc.*, vol. xiii., p. 40.
OWEN, PROF. [SIR] R.—On the *Dichodon cuspidatus*, Owen.   *Ibid.*, vol. xiii., p. 190.
———.—Description of the Lower Jaw and Teeth of an Anoplotheroid Quadruped (*Dichobune ovina*) . . . . . from the Upper Eocene Marl, Isle of Wight.   *Ibid.*, vol. xiii., pp. 254–260.
PRESTWICH, PROF. J.—On the Correlation of the Eocene Tertiaries of England, France, and Belgium.   Part II.—The Paris Group.   *Ibid.*, vol. xiii., p. 89.
SORBY, H. C.—On the Physical Geography of the Tertiary Estuary of the Isle of Wight.   *Edinb. New Phil. Journ.*, ser. 2., vol. v., p. 275–298.

1858.

BRION, J.—Stanford's Relief Map of the Isle of Wight (coloured geologically). *London.*

GERVAIS, PROF. P.—On some teeth of the Anchitherium, recently discovered in the Isle of Wight. *Geologist*, vol. i., p. 153.

GIBSON, T. F.—Notice of the Discovery of a Large Femur of the Iguanodon in the Weald Clay at Sandown Bay, Isle of Wight. *Quart. Journ. Geol. Soc.*, vol. xiv., p. 175.

LA HARPE, DR. P. DE.—Quelques Mots sur la Flore Tertiaire de l'Angleterre. *Bull. Soc. Vaudoise Sci. Nat.*, vol. v., p. 123–143.

MANTELL, DR. G. A.—The Wonders of Geology; or a familiar exposition of Geological Phenomena. 7th Edition. *London.* Revised by PROF. T. R. JONES.

NORMAN, M. W.—Description of the Section of the Upper Greensand at the Undercliff, in the Isle of Wight. *Geologist*, vol. i., pp. 480–484 and 509–513.

OWEN, PROF. [SIR] R.—Notes on the Bones of the Hind-foot of the *Iguanodon*, &c. *Quart. Journ. Geol. Soc.*, vol. xiv., p. 174.

WILKINS, DR. E. P.—On Mammalian Remains from Gravel near Newport I.W. *Geologist*, vol. i., p. 444.

1859.

GOULD, C.—Description of a New Fossil Crustacean from the Lower Greensand (Atherfield). *Quart. Journ. Geol. Soc.*, vol. xv., p. 237.

HEER, REV. DR. O.—Flora Tertiaria Helveticæ. Vol. iii. Fol. *Winterthur.*

NORMAN, M. W.—The "Crackers" and other fossiliferous nodules. *Geologist*, vol. ii., p. 91.

——, ——.—The Flints of High Port, near Ventnor. *Ibid.*, vol. ii., p. 297.

WILKINS, DR. E. P.—Sand-pipes near Swainstone, Isle of Wight. *Ibid.*, vol. ii., p. 175.

——, ——.—A concise Exposition of the Geology, Antiquities, and Topography of the Isle of Wight (with a Relievo Map). 8vo. *London and Newport, I.W.*

1860.

CORNUEL, J.—Lettre sur l'étage Néocomien du département de la Haute Marne. *Bull. Soc. Géol. France*, 2 ser. t. xvii., p. 425.

———.—Notice sur le groupe des grès vert inférieur du bassin de la Seine . . . . et sur les rapports, assise par assise, avec les diverses parties du groupe Wealden et du lower greensand d'Angleterre. *Ibid.*, p. 736.

NORMAN, M. W.—Notes on the Geology of White Cliff Bay, Isle of Wight. *Proc. Geol. Assoc.*, vol. i., pp. 38–46.

———.—On the re-occurrence of Fossil Species at various stratal horizons. *Geologist*, vol. iii., p. 149.

1861.

HEER, REV. DR. O.—Recherches sur le Climat et la Vegetation du Pays Tertiaire. Fol. *Winterthur.*

NORMAN, M. W.—On a Deposit of Recent Shells and Bones in the Cliff of Monk's Bay, Isle of Wight. *Proc. Geol. Assoc.*, vol. i., p. 160.

WILKINS, DR. E. P.—On a newly-discovered Outlier of the Hempstead Strata, on the Osborne Estate, Isle of Wight. *Proc. Geol. Assoc.*, vol. i., p. 194.

WILKINSON, REV. J.—The Farming of Hampshire. [Report on the Agriculture of the Isle of Wight.] *Journ. Roy. Agric. Soc. England*, ser. i., vol. xxii., p. 239 (Isle of Wight, pp. 348–57).

1862.

BECKLES, S. H.—On some Natural Casts of Reptilian Footprints in the Wealden Beds of the Isle of Wight, and of Swanage. *Quart. Journ. Geol. Soc.*, vol xviii., p. 443.

CORNUEL, J.—Essai sur les rapports qui existent entre le grès vert inférieur du pays de Bray et celui du sud-est et du nord-ouest du bassin Anglo-Français. *Bull. Soc. Geol. France*, 2 ser., t. xix., p. 975.

FISHER, REV. O.—On the Brackleshani Beds of the Isle of Wight Basin. *Quart. Journ. Geol. Soc.*, vol. xviii., pp. 65-94.

FOX, REV. W.—When and How was the Isle of Wight severed from the Mainland ? *Geologist*, vol. v., p. 452.

HEER, REV. DR. O.—On Certain Fossil Plants from the Hempstead Beds of the Isle of Wight. (With an Introduction by W. PENGELLY, F.G.S.) *Quart Journ. Geol. Soc.*, vol. xviii., p. 369, plate xviii.

JONES, PROF. T. R.—On the microscopical examination of some Brackleshani Beds. *Geologist*, vol. v., p. 59, and in Dixon's " Geology of Sussex," 2nd Edition (1878), p. 169.

PHIPSON, DR. T. L.—On the Composition of a Specimen of Fossil Wood from the Green Sand of the Isle of Wight. *Chemical News*, vol. vi., p. 194, and vol. ix., p. 28 (1864).

SANDBERGER, PROF. F.—On Upper Eocene Fossils from the Isle of Wight. *Quart. Journ. Geol. Soc.*, vol. xviii., p. 330, plate xii.

1863.

CORNUEL, J.—Sur la limite des deux étages du grès vert inférieur dans le bassin parisien, etc. *Bull. Soc. Géol. France*, 2 ser., t. xx., p. 575.

LANKESTER, [PROF.] E. RAY.—On certain Cretaceous Brachiopoda. *Geologist*, vol. vi., p. 414.

1864.

CORNUEL, J.—Sur l'insuffisance de l'Ostrea aquila pour prouver que la couche à Ostrea aquila du bassin de la Seine serait contemporaine des Perna beds de l'île de Wight. *Bull. Soc. Géol. France*, 2 ser., t. xxi., p. 351.

KOENEN, A. VON.—On the Correlation of the Oligocene Deposits of Belgium, Northern Germany, and the South of England.—*Quart. Journ. Geol. Soc.*, vol. xx., p. 97.

WHITAKER, W.—On some Evidence of there being a Reversal of the Beds near Whitecliff Bay, Isle of Wight. *Geol. Mag.*, vol. i., p. 69.

1865.

ANON.—Note of a new Reptile [*Polacanthus*] from the Wealden Beds. *Geol. Mag.*, vol. ii., p. 432, and *Athenæum*, Aug. 5.

———.—Notice of *Polacanthus* found by REV. W. Fox in Wealden Beds at the back of the Isle of Wight. *Ibid.*, vol. ii., p. 432.

HARRIS, T., & W. DAVIES.—Fossil Jaw-bone of Red-deer found at No Man's Land Shoal, eastward of Ryde, Isle of Wight. *Ibid.*, vol. ii., p. 46.

MITCHELL, W. S.—On some hitherto unrecorded Leaf-forms from the Pipe-clay of Alum Bay. *Ibid.*, vol. ii., p. 515.

PENGELLY, W.—On the Correlation of the Lignite Formation of Bovey Tracey, Devonshire, with the Hempstead Beds of the Isle of Wight. *Trans. Devon. Assoc. of. Sci., Lit., and Art.*, part 4, p. 90.

TATE, PROF. R.—On the so-called Rostellariæ of the Cretaceous Rocks, with a Descriptive Catalogue of the British Species. *Geol. & Nat. Hist. Repertory*, vol. i., p. 93.

WHITAKER, W.—On the Chalk of the Isle of Wight. *Quart. Journ. Geol. Soc.*, vol. xxi., p. 400.

1866.

FOX, REV. W.—On a New Wealden Saurian named *Polacanthus*. *Rep. Brit. Assoc. for 1865, Trans. of Sections*, p. 56.

LEIGHTON, W. H.—On an excursion to the Isle of Wight. *Geol. and Nat. Hist. Repertory*, vol. i., p. 28.

MEYER, C. J. A.—Notes on the Correlation of the Cretaceous Rocks of the South-east and West of England. *Geol. Mag.*, vol. iii., p. 13.

SEELEY, PROF. H. G.—Note on some new Genera of Fossil Birds in the Woodwardian Museum. *Ann. & Mag. Nat. Hist.*, ser. 3, vol. xviii., p. 109.

1867.

CARRUTHERS, W.—On some Cycadean Fruits from the Secondary Rocks of Britain. *Geol. Mag.*, vol. iv., p. 101.

MITCHELL, W. S.—Report of the Committee appointed to Investigate the Alum Bay Leaf-bed. *Rep. Brit. Assoc.* for 1866, p. 146.

RICKETTS, DR. C.—On the Oscillations of Level on the Coast of Hampshire during the Eocene Period. *Proc. Liverpool Geol. Soc.*, Session 8, p. 11.

WHITAKER, W.—On Subaërial Denudation, and on Cliffs and Escarpments of the Chalk and the Lower Tertiary Beds. Part 2. *Geol. Mag.*, vol. iv., p. 483. (Corrections in vol v., p. 47.)

1868.

CODRINGTON, T.—Notes to accompany a Section of the Strata from the Chalk to the Bembridge Limestone at Whitecliff Bay, Isle of Wight. *Quart. Journ. Geol. Soc.*, vol. xxiv., p. 519.

TYLOR, A.—On the Amiens Gravels. *Quart. Journ. Geol. Soc.*, vol. xxiv., p. 103.

WALKER, J. F.—On the Species of Brachiopoda, which occur in the Lower Greensand at Upware. *Geol. Mag.*, vol. v., p. 399.

1869.

CARRUTHERS, W.—On some Undescribed Coniferous Fruits from the Secondary Rocks of Britain. *Geol. Mag.*, vol. vi., p. 2.

FOX, REV. W.—On the Skull and Bones of an Iguanodon. *Rep. Brit. Assoc.* for 1868, Trans. of Sections, p. 64.

MEYER, C. J. A.—On the Lower Greensand of Godalming (*Geol. Assoc.*)

TYLOR, A.—On Quaternary Gravels. *Quart. Journ. Geol. Soc.*, vol. xxv., p. 57.

WOODWARD, DR. H.—Fourth Report on the Structure and Classification of the Fossil Crustacea. *Rep. Brit. Assoc.* for 1868, pp. 72–75.

1870.

CARRUTHERS, W.—On the Fossil Cycadean Stems from the Secondary Rocks of Britain. *Trans. Linn. Soc.*, vol. xxvi., pp. 675–708, plates 54, 57, 63.

CODRINGTON, T.—On the Superficial Deposits of the South of Hampshire and the Isle of Wight. *Quart. Journ. Geol. Soc.*, vol. xxvi., p. 528.

DUNCAN, PROF. P. M.—Second Report on the British Fossil Corals. *Rep. Brit. Assoc.* for 1869, p. 150.

HULKE, J. W.—Note on a New and Undescribed Wealden Vertebra. *Quart. Journ. Geol. Soc.*, vol. xxvi., p. 318.

HUXLEY, PROF. T. H.—On *Hypsilophodon Foxii*, a new Dinosaurian from the Wealden of the Isle of Wight. *Ibid.*, vol. xxvi., p. 3.

JONES, PROF. T. R.—Notes on the Tertiary Entomostraca of England. *Geol. Mag.*, vol. vii., p. 155.

1871.

CARRUTHERS, W.—On some supposed Vegetable Fossils. *Quart. Journ. Geol. Soc.*, vol. xxvii., p. 443.

EVANS, C.—On the Geology of the neighbourhood of Portsmouth and Ryde. *Proc. Geol. Assoc.*, vol. ii., part 1, pp. 61–76.

HULKE, J. W.—Note on a Large Reptilian Skull from Brooke, Isle of Wight, probably Dinosaurian and referable to the Genus *Iguanodon*. *Quart. Journ. Geol. Soc.*, vol. xxvii., p. 199, plate xi.

JUDD, PROF. J. W.—On the Punfield Formation. *Ibid.*, vol. xxvii., p. 207.

JUKES-BROWNE, A. J.—The Valley of the Yar, Isle of Wight. *Geol. Mag.*, vol. viii., p. 561.

LIVEING, PROF.—On a pipe in the Chalk at Alum Bay. *Proc. Camb. Phil. Soc.*, part xii., pp. 194, 195.

MEYER, C. J. A.—On Lower Tertiary Deposits recently exposed at Portsmouth. *Quart. Journ. Geol. Soc.*, vol. xxvii., p. 74.

1872.

CARRUTHERS, W.—Notes on some Fossil Plants. *Geol. Mag.*, vol. ix., p. 49.

EVANS, C.—On the Geology of the neighbourhood of Portsmouth and Ryde. Part II., *Proc. Geol. Assoc.*, vol. ii., p. 149.

HULKE, J. W.—Appendix to a " Note on a new and undescribed Wealden Vertebra." (Vol. xxvi., p. 318.) *Quart. Journ. Geol. Soc.*, vol. xxviii., p. 36.

MEYER, C. J. A.—On the Wealden as a Fluvio-lacustrine Formation, and on the relation of the so-called " Punfield Formation " to the Wealden and Neocomian. *Quart. Journ. Geol. Soc.*, vol. xxviii., p. 243.

1873.

HULKE, J. W.—Contribution to the Anatomy of *Hypsilophodon Foxii*. An Account of some recently acquired Remains. *Quart. Journ. Geol. Soc.*, vol. xxix., p. 522, pl. xviii.

MEYER, C. J. A.—Further notes on the Punfield Section. *Quart. Journ. Geol. Soc.*, vol. xxix., p. 70.

RYLE, T.—On Cretaceous Fossils from the Isle of Wight. *Eastbourne Nat. Hist. Soc.*, pp. 11–13 (1873–74).

WHITAKER, W.—List of Works on the Geology, Mineralogy, and Palæontology of the Hampshire Basin. *Proc. Winchester and Hampshire Sci. Soc.* for 1873, pp. 108–127.

1874.

BARROIS, DR. C.—Sur la Craie de l'île de Wight (1873–4). *Ann. Soc. Géol. du Nord*, t. i., pp. 74–81. *Bull. Soc. Géol. France*, ser. 3, t. ii., pp. 428–435.

BRION, H. F.—Relievo Map of the Isle of Wight. [Scale 3 miles to 1 inch.] Reduced from the Map of the Geological Survey of Great Britain, by H. W. BRISTOW, F.R.S., F.G.S.

HULKE, J. W.—Supplemental Note on the Anatomy of *Hypsilophodon Foxii*. *Quart. Journ. Geol. Soc.*, vol. xxx., p. 18.

———.—Note on an Astragalus of *Iguanodon Mantelli*. *Ibid.*, vol. xxx., p. 24.

———.—Note on a Reptilian Tibia and Humerus (probably of *Hylæosaurus*) from the Wealden Formation in the Isle of Wight. *Ibid.*, vol. xxx., p. 516–520, pl. xxxi.

———.—Note on a Modified Form of Dinosaurian *Ilium*, hitherto reputed Scapula. *Quart. Journ. Geol. Soc.*, vol. xxx., p. 521, pl. xxxii.

KOWALEVSKY, DR. W.—On the Osteology of the Hyopotamidæ. *Phil. Trans.*, vol. 163, pt. 1, pp. 19–94, plates xxxv.–xl.

REPORT.—Sixth Report of the Commissioners appointed in 1868 to inquire into the best means of preventing the pollution of Rivers. The Domestic Water Supply of Great Britain. *Folio. London.*

1875.

BARROIS, DR. C.—L'Age des Couches de Blackdown. *Ann. Soc. Géol. Nord.*, t. iii., p. 1–8.

———.—Ondulations de la Craie dans le Sud de L'Angleterre. *Ann. Soc. Geol. Nord.*, t. ii., p. 85–111.

———.—Le Tunnel de la Manche. *Rev. Sci.*, 2 sér., 4 Ann., pp. 1070–1072, 1192–93.

———.—Description Géologique de la Craie de l'Ile de Wight. *Ann. Sci. Géol.*, Sér. 4, t. vi., livre 2, art. 3.

CARRUTHERS, W.—On the Flora of the London Clay of Sheppey. *Proc. Geol. Assoc.*, vol. iv., no. 5, pp. 318–319.

GARDNER, J. S,—On the Cretaceous Aporrhaïdæ. *Geol. Mag. New Ser.*, dec. ii., vol. ii., pp. 198, 291, 392.

SEELEY, PROF. H. G.—On the Axis of a Dinosaur from the Wealden of Brook in the Isle of Wight, probably referable to the Iguanodon. *Quart. Journ. Geol. Soc.*, vol. xxxi., pp. 461–4.

1876.

BARROIS, DR. C.—Recherches sur le Terrain Crétacé Supérieur de l'Angleterre et de l'Irlande. *Mémoires de la Soc. Géol. du Nord.* Lille. See also *Geol. Mag.*, new ser., dec. ii., vol. ii., p. 513.

GARDNER, J. S.—On the Eocene Floras of the Hampshire Basin. Conferences held in connection with the special Loan Collection of Scientific Apparatus, at South Kensington Museum. Section Physical Geography, Geology, &c., p. 412.

HULKE, J. W.—Appendix to "Note on a Modified Form of Dinosaurian Ilium, hitherto reputed Scapula." (Vol. xxx., p. 521.) *Quart. Journ. Geol. Soc.*, vol. xxxii., p. 364.

1877.

GARDNER, J. S.—On British Cretaceous Patellidæ and other Families of of Patelloid Gasteropoda. *Quart. Journ. Geol. Soc.*, vol. xxxiii., p. 192, plates vii. to ix.

——.—On the Tropical Forests of Hampshire. *Nature*, vol. xv., pp. 229, 258, 279.

JUKES-BROWNE, A. J.—Notes on the Correlation of the Beds constituting the Upper Greensand and Chloritic Marl. *Geol. Mag.*, dec. 2, vol. iv., pp. 350–364.

SOLLAS, PROF. W. J.—On the structure and affinities of the Genus *Siphonia*. *Quart. Journ. Geol. Soc.*, vol. xxxiii., p. 792, plates xxv. and xxvi.

1878.

BRODIE, REV. P. B.—On the Discovery of a large and varied Series of Fossil Insects and other associated fossils in the Eocene (Tertiary) Strata of the Isle of Wight. *Proc. Warwicksh. Nat. and Arch. Field Club*, pp. 3–12.

DIXON, F.—The Geology of Sussex (1850). New Edition revised and augmented by T. RUPERT JONES. 4to. *Brighton.*

HULKE, J. W.—Note on an Os articulare, presumably that of Iguanodon Mantelli. *Quart. Journ. Geol. Soc.*, vol. xxxiv., p. 744.

RAMSAY, [SIR] A. C.—The Physical Geology and Geography of Great Britain. 5th Edition.

WIGNER, G. W.—The Water Supply of Sea-Side Watering-Places . . . . 8vo. *London.* Also in a shorter form, under the title " Sea-Side Water . . . ." 8vo. *London.*

1879.

GARDNER, J. S.—On the British Eocenes and their Deposition. *Proc. Geol. Assoc.*, vol. vi., pp. 83–106.

—— ——. — Description and Correlation of the Bournemouth Beds. Part I. Upper Marine Series. *Quart. Journ. Geol. Soc.*, vol. xxxv., p. 209.

ETTINGSHAUSEN, DR. C. VON.—Report on the Phyto-Palæontological Investigations of the Fossil Flora of Sheppey. *Proc. Roy. Soc.*, vol. xxix., p. 388.

GRIMSHAW, H.—On a Peculiar Feature in the Water of the Well in Carisbrooke Castle, Isle of Wight. *Chem. News*, vol. xl., pp. 310–1.

HULKE, J. W.—Note (3rd) on (Eucamerotus, *Hulke*) Ornithopsis, H. G. *Seeley*=Bothriospondylus magnus, *Owen.*=Chondrosteosaurus magnus, Owen. *Quart. Journ. Geol. Soc.*, vol. xxxv., pp. 752–762.

——, ——.—*Vectisaurus Valdensis*, a new Wealden Dinosaur. *Ibid.*, vol. xxxv., pl. xxi., pp. 421–424. Pl. xxi.

PARKINSON, C.—The Cephalopoda of the Chalk Marl, and Upper Greensand, Isle of Wight. *Science Gossip*, no. 177, pp. 204–205, figs. 155 to 164.

PRICE, F. G. H.—The Gault, being the substance of a lecture delivered in the Woodwardian Museum, Cambridge. 8vo. *London.*

WOODWARD, DR. H.—On the occurrence of *Branchipus* (or *Chirocephalus*) in a Fossil state with *Euspheroma*, and with numerous Insect-remains, in the Eocene Freshwater (Bembridge) Limestone of Gurnet Bay, Isle of Wight. *Quart. Journ. Geol. Soc.*, vol. xxxv., pp. 342–350. Pl. xiv.

1880.

ETTINGSHAUSEN, DR. C. VON.—Report on Phyto-Palæontological Investigations of the Fossil Flora of Alum Bay. *Proc. Roy. Soc.*, vol. xxx., p. 228.

GARDNER, J. S.—On the Alum Bay Flora. *Nature*, vol. xxi., p. 588. See also p. 555.

HULKE, J. W.—Iguanodon Prestwichii, a new Species from the Kimeridge Clay, distinguished from I. Mantelli of the Wealden Formation in the S.E. of England and Isle of Wight. *Quart. Journ. Geol. Soc.*, vol. xxxvi., p. 433, Pl. xvii.-xx.

JUDD, PROF. J. W.—On the Oligocene Strata of the Hampshire Basin. *Quart. Journ. Geol. Soc.*, vol. xxxvi., pp. 137-177. Pl. vii.

LEFÈVRE, TH.—Note sur le Bulimus ellipticus, Sow., fossile des Calcaires de Bembridge, Ile de Wight. *Ann. Soc. Malacologique de Belgique*, vol. xiv. pp. 82-87, and plate viii.

1881.

BLAKE, PROF. J. F.—On a continuous section of the Oligocene Strata from Colwell Bay to Headon Hill. *Proc. Geol. Assoc.*, vol. vii., pp. 151-161.

HULKE, J. W.—*Polacanthus Foxii*, a large undescribed Dinosaur from the Wealden Formation in the Isle of Wight. *Phil. Trans.*, vol. clxxii., p. 653. Plates 70-76.

KEEPING, H., and E. B. TAWNEY.—On the Beds at Headon Hill and Colwell Bay in the Isle of Wight. *Quart. Journ. Geol. Soc.*, vol. xxxvii., pp. 85-127 [plate v.]. See also *Camb. Phil. Soc.*, vol. iv., part 1, p. 59.

PARKINSON, C.—Upper Greensand and Chloritic Marl, Isle of Wight. *Quart. Journ. Geol. Soc.*, vol. xxxvii., p. 370.

TAWNEY, E. B.—On the Upper Bagshot Sands of Hordwell Cliffs, Hampshire. *Proc. Cambridge Phil. Soc.*, vol. iv., part iii., p. 140.

————.—Excursion to the East End of the Isle of Wight. *Proc. Geol. Assoc.*, vol. vii., p. 185.

1882.

DE RANCE, C. E.—The Water Supply of England and Wales. (Isle of Wight Streams, pp. 293-299). 8vo. *London.*

ELWES, J. W.—On the Classification of Oligocene Strata in the Hampshire Basin. *Rep. Brit. Assoc.* for 1882, p. 539.

FISHER, REV. O.—On the Strata of Colwell Bay, Headon Hill, and Hordwell Cliff. *Geol. Mag.*, dec. ii., vol. ix., p. 138.

GARDNER, J. S.—A Chapter in the History of the Coniferæ. *Nature*, vol. xxv., p. 228.

HARRISON, W. J.—Geology of the Counties of England and of North and South Wales. 8vo. *London.*

HULKE, J. W.—Note on the Os Pubis and Ischium of *Ornithopsis Eucamerotus. Quart. Journ. Geol. Soc.*, vol xxxviii., p. 372. Plate xiv.

————.—An attempt at a complete Osteology of *Hypsilophodon Foxii*; a British Wealden Dinosaur. *Phil. Trans.*, vol. 173, pp. 1035-1062.

————.—Description of some *Iguanodon*-remains indicating a new species, *I. Seelyi. Quart. Journ. Geol. Soc.*, vol. xxxviii., p. 135. Plate v.

JUDD, PROF. J. W.—On the Relations of the Eocene and Oligocene Strata in the Hampshire Basin. *Quart. Journ. Geol. Soc.*, vol. xxxviii, p. 461-489.

————.—The Headon Hill Section. *Geol. Mag.*, dec. ii., vol. ix., p. 189.

LUCAS, A. H. S.—On the Headon Beds of the Western Extremity of the Isle of Wight. *Ibid.*, p. 97.

NORMAN, M. W.—The Chloritic Marl and Upper Greensand of the Isle of Wight. *Ibid.*, p. 440. Plate x.

SEELEY, PROF. H. G.—On a remarkable Dinosaurian Coracoid from the Wealden of Brook in the Isle of Wight, probably referable to *Ornithopsis. Quart. Journ. Geol. Soc.*, vol. xxxviii., p. 367.

1883.

ETHERIDGE, R.—President's address on Geology. *Rep. Brit. Assoc.* for 1882, pp. 502-529.

HINDE, G. J.—Catalogue of Fossil Sponges in the British Museum.   4to.
- PRESTWICH, PROF. J.—Notes relating to some of the Drift Phenomena of Hampshire [Elephant Bed, Freshwater Gate].   Rep. Brit. Assoc. for 1882, p. 529.
SEELEY, H. G.—On the Dorsal Region of the Vertebral Column of a new Dinosaur (indicating a new genus, *Sphenospondylus*), from the Wealden of Brook in the Isle of Wight.   Quart. Journ. Geol. Soc., vol. xxxix., p. 55.

1884.

LYDEKKER, R.—Note on the Anthracotheriidæ of the Isle of Wight. Geol. Mag., dec. iii., vol i., p. 547.

1885.

BRODIE, P. B.—Fossil Birds.   Geol. Mag., dec. iii., vol. 2, p. 384.
GARDNER, J. S.—On the Land Mollusca of the Eocene.   Ibid., p. 241. Plate vi.
JONES, PROF. T. R.—On the Ostracoda of the Purbeck Formation: with Notes on the Wealden Species.   Quart. Journ. Geol. Soc., vol. xli., pp. 311-353.
LYDEKKER, R.—Catalogue of the Fossil Mammalia in the British Museum (Natural History).   Parts i. and ii.   8vo.   London.
TOMES, R. F.—Observations on some imperfectly known Madreporaria, from the Cretaceous Formation of England.   Geol. Mag., dec. iii., vol. ii., p. 541.
WOODWARD, A. S.—On the Literature and Nomenclature of British Fossil Crocodilia.   Ibid, p. 496.

1886

HINDE, G. J.—On Beds of Sponge-remains in the Lower and Upper Green-sand of the South of England.   Phil. Trans., vol. 176, Part II., plates 40–45, pp. 403, 412, 418–20, 447.
REPORT of the Committee on the Erosion of the Sea Coasts of England and Wales.   Rep. Brit. Assoc., pp. 428–432.

1887.

JONES, PROF. T. R. — Notes on *Nummulites elegans*, Sow., and other English Nummulites.   Geol. Mag., dec. iii., vol. iv., p. 89.
———, —, and SHERBORN, C. D.—Further notes on the Tertiary Entomostraca of England, with special reference to those from the London Clay.   Ibid., pp. 385 and 450.
KEEPING, H.—[Letter.]   On the Osborne Beds.   Ibid., p. 70.
———, —.—On the discovery of the *Nummulina elegans* Zone at Whitecliff Bay, Isle of Wight.   Ibid., p. 70.
LYDEKKER, R.—Catalogue of the Fossil Mammalia in the British Museum (Natural History).   Parts III–V.   8vo.   London.
———, —.—On certain Dinosaurian Vertebræ from the Cretaceous of India and the Isle of Wight.   Quart. Journ. Geol. Soc., vol. xliii., p. 156.
———, —.—Note on the Hordwell and other Crocodilians.   Geol. Mag., dec. iii., vol. iv., p. 307.
NORMAN, M. W.—Geological Guide to the Isle of Wight.   Map, sections, and 15 plates of fossils.   8vo. Ventnor.
REID, C.—The Extent of the Hempstead Beds in the Isle of Wight.   Geol. Mag., dec. iii., vol. iv., p. 510.
SEELEY, PROF. H. G. — On *Aristosuchus pusillus*, Ow., being further Notes on the fossils described by Sir R. Owen as *Poikilopleuron pusillus*, Ow. Quart. Journ. Geol. Soc., vol. xliii., p. 221.
———, —.—On a Sacrum, apparently indicating a new type of Bird (*Ornithodesmus cluniculus*, Seeley), from the Wealden of Brook.   Ibid., p. 206.
WOODS, H.—[Letter] On the occurrence of Phosphatic Nodules in the Lower Greensand, east of Sandown.   Geol. Mag., dec. iii., vol. iv., p. 46.

1888.

COLENUTT, G. W.—On a Portion of the Osborne Beds of the Isle of Wight, and on some Remarkable Organic Remains recently discovered therein. *Geol. Mag.*, dec. iii. vol. v., p 358.

GARDNER, J. S.—[Letter] On the Correlation of the Grès de Belleu with the Lower Bagshot. *Ibid.*, p. 188.

———, —, H. KEEPING, and H. W. MONCKTON.—The upper Eocene, comprising the Barton and Upper Bagshot Formations. *Quart. Journ. Geol. Soc.*, vol. xliv., pp. 578–635.

HULKE, J. W.—Supplemental Note on Polacanthus Foxii, describing the Dorsal Shield and some points of the Endoskeleton, imperfectly known in 1881. *Phil. Trans.*, vol. 178 (B.) Plates 8, 9.

JONES, PROF. T. R.—Ostracoda from the Weald Clay of the Isle of Wight. *Geol. Mag.*, dec. iii., vol. v., p. 534.

LYDEKKER, R.—Catalogue of Fossil Reptilia in the British Museum, vol. i. 8vo.

———.—Note on a new Wealden Iguanodont and other Dinosaurs. *Quart. Journ. Geol. Soc.*, vol. xliv., p. 46. Plate iii.

McCOOK, H. C.—A new Fossil Spider (Eoatypus Woodwardii). *Proc. Acad. Nat. Sc. Philadelphia* for 1888, pp. 200–202, and *Annals & Mag. Nat. Hist.*, ser 6, vol. ii.

PRESTWICH, PROF. J.—Further observations on the correlation of the Eocene Strata in England, Belgium, and the North of France. *Ibid.*, p. 88, pl. v.

SEELEY, PROF. H. G.—On Thecospondylus Daviesi, with some remarks on the Classification of the *Dinosauria. Ibid.*, p. 79.

STRAHAN, A., and C. REID.—La Géologie d l'Ile de Wight. Printed for the International Geol. Congress. 8vo. *London.*

1889.

BLYTT, A.—The probable Cause of the displacement of beach-lines. An attempt to compute geological epochs [with additional note]. *Christiania Videnskabs—Selskabs Forhandlinger*, 1889, No. 1.

LYDEKKER, R.—On the Remains and Affinities of five Genera of Mesozoic Reptiles. *Quart. Journ. Geol. Soc.*, vol. xlv., p. 41.

———.—On a Cœluroid Dinosaur from the Wealden. *Geol. Mag.*, dec. iii., vol vi., p. 119.

———.—On Remains of Eocene and Mesozoic Chelonia, and on a Tooth of (?) Ornithopsis. *Quart. Journ. Geol. Soc.*, vol. xlv., pp. 237–239.

NEWTON, E. T.—Description of a New Species of Clupea (C. vectensis) from Oligocene Strata in the Isle of Wight. *Ibid*, p. 112. Plate iv.

3. LIST of the MONOGRAPHS published by the PALÆONTOGRAPHICAL SOCIETY, which refer to the ISLE OF WIGHT.

GARDNER, J. S., and C. VON ETTINGSHAUSEN.—Eocene Flora, vol. i. (Filices). 1879–1882.

GARDNER, J. S.—Eocene Flora, vol. ii. (Gymnospermæ). 1883–1885.

MILNE-EDWARDS, H., and J. HAIME.—Tertiary, Cretaceous, ... Corals. 1849–1854.

DUNCAN, P. M.—Supplement to the Fossil Corals. 1866–1870.

FORBES, E.—Tertiary Echinodermata. 1852.

WRIGHT, T.—Cretaceous Echinodermata, vol. i. 1862–1882.

DARWIN, C.—Fossil Cirripedes. 1851–1858.

JONES, T. R.—Cretaceous Entomostraca. 1849.

JONES, T. R.—Tertiary Entomostraca. 1855.

JONES, T. R., and C. D. SHERBORN.—Supplement to the Tertiary Entomostraca. 1889.

BELL, T.—Malacostracous Crustacea. 1856–1860.

DAVIDSON, T.—Fossil Brachiopoda, vols. i., iv., and v. 1850–1884.

LYCETT, J.—Fossil Trigoniæ. 1872–1883.

EDWARDS, F. E., and S. V. WOOD. Eocene Mollusca, Cephalopoda and Univalves, vol. i. 1848–1877.

WOOD, S. V.—Eocene Mollusca, Bivalves, vol. i. 1859–1870. Supplement to the Eocene Mollusca, vol. i. 1877.

SHARPE, D.—Upper Cretaceous Cephalopoda. 1853–1855.
OWEN, R., and T. BELL.—Reptilia of the London Clay [and of the Brackleshain and other Tertiary Beds], vol. i. 1848.
OWEN, R.—Reptilia of the Cretaceous Formations. 1851–1862.
OWEN, R.—Reptilia of the Wealden and Purbeck Formations. 1853–1879.

INDEX OF AUTHORS.

(The figures refer to the dates of publication.)

McCook, II. C., 1888.
McCoy, F., 1849, 1854.
Mantell, Dr. G. A., 1838, 1844, 1846, 1847, 1849, 1854, 1858.
Marcet, Dr. A., 1811.
Marshall, W., 1798.
Meyer, C. J. A., 1866, 1869, 1871-3.
Middleton, J., 1812.
Milne-Edwards, Prof. H. *See* List 3.
Mitchell, W. S., 1865, 1867.
Monckton, II. C. W., 1888.
Morris, Prof. J., 1835, 1848, 1854. *See also* List 1.
Mudie, R., 1841.
Murchison, Sir R. I., 1843.

Newton, E. T., 1889. *See also* List 1.
Nesbit, J. C., 1848.
Norman, M. W., 1858-61, 1882, 1887.

Ogilby, W., 1839.
Owen, Prof. Sir R., 1838, 1841, 1842, 1844, 1846, 1848, 1857, 1858, 1861. *See also* List 3.

Paine, J. M., 1849, 1853.
Parkinson, C., 1879, 1881.
Pengelly, W., 1862, 1865.
Pennant, T., 1801.
Phillips, W., 1822.
Phipson, Dr. T. L., 1862.
Pratt, S. P., 1835.
Prestwich, Prof. J., 1846, 1847, 1849, 1850, 1854, 1857, 1883, 1888.
Prévost, C., 1837.
Price, F. G. II., 1879.

Ramsay, Sir A. C., 1878. *See also* List 1.
Reid, C., 1887, 1888. *See also* List 1.
Renevier, E., 1856.
Ricketts, Dr. C., 1867.
Rickman, W., 1840.
Rivers Pollution, Report, 1874.
Ryle, T., 1873.

Salter, J. W. *See* List 1.
Sandberger, Prof. F., 1862.

Saxby, S. M., 1846.
Scudamore, C., 1820.
Sedgwick, Rev. Prof. A., 1822, 1825, 1831.
Seeley, Prof. II. G., 1866, 1875, 1882, 1883, 1887, 1888.
Sharpe, D. *See* List 3.
Sherborn, C. D., 1887. *See also* List 3.
Simms, F. W., 1845.
Sollas, Prof. W. J., 1877.
Sorby, Dr. II. C., 1853, 1857.
Sowerby, G. B., 1821.
———, J. De C., 1816-1829, 1837, 1840.
Strahan, A., 1888. *See also* List 1.

Tate, Prof. R., 1865.
Tawney, E. B., 1881.
Tomes, R. F., 1885.
Tooke, A. W., 1837.
Townsend, Rev. J., 1813.
Trimmer, J., 1841, 1854, 1855.
Tylor, A., 1868, 1869.

Vancouver, C., 1810.
Verstegan, R., 1605.

Walker, J. F., 1868.
Warner, Rev. R., 1794, 1795.
Waterworth, Dr. T. L., 1813.
Way, J. T., 1849, 1853.
Webster, T., 1813, 1814, 1816, 1824, 1825.
Wetherell, N. T., 1852.
Whitaker, W., 1864, 1865, 1867, 1873.
Whitley, N., 1856.
Wigner, G. W., 1878.
Wilkins, Dr. E. P., 1858, 1859, 1861.
Wilkinson, Rev. J., 1861.
Wood, S. V. *See* List 3.
Woods, II., 1887.
Woodward, A. S., 1885.
Woodward, Dr. II, 1869, 1879.
Wright, Dr. T., 1851, 1852. *See also* List 3.

Young, A., 1794.

INDEX.

LONDON: Printed by EYRE and SPOTTISWOODE,
Printers to the Queen's most Excellent Majesty.
For Her Majesty's Stationery Office.
[17374.—500.—10/89.]

GOSPORT
PORTSMOUTH

Haslacar P[t]
S Sea Castle

OLOG

OF THE

OF

from

of the

Y OF

by

Bris

SPITHEAD

Ryde Roads

NETTLESTONE P[t]
Light Ho
Powers P[t]
Summer Ho
Watch Ho P[t]
Bembridge Limestone
S[t] Helens Ch.
Church

Brambridge P[t]

Brading Harbour
Bembridge

FORELAND

Yaverland Down
White Cliff Bay
S[t] CULVER CLIFF
Red Cliff

BASE OF TERTIARY

Sand down
SANDOWN BAY

Section A

Shanklin Chine

Luccomb Chine
Shine Head
DUNNOSE

UPPER SECONDARY

Bonchurch
Niton

North

Osborne
Norris
Sea Wall

BEMBRIDGE
HEAD SERIES
OSBORNE & HEADON
Sea Level

Engraved by J W Lowry

Judd & C[o] Lith 73 & 75 Farringdon R[d] & Doria, e'Cromicni

INDEX GEOLOGICAL MAP
of the
ISLE OF WIGHT,
GEOLOGICAL SURVEY OF ENGLAND & WALES.

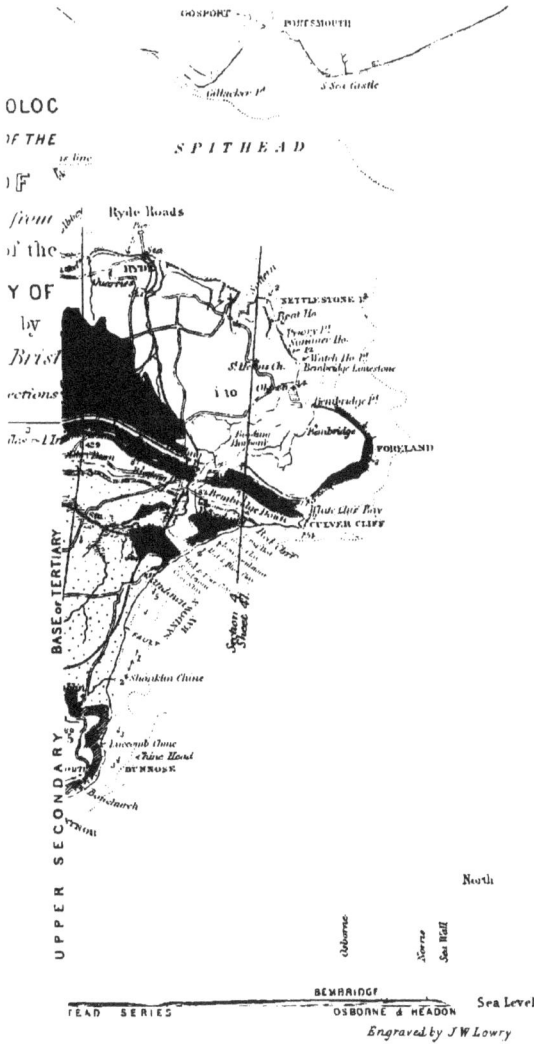

GOSPORT
PORTSMOUTH

Gilkicker P.
S Sea Castle

OLOC

OF THE

OF

SPITHEAD

Ryde Roads

from

of the

Y OF

by

Brist

ections

NETTLESTONE P.
Bent Ho.
Priory Ho.
Nunney Ho.
Watch Ho. P.
Bembridge Limestone
S. Helens Ch.

Bembridge P.

Bembridge

FORELAND

Whitecliff Bay
CULVER CLIFF

BASE or TERTIARY

Shanklin Chine

Luccomb Chine
Chine Head
DUNNOSE

Bonchurch

UPPER SECONDARY

North

Osborne
Nurse
Sea Wall

BEMBRIDGE
OSBORNE & HEADON
Sea Level

EAD SERIES

Engraved by J W Lowry

Judd & Co. Ld. Lith 73 & 75 Farringdon Rd & Dorian's Crescent

Brook Chine

WEALDEN
BEDS

k

WEALDEN
BEDS

i

h

WEALDEN
BEDS

l   *k*

Blown Sand

LOWER
GREENSAND

YPHÆA BED   S C A P H I T E S   B E D S

Southern end of
S.t Catherine's Hill

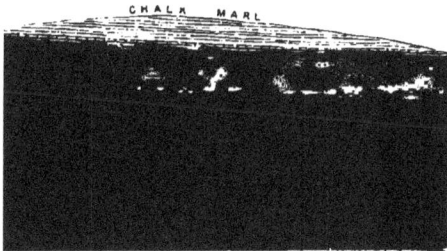

CHALK   MARL

CHLORITIC MARL

UPPER
GREENSAND

GAULT

CARSTONE

SANDROCK
BEDS

Ordnance Datum

FERRUGINOUS BEDS   CLAY - BED
OF BLACKGANG CHINE

I   F   F

Engraved by W.M.Redaway

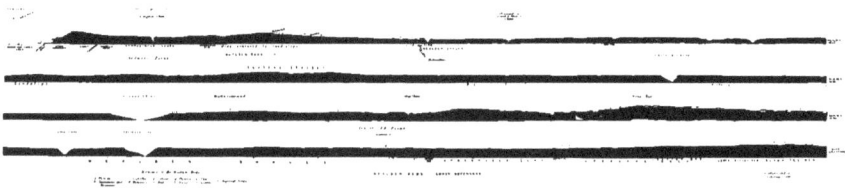

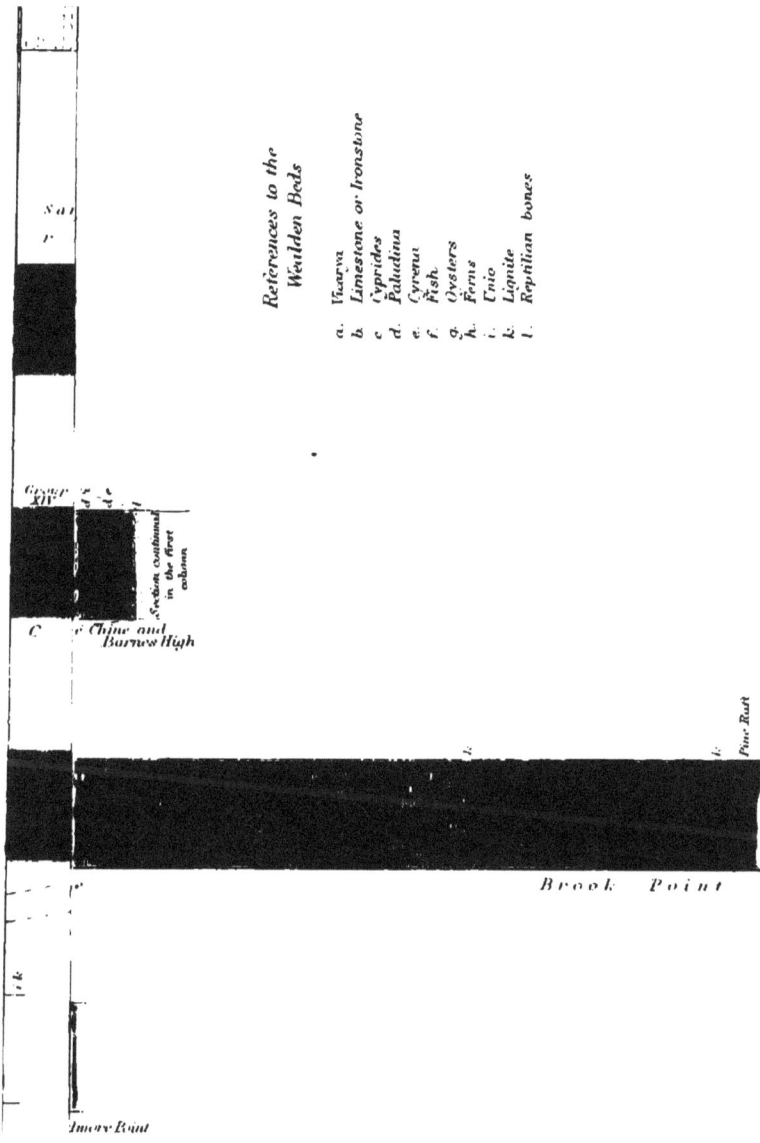

References to the
Wealden Beds

a. Vivarva
b. Limestone or Ironstone
c. Cyprides
d. Paludina
e. Cyrenu
f. Fish
g. Oysters
h. Ferns
i. Unio
k. Lignite
l. Reptilian bones

Sar
r

Grear
XIV

Section continued
in the first
column

C   Chine and
Barnes High

Brook   Point

Pine Raft

k

Imore Point

S.5°W.

The English Channel

Sea level

Chalk Rock

Layers of Flints

containing numerous

surface

marine remains

1.

1500 Feet

40 Chains or ½ Mile

Bench

S.11°W.

The English Channel

Sea level

Middle and Lower Chalk

F.L.I.N.T.S.

or

numerous

layers

with

(Chalk)

Reading Beds (Plastic Clay)

overlap with layers of Plastic

Chalk Rock

Beacon 483 feet

Faux

& Footpath

High Down

...d to
...e in Bay.
...d pit

LIGOCENE, OR FLUVIO-MARINE SERIES

F WIGHT.

nd H.W. Bristow, F.R.S.

Clement Reid, F.L.S.,F.G.S.

inch.

60   80   100 Feet

4

High Down

Beacon.
489 feet

S.II°W.

ad to
ve in Bay
nd pit

Fence
& Footpath

The English Channel

Chalk Rock

Sea level

peclay, with leaves of plants

Reading Beds (Plastic Clay)

Chalk, with numerous layers of FLINTS.

Middle and Lower Chalk

Bench

40 Chains or ¼ Mile
2500 Feet

High Down

S.5°W.

The English Channel

Sea level

Chalk Rock

uneven cracked

surface and

containing numerous

layers of flints

ENGRAVED BY J.W. LOWRY

LIGOCENE, *or* FLUVIO-MARINE SERIES

F WIGHT.

nd H.W. Bristow, F.R.S.

Clement Reid, F.L.S.,F.G.S.

[illegible]

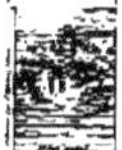

[illegible]